THE BRAIN
THE LAST FRONTIER

Richard M. Restak, M.D.

WARNER BOOKS

A Warner Communications Company

WARNER BOOKS EDITION

This Warner Books Edition is published by arrangement with
Doubleday & Co., Inc., 245 Park Avenue, New York, N.Y. 10017.

Cover design by NEW Studio
Cover photos by Georgiana B. Silk

Warner Books, Inc., 75 Rockefeller Plaza, New York, N.Y. 10019

Ⓦ A Warner Communications Company

Printed in the United States of America

First Printing: July, 1980

10   9   8   7   6   5   4   3

- Will an infant deprived of close physical contact develop psychopathic tendencies?
- Are the behavioral differences between the sexes founded in neurological fact?
- Can senility be diagnosed before its symptoms appear?

Through the new approach of psychobiology, which combines brain science with behavioral science, Dr. Richard M. Restak considers questions such as these. He takes you on a revolutionary exploration of the human mind, the ultimate intellectual challenge.

He explains how the brain works.

> "Restak is most impressive in describing specific phenomena. Some of his discussions . . . are among the best I've seen in an introductory treatise."
>
> —*Psychology Today*

He explains how it perceives and reacts to stimuli.

> "Restak effectively uses a light, conversational style to probe into such issues as brain chemistry and violence, microcircuit and holographic models of brain function, and the extent to which brain structure mirrors the nature of reality . . . fascinating."
>
> —*Library Journal*

He describes the new discoveries and what is to come.

> "Comprehensive . . . optimistic . . . Restak makes clear the border between accepted findings and theories or interpretations. He is particularly good in discussing some of the newer work."
>
> —*Kirkus Reviews*

"Having explored the dark heart of the atom and having paid several visits to the moon's wrinkled old skin, we may be ready at last to find out what's going on in the inner space between our ears," says *The Cleveland Plain Dealer*, inviting you to explore . . .

THE BRAIN: THE LAST FRONTIER

To Ann Buchwald, Lisa Drew, and Carolyn Restak

Contents

An interest in the brain requires no justification other than a curiosity to know why we are here, what we are doing here, and where we are going.

PAUL MAC LEAN

PART
1

1

The Philosopher's Myth

On the contrary, it is impossible to obtain an adequate version of the laws for which we are looking unless the physical system is regarded as a whole.

M. PLANCK, 1931

Two former Nobel Prize laureates in physics were recently asked to guess what area of research would win the Nobel Prize for physics in the year 2000. Both of them, without prior consultation and with hardly a hesitation, said brain research. The human brain, they concluded, is our ultimate intellectual challenge in the last quarter of the twentieth century.

There is a double irony in the choice of the brain as an area of research worthy of Nobel Prizes in physics. For one thing, physics is one of the "hard sciences," a model of the "objective" view of the world. Brain research, in contrast, has been modeled on biology, a "softer," more fuzzy discipline where disagreement abounds. (Articles are still appearing from time to time attempting to disprove even such well-established biological principles as evolution.)

Secondly, brain research is taken up with questions even more fundamental than those which challenge the theoretical physicists: "How do we know what we know?" "What is the *real* world?" "Who am I?" Until very recently such questions were referred to theologians or philosophers, whose speculations provided the foundations for complicated and impressive philosophical systems. But today, philosophy and theology exert a far less compelling influence on our lives. When was the last time you saw a philosopher or theologian on a late-night talk show? And how many of us can get out more than a few mumbled generalizations about the major philosophical thinkers of the last five hundred years?

With the collapse of philosophy and theology as major influences on our lives, questions such as "How do we know what we know?" or the ringing "What is truth?" have remained unanswered and, even, largely unasked. In their place, we've focused on questions of a more "practical" nature. Food, energy supplies, pollution, crowding, the breakdown of needed social services—these are some of the areas which have received and continue to receive the most attention. Already some of them are reaching crisis proportions.

The world's population, for instance, is expected to double in the next thirty to forty years. In addition, population growth now indicates that each doubling of the population will take place in half the previous time. Instead of thirty to forty years, we move to fifteen to twenty, with our children witnessing a population doubling spaced over one or two decades.

Realizations such as this tend to encourage even greater emphasis on the external environment. Birth-control techniques are expanded and, at least in India, sterilization procedures are forced on groups of citizens without their consent. Energy policies are formulated which concentrate on how best to cope with a dwindling supply of oil and energy reserves. Everywhere we are concerned with how we manipulate or manage one crisis after another.

In recent years, however, we've experienced a change in the types of crises that immediately threaten us. No one can now board an airplane without a gnawing fear of a hijacking or a bombing. In less than five years, terrorism has become one of the most decisive influences in our lives. It is also one of the most conspicuous examples of an *internal* as well as an external threat to our continued existence.

14

One understandable response to all of this is to turn to the behavioral sciences to provide the answers to the questions that plague us. Although initially appealing, this suggestion overlooks the rather poor track record that the behavioral sciences have shown in the past. Different authorities hold radically different views on why we act the way we do. Things have now become so complicated, in fact, that psychologists trained in one orientation are often unable to agree with other psychologists on such basic concepts as what psychology is about. Those trained in behavioral methods, for instance, operate as if consciousness doesn't exist, while scoffing at their counterparts who value and rely upon subjective experiences.

In the last several years, a new field has emerged which may offer us a better means of understanding and controlling some of our internal and external threats. Known as psychobiology, this new discipline is a combination of the behavioral sciences and the brain sciences. It differs from other studies of behavior in some of its initial assumptions. While the working of the brain is often peripheral to most theories of psychology, psychobiology depends primarily on what we have learned about the brain and how it works. The emphasis is on how the brain influences our perceptions of the world, how we know ourselves, the nature of reality—in essence, the questions we mentioned earlier as formerly asked by philosophers and theologians.

Basically, psychobiology is concerned with the mind's attempt to know itself through the study of the brain. Today we accept as a truism that the brain is the physical basis of the mind, although this is not quite the same thing as stating that the brain *is* the mind.

Throughout history, almost every major part of the body has been credited at one time or another as the seat of the mind or the soul. In civilizations such as the Sumerian and the Assyrian, the liver was considered the repository of the soul and the physical basis for the personality. To Aristotle, the heart was the central organ, while the brain existed as a sort of cooling mechanism for the blood as it left the heart. This view of the importance of the heart survives today in our popular images of "heartbreak" as a description of the effect of unrequited love, and "bleeding hearts" as a contemptuous term for people who are ruled by sympathy and sentimentalism.

Today we look to the brain rather than to the heart for an understanding of the mind. While this is largely an advance over the superstition and ignorance of the past, it presents us immediately with some rather knotty problems. The first of these is an organizational one.

The number of neurons in the human brain is almost equal to the number of stars in our Milky Way—over fifteen billion. What are the relationships of the nerve cells to each other? To understand the brain, is it necessary to deal with all possible interactions between nerve cells? or with only some of them? Put another way, What level of organization offers the best hope toward understanding how our brain works?

One of the stumbling blocks to understanding the brain can be traced, I believe, to selection of the wrong level of organization. Let me illustrate what I mean.

Forget about the fifteen-billion figure I gave you a moment ago, because that is a staggering number to deal with. Instead, let's work with something more manageable: the number of people presently inhabiting the earth. Imagine each person with a telephone capable of calling any other person in the world. In addition, assume that in our model the important information about people's activities and behavior is always discussed over the telephone. Our job will be to keep track of all the calls and to correlate telephone calls with behavior. Some of this may be very easy, as when a person in London calls a friend in New York and invites him to come over for a week. Within a day or two of such a call, the friend in New York can be observed carrying his bags while leaving his apartment on the way to the airport. At such a time we may feel quite confident that our telephone monitoring system is working well and giving us an accurate and useful correlation between person-to-person communication and behavior. As the number of people increases, however, our method will soon break down. Imagine all the telephone calls we would have to monitor to enable us to predict the identity of the passenger list of a jumbo jet leaving for Athens six

16

months from now. Or imagine the number of telephone calls that would be needed to estimate the population of Schenectady, New York, between the years 1979 and 1980.

Perhaps you think such a task could be handled very well by computers. To see why this is impossible, let's simplify the situation a bit by reducing the number of variables from fifteen billion to a mere thirty-two. Thirty-two just happens to be the maximum number of pieces we can play on a chessboard at any one time. At a typical point in the game, thirty permissible moves are open to each player. If it's white's turn to play, for instance, each of his thirty moves can be countered by thirty moves on the part of black, which leads to about one thousand variations at the end of one round of play.

With white's next move and black's response, everything is computed again, yielding one million positions. By the third move we're at one billion, and so on. In a very short time the number of possible calculations becomes, for all practical purposes, infinite.

In the case of our telephone system we are helped by the fact that every person isn't likely to call everyone in the world. Barriers of language, common interest, and acquaintance limit somewhat the number of possible calls. In chess a similar situation exists, since not every move is equally good and some are downright disastrous, leading to a quick checkmate. In the brain, however, we have no rules that would enable us to know beforehand with absolute certainty whether one brain cell is influencing the activity of another one. This gets us back to the stars-in-the-galaxy-neurons-in-the-brain situation, which represents a level of complexity no human mind or computer can ever be prepared to deal with. In short, if we focus our attention on the neurons and their interconnections, we're selecting a level of organization that can never satisfy our efforts at understanding.

On the other hand, if the level chosen is too sweeping, we come out with generalizations that are useless. "The brain works as an information-processing system" is intended to be informative, but it really doesn't tell us anything at all. The key is to focus on the correct level of organization.

Another related problem in our attempts to understand brain functioning is a more philosophical one. We live in a world of *things*. Our perceptions are geared to encountering objects and people who change very little from day to day. "A rose is a rose is a rose," according to Gertrude Stein, and she might have selected, with equal validity, any one of the tens of thousands of objects and people we encounter every day.

The alternative view, never a popular one, holds that the world consists of *processes* and that what we perceive around us are only frames in a movie. The Greek philosopher Heraclitus said you never step into the same river twice, since the water continues to flow and tomorrow you will encounter different water than today. Things are always in a process of change, like a flame converting combustible substances into heat, light, and hot gases. A fire is not a thing but a process of combustion.

Modern physics is very much a *process* science and poses the greatest challenge to our *things* view of the universe. Quarks, black holes, and particles of antimatter are neither accessible to our direct experience nor can they be considered fixed entities. Despite this, each of these strange-sounding words has already become part of our everyday vocabulary. With the discovery of the atom and the subsequent exploration of the inner world of subatomic physics, the mechanistic *things* view of the universe began to crumble. Even before that, nineteenth-century scientists demonstrated that the movement of a magnet near a coil of copper created electrical energy described by the experimenters as a "disturbance" or a "condition" rather than a thing. By the third decade of this century the mechanistic *things* view of space and time yielded to the concept of a space-time continuum. All measurements involving space and time thus lost any absolute significance.

Today, physics is the study of interactive processes rather than discrete particles. In such a model the atom is not solid but a tiny universe of energy, with the nucleus and its orbiting charges separated by vast space. If the actual mass of our brain were condensed minus that energy space, it would occupy an area smaller than the head of a pin.

My purpose in introducing physics at this point is to contrast it with our approach to the study of the brain. While physics has become very much a *process* science, our ways of thinking about the brain have been locked into looking at the brain as a thing. This creates some immediate, and occasionally subtle, difficulties.

Is the brain the mind? This question at first seems sensible and capable of verification. It has in fact stimulated the imaginations of scientists and philosophers for centuries. Some now feel that the answer is obvious: The mind is nothing but the action of the brain and is a meaningless concept without reference to a brain. As a neurologist, I felt for the longest time that this view was correct. But the question is actually a trick, of the "Have you stopped beating your wife?" type, where either a yes or no carries with it undesirable implications. In both instances, the proper response is to focus on the question itself and show how the form of the question results in a "loaded dice" situation.

The philosopher Gilbert Ryle once described the mind-brain dilemma as a "philosopher's myth" based on what he called a "category mistake."

Imagine an eight-year-old boy taking his first trip to Washington, D.C. He's been told in school that Washington is the seat of the nation's government. On the first day, the boy visits Congress. On day two, he takes a tour of the White House. On the third day, he's shown the Supreme Court building. At this point the boy seems puzzled and asks: "But where is the government? I've seen Congress, the White House, and the Supreme Court, but I still haven't seen the government." Ryle would explain our young man's puzzlement as resulting from a category mistake. Congress, the White House, and the Supreme Court are things—at least each of them can be physically located in space and time. The government, in contrast, is quite another category altogether. It is, in fact, a supercategory which describes the interactions of the other three.

In a similar manner the brain can be dissected, electrically stimulated, or even placed in a blender and homogenized to a few ounces of froth. The mind, however, remains as elusive as the "government" our eight-year-old was trying to find in Washington.

When we're trying to relate the brain to the mind, we're dealing with a category mistake, since the brain is best conceived as a process rather than a three-pound lump of protoplasm. Think of the problems that resulted from our telephone model of human communication, and that model involved only the population of the earth instead of the fifteen-billion interactions that are possible within our heads! Understanding ultimately depends on our ability to concentrate on the process. Once a process is understood, the actual mechanisms of how it is carried out are of less importance. To correlate brain with mind, or vice versa, is a category mistake that insists on equating two different processes. The trick is to exercise care in distinguishing a prerequisite from a cause.

On any given morning, the Los Angeles Freeway is a prerequisite for thousands of commuters getting to work. It is hardly a causative explanation, however, of how a particular commuter, say a stockbroker, can be found at nine o'clock propping his feet on his desk while perusing a list of client telephone numbers.

The question "Is the brain a sufficient explanation for the mind?" was anticipated by the biologist Sir Julian Huxley: "The brain alone is not responsible for mind, even though it is a necessary organ for its manifestation. Indeed, an isolated brain is a piece of biological nonsense as meaningless as an isolated individual."

For the sake of argument, however, let's assume that we've reached a point where every mental event, every product of "mind," can be correlated with something going on within our brain. This is yielding a lot, incidentally, since we're not anywhere near the point of making such correlations. But for the moment, let's imagine it's possible that we can. Since we think of the world as being governed by physical laws, the explanation for such a correlation seems obvious: The brain events are the cause of the mental phenomenon of thinking, perception, whatever. In essence, the mind is dependent on the brain. But logically, isn't just the opposite explanation—The brain is dependent on the mind—equally likely? From a strictly logical point

20

of view, this is an equally valid stance to take, given the postulated parallel between mental events and brain processes.

Despite the seeming logic of all this, most people have a terribly strong hunch that such a view ultimately doesn't make sense. The mind created the brain? I don't believe it for an instant, and I hope you don't. I am only bringing it up to show why the mind-brain controversy may never be resolved. Category mistakes result from our equating the brain and mind as *things* when they are actually *processes*. This confusion immediately leads to mistakes in the way we think about behavior. Naturally, if we start from wrong premises, we are going to end up with theories that are useless and don't work. Let's take a moment to consider some of the ways we traditionally "explain behavior."

2
Dr. Punishberg and
Dr. Rewardnik

Animals manifestly enjoy excitement and suffer from
ennui . . . and many exhibit curiosity.

CHARLES DARWIN,
The Descent of Man, 1874

At this moment your eyes are moving from left to right as
you read this sentence to conclusion. Your understanding
of what I'm saying depends on the scanning movements
your eyes make as they move across the sentence and then
down to the next sentence, where the process is repeated.
Even with something like reading, where you may remain
perfectly motionless, muscle movements (in this case, eye
movements) are taking place all the time. If I were to
completely immobilize your head in a halter and then
cause your eye movements to cease altogether, everything
around you would become a blur and you'd see nothing.
You can test this for yourself by staring fixedly at one spot.
Your visual field will gradually become blurred and all
visual perception will gradually fade away. After stabilization of an image, clarity is replaced by an impression of
mistiness and indefinite depth. Everything we do, even
something as seemingly passive as reading a book, requires
the active exploration of our environment through some
form of movement.

The original explanation given for human voluntary
movement was "an effort of the will." One might just as
easily postulate a tiny man, living inside everyone's head,

giving orders and pushing buttons which result in the "desired" movement. If a person touches his hand to a hot stove, the tiny man barks out an order to the muscle of the arm and the arm withdraws. If a beautiful girl appears over the horizon, the order might invoke a wide range of programed approach tactics. If the girl turns out later, on closer inspection, to be a transvestite, the tiny man can immediately call for a nerve-jangling strategy reversal culminating in a quick exit. The problem with such a view of human behavior stems from the fact that, as examples become more complex, the population of miniature people within our heads continues to grow. Was there another little man who told the first one that he didn't fancy transvestites?

The second approach—the one that predominates in contemporary psychology—is based on the idea that voluntary movement as well as behavior is only apparent, and that both processes consist of nothing more than automatic responses to external stimulation. This, in essence, is the tenet of modern behavioral psychology. In place of the tiny man, we wind up with a model in which "will" or "desire" will be replaced by stimulus and response.

I am introducing behaviorism at this point in order that we might examine it in terms of the process approach we discussed in the last chapter. In addition, it is currently the most influential theory that claims to explain human behavior—and, on superficial examination, for good reason. Instead of endless speculation about imaginary little men in our heads, it presents us with a scientific way of studying organisms ranging from one-cell bacteria to creatures as complicated as ourselves, with over three trillion cells.

But is behaviorism a sensible level of organization in which to explain the complexities of human behavior? There are several reasons that make this unlikely.

For one thing, behaviorism dodges the most difficult questions about behavior. How do our brains inform us of reality, guide our perceptions, formulate our desires, and help us to reach our goals? The behavioral psychologist reading the previous sentence would be tearing his hair out at this point in response to some of the words I used. "In-

24

form," "guide," "desires," "goals"—to the behavioral psychologist such words are meaningless; in fact, more than meaningless, since they represent processes which either don't exist or are unnecessary in explaining behavior. I think such an orientation is more appropriate for studying machines rather than living organisms.

If you study the gearing mechanisms of a three-speed bicycle, you can quickly see how shifting the gears changes the ratio and makes pedaling easier or harder. With a ten-speed bike, the derailler mechanism makes a wider selection of gear ratio possible, resulting in even more variation in the effort required for pedaling.

If you accidentally snap off the back of your watch while playing tennis, you might later take a moment to study the wheels, springs, and escapement anchor with its pallets and gears. This requires a little bit more in the way of mechanical insight than do the previous examples, but with patience and the use of common sense most of us can understand something about how the winding mechanism drives the gears.

Each of these examples might be compared to the behavioral psychologists' method of studying behavior, with stimulus and response as the motif. Bicycles and watches are mechanical systems in which response can be predicted beforehand by studying the parts of the bicycle or the watch. In most instances the individual parts can even be ignored (until something breaks down and the watch or bicycle has to be returned to someone with a more practical and less philosophical interest in watches and bikes). For all practical purposes, though, function follows form, and knowledge of a stimulus suffices to guarantee a predictable response. According to such a view, a rat pressing a button to obtain a food pellet can be thought of as a series of levers, springs, and cogwheels.

In contrast to this, imagine yourself engaged in an amateur archaeological dig somewhere in the Sahara Desert. To your great surprise and delight, your party succeeds in finding a previously undiscovered crypt in which are set out the belongings of a long-dead monarch. Among the possessions is a chessboard with the pieces arranged as they were left in the middle of a game that started two thousand years ago. At this point your understanding of what is going on demands an entirely different process than understanding the workings of a watch or a bicycle.

The chess pieces stand in symbolic rather than mechanical relationship to each other. The pieces are only so much wood and ivory, and their function cannot be predicted from their form. Only the knowledge of the process of chess can provide an understanding of what is going on.

Think of behaviorism as most appropriate to the bicycle and watch examples: Behavior can be predicted from studying the inner actions of gears and flywheels. Psychobiology, in contrast, concentrates on describing, as best we are able with available technology, how behavior stands in a symbolic relationship to events going on in the brain. Let's take a close look at some classic behavioral psychology experiments and see if their results can truly be explained in terms of stimulus and response.

One of the key notions of behavioral psychology concerns the role of reward. In a typical experiment an animal is placed in a box equipped with a bar-pressing apparatus (the Skinner box, named after psychologist B. F. Skinner). Pressing a lever causes the arrival of a pellet of food. Repeated bar-pressing causes more food pellets until eventually the animal is filled and stops the bar-pressing. What accounts for such an experiment's appeal is its utter simplicity: The act of pressing a bar (the stimulus) is followed immediately by the arrival of the food pellet (the response). In such a situation there is no need to postulate variables such as a "decision," the existence of the "mind," or the exercise of a "will"—in short, the little man inside our heads has been done in.

Such experiments derive their inspiration from the Russian physiologist Ivan Pavlov, who in the early twentieth century strapped dogs to a laboratory table at Moscow University in order to precisely record such responses as their salivation, without interference from more complex behaviors. As we all know, the dogs salivated when a bell was sounded at the time of their feedings. Later, the sound of the bell alone stimulated salivation. An over-

26

simplified understanding of conditioned reflexes, based on Pavlov's work, led to the explanation of behavior in mechanistic terms. For simple tasks such as salivation, total immobility on a laboratory table was used. With more complex tasks such as bar-pressing, the Skinner box was devised. In each case "behavior" is the object of study.

Anyone who has ever studied an animal in a Skinner box will note first of all that bar-pressing occupies a minimal amount of the animal's time. In addition, other highly characteristic forms of behavior regularly occur: running, rearing up on the hind legs, sniffing, and exploration of the box. Each of these separate behaviors is treated by behavioral psychologists as, for all practical purposes, irrelevant. The only behavior worth noting seems to be how many times the animal will press the bar and receive its reward.

What is actually occurring in such a highly artificial situation is a selective definition of "behavior" on the part of the experimenter. Since he or she is interested in bar-pressing and pellets as a measure of stimulus and reward, those aspects of the animal's behavior that do not fit into a stimulus/reward paradigm are ignored. Now let's take the rat out of its cage for a minute and see if the animal's behavior can be explained simply in terms of pellets and bar-pressing.

If the laboratory rat *(Rattus norvegicus)* is removed from its cage and put on a table, it begins by sniffing and investigating each object it encounters. At any given moment its general behavior is unpredictable, but its movements will eventually cover the whole table. When these exploratory movements are photographed and analyzed with the help of a computer, we find that they are not random but ordered: They tend to take the animal to unfamiliar areas.

To prove this, the rat can be placed at the stem end of a Y-shaped maze and released. He first moves to one end of the maze, say the left, and travels to the end of it. The rat is then brought back to the starting point and released again. This time, like a curious shopper, the rat walks into the right arm of the maze. Such behavior doesn't fit into the pattern of stimulus/response: There is

27

no cheese at the other end of the passage, no pellets to be gotten from pressing a bar, no electric shock to be avoided. In essence, such behavior doesn't fit the explanations of behavioral psychologists and, as a result, is ignored.

An alternative explanation of the rat's behavior replaces stimulus/reward concepts with an unbiased look at the adaptive value of novelty. Is it possible that novel stimulation has survival value?

To find out, psychologist David Hebb offered rats two routes to food—one direct and familiar, the other indirect and new. Contrary to behavioral psychology patterns, the rats did not always choose the familiar path, but at least as often chose the alternative, unexplored route. Such behavior maximizes the animal's range of movement independent of any special incentive.

Biologically, novelty-seeking behavior would have no point if the behavior did not lead to some useful consequences. In a stimulus/response model the consequence is an immediate one (food)—hence, its appeal. From here, behaviorism assumes that learning always depends on reward: food, sex, shelter, etc. But is it possible that novelty is itself rewarding? There are suggestions that it is.

To prove it, laboratory animals are allowed to move freely about in an environment such as a maze. They are then taught to run from one point to another in the maze to obtain food. In such a test, those animals with prior experience of moving about in the maze learn the task more quickly than animals without prior exploratory experience. Since none of the rats could possibly know beforehand that the maze was later going to be used for food-seeking behavior, the results suggest that the opportunity to explore is itself rewarding. Prior exploration allows an animal to develop new patterns of responses that can later be used to meet future demand situations—e.g., finding food. The biological utility of novelty can also be demonstrated on the behavioral psychologist's home court, the Skinner box.

Dr. G. B. Kish and J. Antonitis placed laboratory mice in the Skinner box. In this case, however, instead of rewarding bar-pressing by food pellets, these researchers offered nothing more complicated than a clicking sound or a change in the level of illumination. In all cases the mice developed the bar-pressing habit.

Similar behavior can be observed in primates. Rhesus monkeys will go to great trouble in order to watch a scene filled with action and movement. Bar-pressing can be used to demonstrate preferences for certain colors and levels of brightness. One researcher has dubbed this "the factor of pleasure." In addition, the monkeys will bar-press to create more complex situations—for example, movies instead of a still photograph or a plain field of light. In each case the determining factor is believed to be the information content of the stimulus.

In introducing these observations I am not suggesting that complicated human behavior can be understood in terms of Skinner boxes. In fact, I'm suggesting just the opposite. The experiments with light and sound underscore the oversimplified and highly selective kinds of information that behavioral psychologists have been collecting for years. After proving that laboratory animals will bar-press for reward, they neglected to observe that the animals would also bar-press for novelty. In essence, novelty itself is rewarding.

In recent years, psychobiologists have isolated the portion of the brain that responds to novelty. The hippocampus, named after the Greek word for seahorse (which this part of the brain resembles in shape) has been shown to be critical in memory formation. Naturally, an animal could not survive if it were incapable of registering which part of the environment had already been explored, or learning from experience those areas inhabited by predators. To do this the animal must first form a memory of its exploration, and this task is performed by the hippocampus.

Experiments by Dr. O. S. Vinogradova suggest that the hippocampus is necessary for information to be registered independent of any reward or punishment considerations. Instead, hippocampal function is closely related to the search for novelty. If the hippocampus is destroyed, laboratory animals will return monotonously to areas of the cage where they have already visited, while seeming to ignore unexplored portions of their environment.

If recordings are made of single cells in the hippocampus of a normal rat or mouse, they are found to fire selectively in response to novel changes in the environment. EEG recordings demonstrate slow, or even absent, hippocampal discharges when animals are immobile, with gradually increased firing during movement reaching the greatest discharge rate during large-scale exploratory excursions into completely new environments.

In other words, the behavioral psychologist's explanation for behavior based on stimulus and reward is incorrect, even at the level of rat and mouse behavior. It resulted from incomplete, and sometimes incorrect, behavioral observations. In addition, neurophysiological techniques, such as single-cell recordings of the hippocampus, enabled psychologists to isolate novelty rather than reward as the factor that "turns on" the hippocampus to form a memory.

For over twenty-five years, human behavior has been explained on the basis of incorrectly interpreted observations of rats and mice. Operative conditioning has spawned a new field of behavioral psychology that explains human personality in terms of increasingly alluring and sophisticated rewards. Aversive therapy in our prisons, token economies in our mental hospitals, hierarchies of rewards in business—all are spinoffs from the operant-conditioning experiments based on stimulus and reward.

Such explanations have never been satisfactory to anyone who closely observes the factors that motivate human behavior. Mountain climbers and hang gliders are only two examples of people who demand stimulation and novelty even at the risk of danger or physical discomfort.

Why, then, has American psychology been dominated for over twenty-five years by a behavioral paradigm that ignores the greater part of animal as well as human behavior? Modern psychobiology provides the answer to this technical question based on a re-examination of the neural basis of action and movement.

If your eyes weren't moving right now, you wouldn't be able to visualize the words on this line. As you reach the end of the page your hand reaches down and turns the page, your head turns slightly upward and to the left, and you begin to read the next page. The traditional behavioral explanation for this series of events goes something like this: Light is converted via electrochemical impulses which, after relay into the brain, stimulate part of the motor cortex to direct the action of the muscles of the right hand to reach out and grasp the page and turn it. The muscles of the neck and shoulders then bring about a slight shift to the left and you're ready to begin reading the next page. In such a model the emphasis is on observed behavior, the motor act of grasping the page, lifting it and finally turning it. Although seemingly reasonable, this approach ignores the most important components of the act of page-turning: what you are reading, whether or not you are enjoying it enough to turn the page and read on.

In addition, the actual behavioral observations are static and unimaginative.

As you read this page you're not sitting in the same position as you were yesterday. Perhaps then you were sitting home in your living room by your favorite lamp, balancing the book in your lap while you sipped a martini. Today you may be on a crowded airplane with the book propped on the tray attached to the seat in front of you. In these two instances your page-turning movements vary widely, with your hand traveling different distances to reach the page. In both cases the results are the same but the movements are not. This variation in detail between two descriptively similar movements ("He turned the page of his book") is not accidental but an essential part of every move you make.

Page-turning requires an ongoing comparison involving information about the position of your hand in space and its relationship to the position of the book. Rather than an automatic task which is done only one way, page-turning is a movement of almost infinite complexity. You can use your right hand or your left hand; if both are engaged, you can lean forward and do it with your tongue; if you are reading along with somebody else, you might even take advantage of their movements and have them turn it for you. In each case the task of page-turning is carried out using different muscles and starting out from different positions (Will you use your forefinger or your middle finger?). In addition, a constant supply of information is required: checking the hands' position in space, making corrections for any over- or underreach based on the relative positions of the hand and book at any one moment. If you've ever watched a blind man tap his way along the pavement with his cane, you've seen a similar process. His movements depend on what his ears and hands tell him about the contour of the ground below.

Movements, along with behavior, can never be adequately explained in terms of stimulus and response, since both are constantly changing. Reaching for a coffee cup on the dining-room table involves an entirely different motor program than climbing up on a stool and using a ruler in order to reach a cup at the back of the highest shelf in the cupboard. In both cases the response is the same, but the behavior and movements necessary to carry it out are variable. Put another way, the motor act of reaching for the coffee cup is inseparable from the sensory information that is received about the position of the cup in relation to your hand. Both input and output are part of a behavioral unit in which one cannot be intelligently described without referral to the other. If the brain is deprived of information about the location of your hand—let's say by the injection of novocain into the nerves of your arm—reaching for the cup becomes a laborious, clumsy task which will occupy all of your attention. You'll be literally forced to watch your hand and guide it visually as it approaches the table. A similar situation can result from a stroke in a part of the brain which receives and integrates sensory information. But in this case the situation is relatively permanent and extremely handicapping.

Behaviorism's failure stems from its selection of the wrong conceptual framework and level of organization. As the nervous system becomes more complex, increasingly sophisticated models are required to explain movement and behavior. To rely on stimulus/response models is like claiming that there is no difference between the inner structure of a pocket watch built in 1910 and the quartz crystal digital watch of today. While it is true that both serve the same function of telling time, they do so by entirely different mechanisms. As we have said, if you remove the back of the pocket watch, you can figure out how it works by observing the mechanism of the gears, wheels, and springs. The quartz watch, on the other hand, is not a mechanical system but an electrical one, which does not depend directly on the shape of its components. A transistor or a condenser can take on a variety of shapes and work just as well. No doubt this is equally true of nerve cells. In short, "Function follows form" and "Response can be predicted from stimulus" cannot be applied to nervous systems. Even the simplest examples of behavior are more complex than behavioral psychologists have led us to believe. Let's look now at some of the things we know about the brain's functional organization.

3
The Working Brain

The Working Brain

> We often think that when we have completed our study of *one* we know all about *two,* because "two" is "one and one." We forget that we still have to make a study of "and."
>
> A. EDDINGTON,
> *The Nature of Physics*

To be reading this you must be awake. Reading, along with every other brain activity, except dreaming, depends on wakefulness. You might think of wakefulness as similar to the maintenance of a degree of muscle tone without which a muscle becomes weak, flabby, and eventually paralyzed. Wakefulness has in fact been described as the maintenance of a state of optimal "cortical tone" necessary for the carrying out of all organized mental activity. When the tone is lost we may be asleep or in a coma. In either case, our response to our environment is reduced and there is a dramatic fall-off in our brain's functional capacity.

Wakefulness depends on a special formation in the brainstem known as the reticular formation. If a cat is sleeping quietly on the back porch, its reticular formation is at rest, sustaining only a minimal background activity, which can be recorded electronically. When someone calls the cat's name, an excitation wave begins in the reticular formation, spreads upward in the brainstem, first activating the auditory portions of the brain and then spreading to

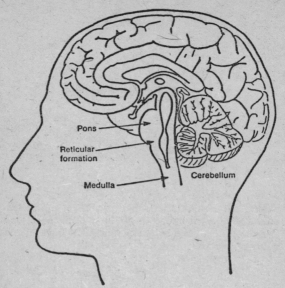

FIGURE 1.

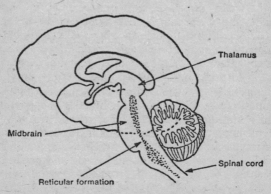

FIGURE 2. Location of the reticular formation in the human nervous system.

36

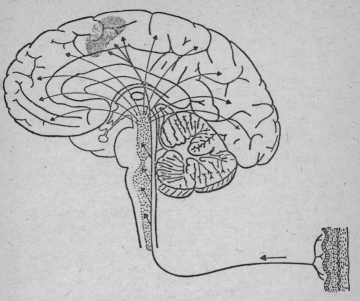

FIGURE 3. The reticular formation, a tiny nerve network approximately the size of the little finger, lies deep within the brainstem and is responsible for the maintenance of alert wakefulness.

other parts of the brain until eventually it activates the motor cortex, resulting in the cat's springing up and heading toward its bowl of milk. The organization of cells in the reticular formation is designed for the rapid spread of excitation throughout the whole brain. The cells are small and connected to each other by short processes. If the reticular formation were spread out on a flat surface, it would resemble a net. Think of a net in three dimensions and you have a mental picture of the reticular formation.

Impulses can go both ways in the reticular formation. (See Figure 3.) If you think about an ice cream sandwich right now, the idea of it—primarily a product of your cerebral cortex, the most recently and highly developed part of the human brain—will stimulate a wave of excitability which will descend into the reticular formation, activating the muscles of your legs and hands and eventually lips and jaw as you proceed to the refrigerator, return to the living-room couch, and continue reading about the reticular formation while you eat your ice cream sandwich.

One of the ways we learned about the reticular formation's activity came from deliberately destroying it in experimental animals. After such an operation, a cat will remain asleep in his cage. He will not respond to his name, even to a loud noise or a painful pinch. Nothing, in fact, is capable of arousing him.

From such observations we have learned that the reticular formation is a powerful brain regulator that determines the level of wakefulness and controls the functional state of the brain from moment to moment. So far, though, no one has shown what differences, if any, exist within the reticular formation that explain why some people require more sleep than others or possess more energy. Its function seems to be that of an overall alerting center responsible for the maintenance of wakefulness. The great Russian psychologist Alexander Luria, whose work provides the basis for this chapter, has called the reticular formation the first functional unit of the brain.

With the discovery of the reticular formation a new principle of brain organization was introduced. Up until that time, most psychobiologists concentrated on the cerebral cortex. It seemed reasonable to assume that we could discover the principles of brain organization by concentrating on the cortex and ignoring the lower brainstem areas, including those portions containing the reticular formation. We now know from animal experiments, as well as from observations of people with brain tumors in the area of the reticular formation, that wakefulness depends on the brainstem more than it does on the "higher" centers of the cerebral cortex. It is possible, in fact, for a small brain tumor or blood clot the size of a match head (if it is located near the reticular formation) to bring about a state of irreversible coma even when the cerebral cortex remains completely normal. This was probably the initial cause of Karen Quinlan's puzzling coma, despite an originally intact cerebral cortex. The principle involved here concerns the way the brain is organized in a vertical dimension. Wakefulness depends on a vertically arranged two-way system capable of conducting information from outside our bodies inward, as well as from the cortex downward to the nerves and muscles controlling the body.

Naturally, more is required for reading and understanding this chapter than just a state of wakefulness. As you proceed from the beginning of this sentence to the end, your alertness may enable you to detect the misspelled word I've hidden in the sentence. Your ability at this point to recall what that word is, however, depends on the brain's second level of functional organization: the reception, analysis, and storage of information. This involves a much more specific task than remaining awake. The part of the brain involved in carrying out the process is different as well.

While the reticular formation is located in the brainstem, the information capacity of the brain is located principally in the cortex. Instead of a nerve net such as was seen in the reticular formation, the cerebral cortex contains isolated nerve cells connected to each other by long threadlike processes which are capable of conducting an electrochemical impulse from one cell to another. Here, too, the brain displays a hierarchical structure. Vision, for instance, depends on impulses transferred along the optic nerve to visual analyzer cells located in the primary visual area at the back of the brain. Surrounding these cells are visual-association cells, which are important for organizing what is seen into meaningful visual patterns. As the final step in the hierarchy, a third group of cells, surrounding the first two, is responsible for co-ordinating what we see and hear into a single perception. When we encounter a friend coming toward us on a street, we may recognize him from a block away. Later, as he approaches, he may call out to us and we recognize his voice. At that moment the brain cells responsible for hearing and vision, co-ordinating together, give us the composite perception of a "friend." Such co-ordinative efforts can be disrupted, however, when the nerve cells in, let's say, the visual area are prevented, after a stroke, from spreading toward the association cells for

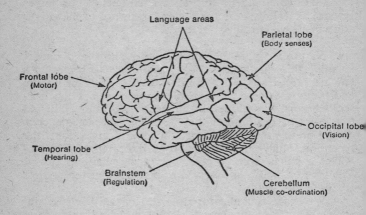

BRAIN SURFACE

Language areas

Parietal lobe
(Body senses)

Frontal lobe
(Motor)

Occipital lobe
(Vision)

Temporal lobe
(Hearing)

Brainstem
(Regulation)

Cerebellum
(Muscle co-ordination)

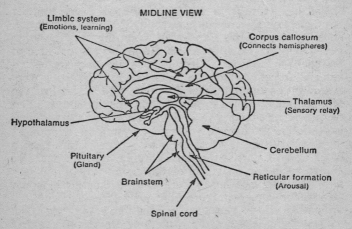

MIDLINE VIEW

Limbic system
(Emotions, learning)

Corpus callosum
(Connects hemispheres)

Thalamus
(Sensory relay)

Hypothalamus

Cerebellum

Pituitary
(Gland)

Reticular formation
(Arousal)

Brainstem

Spinal cord

FIGURE 4. The cerebral cortex with its major functional areas.

hearing. In such cases, the victim of a stroke in the visual-association area may not recognize his friend until he actually hears the friend speak. Recognition in such cases becomes dependent on hearing.

The second functional unit of the brain conforms to several basic laws. First, as with the reticular formation, we have a hierarchical structure that becomes less specific as we proceed upward. The primary visual cells respond only to visual input. At the second level of the hierarchy, vision is co-ordinated and organized. Finally, at the highest level, the nerve cells located in the visual and auditory as well as other perceptive areas overlap, leading to an integrated perception of the friend we spot from a block away.

In addition, different functions tend to become localized principally on one side of the brain or the other. Speech in most of us is in the left hemisphere, while we organize space in our right hemisphere. We will say more, in Chapter Ten, about the right- and left-hemisphere phenomenon; but for now, think of it as an inborn tendency to localize highly specific brain functions in one hemisphere. This is based on the well-known principle that a job is likely to be done better by one tightly knit group of skilled workers than by several fragmented groups working independently of each other.

Reading this chapter demands, as we discussed, wakefulness and the capacity to receive, process, and remember information for later retrieval. But mental processes hardly stop at that. Perhaps at this point you may have decided that this discussion has become too technical and you are considering moving ahead to the next chapter. If so, you are criticizing and reacting. This may result in a series of actions such as turning the pages forward or checking the chapter titles in the Table of Contents. If you do so, you are engaging the third and final functioning unit of the brain, which is concerned with willed action.

The outlet channel for action is the motor cortex, which contains giant nerve cells with long processes running down the spinal cord and eventually connecting with all the muscles in the body. These giant cells, when stimulated with an electrical probe, cause solitary twitches of muscles, which are not of much use in terms of co-ordinated activity. Slightly in front of the giant cells are the premotor cells, which are the next step in the hierarchy and prepare motor programs for transmission to the giant cells. The motor cortex can be thought of as a computer; the premotor cortex, its computer program. The result is an increasingly controlled sequence of muscle movements rather than an isolated and chaotic discharge.

But for every computer and computer program there must also be a programer. At the most anterior part of the brain are the prefrontal lobes, which play a decisive part in firming up intention, deciding on action, and regulating our most complex behaviors. We've learned about the action of the prefrontal lobes from the results of psychosurgical experiments in the 1940s and 1950s in treating certain forms of mental illness. The operations, aimed at reducing the patient's symptoms, exacted a dreadful price. Loss of initiative, placidity, a desire for sameness in the environment—these were only some of the unfortunate results of cutting the fibers of the prefrontal lobes.

The principal role of this area of the brain seems to be the regulation of background tone necessary for behavior. It is, in fact, a superstructure that energizes all the other parts of the cortex and is an overall regulator of our behavior. If we had to describe the prefrontal lobe in one word, it would be *purpose*. Goal-directed behavior is lost after destruction of the prefrontal areas.

Observation of patients with destructive lesions in the frontal lobes—either deliberately induced via psychosurgery, or as the result of strokes or tumors—reveals an apathetic, listless person without interest in his surroundings. Easily distracted by trivial events around him, the prefrontal-lobe patient will break off whatever activity he is engaged in at the moment and direct his full attention to distractions in the environment. Such a person moves slowly and deliberately and often fails to grasp the essentials of a situation. In addition, since he lacks long-term goals and is absorbed completely in the events of the moment, no plans are made for the future. If, for instance, the last portion of milk in a container has just been poured out, the prefrontal-lobe patient is likely to replace the empty container in the refrigerator, seemingly incapable of appreciating that it's now time to buy another container of milk.

Complex goals are replaced by simpler ones, interests dwindle and horizons shrink. The verbal contributions of a person with prefrontal-lobe disease are meager, trivial, and uninspiring. At a later point, thanks to narrowing intellectual horizons and the absence of future-directed behavior, such a person can be extremely puzzling to friends and acquaintances. Eventually, conversation takes on a wooden "here and now" quality that admits of little deviation, either reminiscences about the past or projections into the future. At a certain point, the emphasis on trivia, easy distractibility, and loss of initiative leads to an almost complete communication breakdown. Conversation becomes reduced to a series of grunts and sighs irascibly offered in response to only the most persistent efforts on the part of friends and relatives. Accompanying this, the patient, seeing no purpose in conversation which to him appears senseless and incomprehensible, drifts easily into fits of bad temper. Tantrums of childish stubbornness alternate with periods of complete mutism until, finally, the prefrontal-lobe patient is left to live out his days in a narrowly restricted intellectual world.

From studies of patients with prefrontal-lobe disease, psychobiologists arrived at an understanding of the prefrontal lobes as the coordinators of higher mental activity, particularly activity which depends on speech as a motivating factor. Anticipation of future events and necessities (the purchase of a container of milk) requires the formulation of complex behavioral programs which are heavily dependent on speech. A child is told not to return a used milk carton to the refrigerator and soon learns that, in this case, a more complex behavioral repertoire is required: a trip to the store to purchase a new one. In patients with prefrontal-lobe disease, however, all complex behavioral patterns are replaced by simpler ones. Even if the person can be motivated to go to the store for a new container of milk, he may become distracted on the way and return empty-handed several hours later. In action as well as conversation, the prefrontal-lobe victim is limited to the repetitive carrying out of behavioral stereotypes.

In summary, the brain is best understood in terms of three functioning units: alertness, information processing, and action. While each has its own role in brain functioning, the harmonious interaction of all three is required for optimal functioning. If we are asleep, we obviously can neither register impressions nor act. If our actions are interfered with—say, by a prefrontal lobotomy—then our capacity for handling information is seriously impaired and we lack the necessary "verve" to act out our wishes.

Each form of conscious activity depends on the combined action of all three functional brain units. Mental activity thus takes on the quality of a dynamic process. In such a model it doesn't make sense to ask where a particular brain activity is located. We don't perceive, for instance, in one part of the brain and act in another, while comparing perception and action in yet a third part. Such incorrect models compare the brain to a camera or a tape recorder which passively registers the events in the environment around it. As we discussed in Chapter One, a *thing* explanation is being sought to describe activity which is essentially a *process*.

Even something as simple as sitting quietly looking out a window requires a delicate interplay of all three brain units. Scanning movements of the eyes are taking place all the time as we look for essential clues about what's going on in front of us. At the same time, we are registering and synthesizing sights and sounds which enable us to realize, for instance, that we are looking out on a playground. If we observe a toddler fall off a swing, the combined action of all three functioning brain units will result in our hurrying out to comfort him. Trying to pin down what part of the brain is involved in the perception of the child's fall is like trying to describe the precise location and movement of a subatomic particle.

In addition, the brain operates in a hierarchical pattern. Vision at first consists of lines and patterns of light. When these are co-ordinated in the visual association area, a meaningful visual pattern emerges which can then be

44

linked up with other association areas in distant parts of the brain to produce perception.

At this point, let's discuss some aspects concerning the brain's origin and, in the process, I will try to provide some partial answers to questions that have not yet been raised.

PART
2

4

Answers to Questions Which Have Not Yet Been Raised

There are works which wait, and which one does not understand for a long time; the reason is that they bring answers to questions which have not yet been raised; for the question often arrives a terribly long time after the answer.

OSCAR WILDE

An American professor teaching at a Japanese university was leaving a faculty meeting after taking part in a lengthy and detailed discussion. On the way out of the meeting, the professor remarked to some of his Japanese colleagues that several of the proposed changes were almost certain to be enacted. "Why do you say that?" one of them asked. The professor proceeded to enumerate a long list of distinguished teachers who had spoken in favor of the proposal. His colleagues agreed but remarked: "While it is true that the people you mentioned spoke in favor of the motion, the meeting arrived at just the opposite conclusion and the motion was defeated. You correctly noted and understood the words spoken, but you didn't understand the silences between them."

It's an everyday experience that important events are often decided on the basis of the "silence" between words. Who could feel confident in trying to identify the nonverbal factors involved in such decisions as the choice of our career, the person we marry, the political candidates we vote for, the identification of one colleague as "sincere" and another as a hypocrite?

Even though we like to think of ourselves as rational creatures, we can't help but be impressed by the number of irrational things we do every day. Although we may attempt to make our choices more acceptable by introducing such terms as "intuition" or "gut reaction," the fact remains that when we try to explain many of our actions, we are often reduced to a few platitudes followed by an embarrassed silence. In addition, there are great differences in people's susceptibilities to irrational processes. Certainly anyone who has come into contact with an obsessive-compulsive has learned firsthand the powerful influence an unlocked door or an unclean hand towel can exert on a person's behavior.

At the other extreme, some people seem so much at the mercy of impulse and wishful thinking that they engage in behavior that is not only self-destructive but stupid. An accountant who was arrested for allegedly writing six hundred company checks to her sister over an eight-year period remarked casually that she didn't think she'd be caught. Apparently she was able to sustain this irrational belief despite daily contact with computers and other information-storing devices that have become part of everyday business practices.

In the past, such puzzling behaviors have been written off by facile references to "unconscious processes"— the criminal really wanted to be caught, a "compulsion to confess," etc. This, of course, provides an explanation for seemingly inexplicable behavior while at the same time doing away with any need to modify our views of people's responsibility.

Recent brain-research findings provide an underpinning for our intuition about the powerful influence that irrational forces appear to have in shaping our behavior and our lives. The brain scientist who has been most influential in this area is Dr. Paul MacLean, Director of the Laboratory of Brain Evolution and Behavior of the National Institute of Mental Health. MacLean's laboratory at the Institute is located some twenty miles from the main campus, deeply secluded in the rolling Maryland hills. During my first visit, on a lovely sunlit May morning,

MacLean spoke of his work while conducting a tour of his laboratory.

"In its evolution the human brain expanded in a hierarchical fashion along the lines of three basic patterns. These three formations are markedly different in chemistry and structure and, in an evolutionary sense, are eons apart."

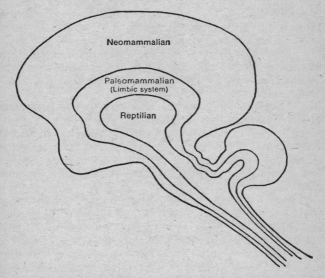

FIGURE 5. The hierarchy of three main brain types which, through evolution, have become part of our biological inheritance.

According to MacLean, the brain is somewhat like an archaeological site, with the outer layer composed of the most recent brain structure, the cerebral cortex, which is highly developed in primates and reaches its greatest level of complexity in humans. Deeper layers of the brain contain structures of our earlier evolutionary forebears, the reptiles and mammals. According to MacLean, we are the possessors of a triune brain—not one brain but three, each with its own way of perceiving and responding to the world.

As a further development of the archaeological model, MacLean, whose earlier premed training included geology and philosophy as well as biology and chemistry, has spent a lifetime searching for and describing *paleopsychic processes,* a term of his own coining that refers to ancient forms of animal mentation that we have inherited from our evolutionary ancestors. At the basis of MacLean's research is the assumption that observations of animals, particularly reptiles and subhuman mammals, are relevant to understanding human behavior.

"The three brains amount to three interconnected biological computers, each having its own intelligence, its own subjectivity, its own sense of time and space, and its own memory and other functions."

The first, most primitive, of these three brain computers MacLean has dubbed the R-complex, a deeply placed series of brain structures that make up almost the entire mass of the brain in lizards and reptiles. As we climb the evolutionary ladder, the R-complex becomes less conspicuous, until in humans it is greatly overshadowed by the cerebral cortex.

Now let's take up the first portion, which corresponds to the brain in reptiles. "Lizards and other reptiles provide examples of complex prototypical patterns of behavior commonly seen in mammals, including man," MacLean comments, as we enter a small laboratory filled with dimly lit terrariums. He points to a terrarium containing a small lizard similar to a chameleon. "Watch him!" Slowly the lizard begins to bob its head up and down as it remains perched on a twig. In another few seconds, a second lizard catches the eye of the first, and both begin a ritual display which includes further bobbing and push-up movements accompanied by a puffing out of their neck scruffs. "That's the beginning of a rather aggressive greeting," MacLean chuckles, "a sort of how-do-you-do in lizard language."

Using the latest electronic observation devices, MacLean and his coworkers have recorded in videotape thousands of examples of lizard behavior. From this and other data, he has dissected out over twenty repeatedly observed behavior patterns concerned with either self-preservation or the survival of the species. Included are such activities as the establishment of territory, growling, foraging, hoarding, greeting, and the formation of social groups.

"Except for altruistic behavior and most aspects of parental behavior, it is remarkable how many behavior patterns seen in reptiles are also found in human beings."

As MacLean is quick to point out, this is not quite the same thing as saying that humans are no different from reptiles (although we each can probably think of human examples that might support such a thesis!). Rather, reptiles display countless behavior patterns commonly observed in human beings. This should not be surprising, according to MacLean, since we still carry around with us, like forgotten luggage, the brain structures of our reptilian ancestors. Reverting to our computer analogy, it seems reasonable to suppose that this ancient brain structure is contributing its own "program" for influencing our behavior. Thus, MacLean postulates that such human characteristics as ritualism, awe for authority, social pecking orders, even obsessive-compulsive neuroses, may be partially caused by our reptilian brains.

While all of this is very pat in theory, how does one prove it? One way comes from classical ablation studies—literally cutting out the parts of the brain in question and observing what happens. But there are fallacies in this, as MacLean is quick to point out. For one thing, if too much brain tissue is cut out, other aspects of the animal's behavior may be affected. For instance, if an ablation destroys an animal's ability to move, or blinds him, then any studies on grooming or the tendency to establish homesites are going to be worthless. For such experiments to be useful, they must affect a comparatively isolated piece of behavior without changing the animal in other significant ways.

MacLean's observations of the squirrel monkeys provided the opportunity to test his theory on the function of the R-complex. In a social situation squirrel monkeys demonstrate a complicated form of "display" which serves in both courtship and aggression; it is a coin in the commerce of social domination. A more powerful squirrel monkey signifies his dominance by display, which includes a high-pitched cry, the spreading of its thighs, and the thrusting of an erect penis toward the other animal. Despite the overt sexual tone of much of this (an amused MacLean recalls expressions of embarrassment on the part of general audiences when shown slides of his squirrel monkeys), the act has little to do with sexuality. Rather, it is a measure of social dominance, seen in its most dramatic form when a new monkey is introduced into an established colony. Within a few minutes, all of the monkeys begin to display to the new animal, which must be quietly submissive and bow its head, or suffer vicious attack, even death.

The tendency for squirrel monkeys to "display" is an innate characteristic observed within two days after birth, even in monkeys exposed only to their mothers. In the somewhat bitterly contested nature-nurture controversy, display in squirrel monkeys is admitted by almost all students of animal behavior as a programed preset pattern which is relatively independent of the environment.

If any lingering doubts still remain, MacLean's next observations put them at rest. One variety of squirrel monkey, the gothic variety (named after the pattern of hair on the monkey's eyes, which resemble a gothic arch), displays to its own reflection in a mirror. Moreover, the display can be measured and graded according to such criteria as loudness of vocalization, the extent of thigh-spreading, etc. Using such measurements, MacLean set out to discover the effects on display of ablating different parts of the monkey's brain, being careful to avoid destructions that would result in paralysis, blindness, or any other defect that might interfere with motor power or sensory perception. In many cases, large destructive lesions had no perceptible effect on display. When MacLean cut into the R-complex, however —specifically a small nest of cells known as the globus pallidus—the animal stopped displaying.

"This effect suggests that the R-complex may be part of a neural repository for behavior that is specific for particular animal species," MacLean comments.

Peering into a microscope where Dr. MacLean had placed one side of a hamster brain, I see two darkly stained walnut-shaped objects that appear to be in the grip of a huge nutcracker of paler surrounding tissue.

"The dark areas are the globus pallidus and the caudate, important parts of the R-complex. As you can see, they can be set off by selective staining, which shows a remarkable chemical contrast between the R-complex and the rest of the brain. That nutcracker you refer to is obviously what would correspond to the cerebral cortex in man."

MacLean's present research is directed to observations in animals as far-ranging as hamsters, monkeys, and turkeys. Already he has built up a telling argument that behavior in lower animal forms is programed via the R-complex. Although the trigger for the behavior may be something in the environment (another squirrel monkey beginning a display, for instance), the neural machinery for carrying it out is preset. In some cases the necessary stimulus may be extremely specific.

"On a recent visit to the zoo, I was unable to attract by any means the attention of a common lizard until I sketched a shadowgram (of a lizard) and held it up to the window. Immediately it came over and gave its full display." MacLean's explanation for such results suggests as much about some aspects of our perceptions and psychic pasts as it does about lizards. "Here, it seems, we are almost next door to the ink blots of a Rorschach test! Is it possible that cubistic art—depicting, for example, the entire human figure in two dimensions (eyes and buttocks in the same plane)—owes some of its appeal to the portrayal of the archetypal patterns and partial representations?"

When we speak of archetypal patterns and partial representations we enter into a world not usually associated with brain scientists and white coats, but a world of Jungian speculations and Plato's cave (whose human inhabitants see "only their own shadow, or the shadows of one another"). MacLean is well aware of the implications of some of his findings. "Walking as we do through a world of shadows, we shall be lucky if we ever see even the broad outlines of neural mechanisms underlying cryptopsychic processes that are the very guts of everyday forms of human behavior."

When I remarked that such a comment was more typical of a philosopher, MacLean smiled. A trim, scholarly appearing man now in his early sixties and a veteran of over thirty years of brain research, MacLean went on to explain that his earliest interest was in philosophy. It was a course on the philosophy of science, taught by A. Northrup of Yale University, that "converted" MacLean to medicine and science.

Although firmly grounded in the principles of vigorous scientific proof, MacLean's interests obviously extend beyond those of a chronicler of lizard behavior. He points out, for instance, that formalities and rituals are an integral part of our daily lives. "It is traditional to belittle the role of instinct in human behavior, but how should we categorize those actions that stem from a predisposition to compulsive and ritualistic behavior, a proclivity to prejudice and deception, a propensity to seek and follow precedent as in legal and other matters, and a natural tendency to imitation?"

Left at this, MacLean's research would be little more than a novel way of thinking about some forms of human behavior. The next time we watch "All in the Family" we might suddenly envision a lizard ensconced in Archie's favorite chair. Or we could carry on about precedent and pecking orders, reducing the United Nations to a mini zoo of displaying ducks, turkeys, and hamsters. But that's only part of the story—in fact, only one third, since we are the possessors of a triune brain, and the R-complex describes only the most primitive part of it.

Our second brain is a bit easier to get into than the R-complex, since few of us have close rapport with lizards or spend much time observing their behavior. Would such an empathy be of any value? or would it tell us anything about the personality of someone who displayed such expertise? In MacLean's case I think not, since his humor, friendliness, and all-round helpfulness to an author are distinctly "warm-blooded." On the other side of the question, I recall an animal trainer in Jamaica who put on a display of alligator wrestling and alligator hypnotism. Later, in a discussion with him at a bar, I noticed he rarely blinked and sat motionless for long periods of time. It was hard to believe that a few hours before, this same man had clamped a lightning-fast hammerlock on the jaws of a deadly alligator. Other examples of this are my Parkinsonian patients, who can sometimes be diagnosed by their infrequent blinking, the "reptilian stare" described by nineteenth-century neurologists. Certainly in the case of Parkinson's disease, MacLean's logic holds up: The defect is in the R-complex and consists of a deficiency of the neurochemical dopamine usually found in the part of the R-complex called the substantia nigra.

But to return to our second brain. Obviously such things as display behavior, homesite building, and foraging for food are all right and necessary in their place. But, on the whole, such a world would be pretty boring after a while, with most of us demanding something beyond pre-programed behavior. We'd insist, in fact, on some way to diversify life's experiences, upping the ante in the game of life. The evolutionary development in lower animals of a cerebral cortex accomplished this and was nature's way of providing the reptilian brain with an "Emancipation Proclamation." Henceforth, higher animals would be free from the dictates of that unreasoning master, the R-complex. "The reptilian brain is filled with ancestral lore and ancestral memories," according to MacLean, "and is faithful in doing what its ancestors say, but is not a very good brain for facing up to new situations. It is as though it were neurosis-bound to an ancestral superego."

We can also think of brain evolution as the development of increasingly complex electronic devices to record the environment. If the reptilian brain provided a radar, the next step in brain development, the mammalian (or limbic) brain was more like one of our first black-and-white television sets. (See Figure 5.)

Both the reptilian and limbic brain are concerned primarily with self- and species-preservation. The brains of mammals contain the neural circuitry necessary for emotional experience. "Tell me where is fancy bred, in the heart or in the head?" can now be definitely answered: not only in the head, but in a particular part of the brain known as the limbic system.

The limbic system is a series of brain structures surrounding the reptilian R-complex. It forms a cap, or limbus, around the brainstem. Although it is not necessary to become a neuroanatomist to understand how the limbic system works, a tiny dose of neuroanatomy at this point will make things a lot easier later on. In addition, it is just possible that some implications of this will occur to you that haven't been thought of before. I can't mention often enough that psychobiology is still in its infancy. In fact, many of the "pioneers" are still living. Think of it as an opportunity to study Renaissance art at the time that Michelangelo worked on the Sistine Chapel.

The topics we're now discussing are more philosophical than psychobiological. In addition, very few psychobiologists know any more about philosophy than philosophers know about psychobiology. So if you read in both disciplines, you might come up with stunning insights. But, unless you are awfully clever, don't count on it.

Now for a mini-neuroanatomy lesson. The earliest descriptions of the brain date from a time when the ultimate scientific precision consisted in describing phenomena in terms of common everyday natural products. Thus, there was a great run of what I call "nut and fruit" descriptions. The brain was compared with fruit such as an orange: The central pulp was named the medulla, while the rind was dubbed the cortex. The medulla is actually an extension of the spinal cord, with the addition of certain areas for control of respiration, heart rate, even wakefulness (the reticular formation we mentioned in Chapter Three). Parts of it, when sectioned (literally cut in a series of slices the way you'd slice up a watermelon), resemble an olive, and are still known as the "olivary complex." Further up the medulla is the pons, named after its resemblance to a bridge. As you can see, the early anatomists weren't uncomfortable with mixed metaphors, and if a fruit or nut couldn't be found, a bridge would do just as well!

Close to these structures is the hypothalamus, which regulates temperature, blood pressure, breathing, and other parts of what the physiologist Claude Bernard referred to as the "milieu intérieur." Close to this is the basal ganglia, which are the oldest parts of the brain and roughly correspond to the R-complex we described earlier. All of these structures are part of the pulp (to return to our orange analogy) and are responsible for the innate species behavior seen in reptiles. Also, our sensations arrive via specialized cables which run through the pulp all the way from our big toe up to the very top of the brain, the cerebral cortex. If railroads existed when the nerve tracts were named, we might be describing the brain in terms of depots, fast passenger lines, and slower freight hauls. But since most of the early descriptions can be traced back to Renaissance times, the anatomists had to be satisfied with fruits and nuts. You can think of the brain of the most ancient reptiles and lower mammals as possessing nothing else—in fact, an orange that is just a pulp. The rind, or cortex, emerges at the point in evolutionary history when amphibians began to turn into reptiles. The cortex, the outer layer of the brain, grew out of the brainstem and folded around it like a shell. It continued to mushroom (my own tribute to nut and fruit anatomy) until we arrived at the situation in humans where the cerebral hemispheres completely enveloped the early mammalian brain structures.

Naturally, the continued enlargement of the cerebral hemispheres would have resulted in a world of top-heavy humans with heads too large for their bodies. (Perhaps when we say that someone has a big head, we're speaking on two planes at once. Neither exaggerated self-estimation nor head size is of evolutionary advantage.)

Another obvious problem with exponentially expanding cerebral hemispheres is the modification of the female pelvis that would have been required. At a certain point, head size and pelvic size would intersect, and either women would no longer be able to walk, or babies could only be born by Caesarian section. Since the survival value of Stone Age obstetrics was probably pretty low, an ingenious architectural change enabled the cerebral hemispheres to continue to grow yet cease to expand. Think of it as a family which has outgrown its house and won't move from a favorite neighborhood and can't purchase more land. The obvious solution—which, incidentally, is seen more frequently in choice but crowded sections of American cities—is to go vertical. Either build another story on top,

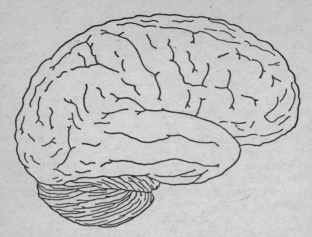

FIGURE 6. The surface of the brain, of which over half the area is enfolded and hidden from view.

or dig down and modify the foundation. In the case of the cerebral hemispheres, the moves were downward, and as a result, the greatest surface is now concealed within the depths of the cerebral hemispheres. If you look at a Chinese fan, for instance, the most artistic parts are hidden beneath the surface. In the same way (see Figures 6 and 7), over 98 per cent of the brain is hidden within the depths of the sulci, or invaginations. Its outer surface, the gyri, are often named in reference to these invaginations. In other instances the gyri are actually biological tombstones in honor of the brain scientists who first described them or made significant discoveries about their function. Paul Broca, a French neurologist who first described an important speech center, is immortalized forever in Broca's center, located in the superior frontal lobe.

The time-honored fruit and nut analogies for brain descriptions are all very well as far as they go. Unfortunately, they don't go far enough and are sometimes misleading. Comparing the cerebral cortex to an orange rind, for instance, is misleading because the cortex, unlike the

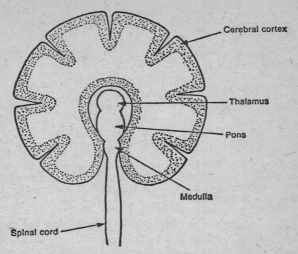

FIGURE 7. A diagram of how the enfolding and overlapping of the brain's surface greatly increases the amount of brain area that can be contained within a fixed space.

rind, is not a homogeneous structure. There are seven layers of cortex, each with a different number and type of cell. In addition, when we speak of "brain cells" we must be very careful to distinguish neurons from the supporting glial cells which surround them. "Glial" is from the Greek word meaning glue and is actually an unfortunate term to describe the largest number of cells in the brain, especially since they do more than serve as struts for the elaborate psychic light show put on by the neurons. We now know that the glial cells serve a nutrient function, they can be responsible for the initiation and termination of seizures, and there is even some evidence that they form a communicating network of their own.

In addition, there are other important differences in brain structure that are concealed by analogies to such things as oranges. Not only are there different kinds of nerve cells within the layers of the cerebral hemispheres, but three basic subdivisions can be distinguished on the

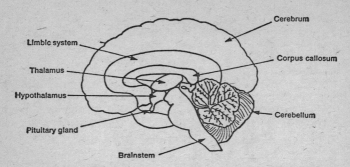

FIGURE 8. The limbic system, corresponding to the second (old mammalian) brain in the three basic brain types, is responsible for certain automatic bodily functions as well as the experience and expression of emotion.

basis of evolutionary development. The names applied are archipallium, paleopallium, and neopallium, but an easy way to think of them is reptilian, old mammalian, and neomammalian. (See Figure 5.) As we discussed, the reptilian complex (or R-complex) seems to be the repository for unlearned preprogramed sets of behavior. As we proceed to the old mammalian and neomammalian brains, the older parts of the cortex fold into two concentric rings which are eventually completely covered in humans by the fully expanded cerebral cortex. To return to the archaeological model, the oldest brain structure is deeply buried beneath the surface, like an unused toy that has been relegated to the cellar.

If you look at this folding-in process from the side (see Figure 8), you'll see an incomplete ring, almost like a doughnut someone has nibbled a small bite out of. The Latin word for *ring,* or "forming a border around," is *limbus*—hence, the name *limbic lobe.* Since the limbic lobe surrounds the brainstem and, in addition, makes connections with the newer enlarged cerebral hemisphere, a more dynamic way of naming it is in terms of its function—hence, the term *limbic system,* first applied by Paul MacLean in 1952.

Originally the limbic system was thought to be important in smelling. Its old name, rhinencephalon (or nose brain), referred to its extensive connections with the olfactory structures, and for the longest time brain scientists seemed satisfied with that. Although smelling plays a relatively minor role in the lives of most humans, the structures underlying it seemed to occupy an astonishingly large brain area. Experimental work on parts of the limbic system suggested a much more intriguing role for the limbic system than just smelling.

For one thing, stimulation or destruction of the limbic system in animals can bring about changes in their behavior. An electrical stimulus to the amygdala, a part of the limbic system, stops the animal in its tracks and stimulates him to display signs of excitement, rage, or fear. A cat will hiss, its pupils dilate, its mouth salivate, its back arch, and it seems prepared to attack. Such demonstrations led psychobiologists to speculate about the effects of destroying the amygdala altogether. This critical experiment, performed in 1939, provided conclusive evidence that the limbic system is the area of the brain most concerned with emotion.

The Kluver-Bucey monkey, named after the two University of Chicago researchers who carried out the experiment, has one of its amygdala severed (there are two of them, one on each side of the brain). Although such a lesion includes fibers connecting the limbic system with both the brainstem and the cerebral cortex, the effect achieved is attributed primarily to amygdala destruction.

The first thing you would notice about the Kluver-Bucey animal is its extreme docility. You can poke at it, squeeze it, and do just about anything that under ordinary circumstances would guarantee an armful of scratches. In addition, the animal engages in some very peculiar and bizarre behavior. For instance, it seems not to recognize such common objects as its food dish. It often puts nonedible objects into its mouth, and in some cases even swallows them. It touches every object it sees and engages in sexual behavior almost constantly.

Soon, other investigators noted behavioral changes resulting from either destruction or stimulation of other parts of the limbic system. The most telling argument for the limbic system's role in emotional behavior, however, came from studies of epilepsy. Epileptic discharges in or near the limbic system were demonstrated to elicit a broad range of vivid emotional feelings, such as this description by a fifty-two-year-old housekeeper.

"Every week I'll feel a 'breeze' come by me, almost as if somebody is brushing past me. It's a very pleasant, smooth, and silky feeling that I can't describe in any other way. I don't pass out or anything and I know it's only my imagination, but I have to look around to see who it is. Even though no one is there, the *feeling* is so real I have to look. If this is bad nerves, then maybe I'm happy to have them." Investigation of this patient revealed epileptic discharges confined to the limbic system in the nearby temporal lobe. With medication, the brain-wave abnormalities as well as the "breeze" and the eerie visitation ceased.

Included in the seizures of other patients with limbic types of epilepsy are peculiar forms of familiarity, a *déjà vu* experience, as if the patient may have existed in another

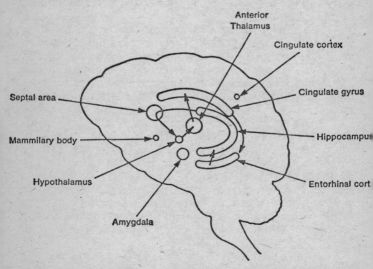

FIGURE 9. A dynamic view of some of the connections within the limbic system responsible for the integration of emotion and thought.

life, or as if something present at the moment had occurred in some distant and elusive past. In other cases the feeling is exactly the opposite. The person may have the feeling that his wife, for instance, seems different and inexplicably changed. In still other cases, the seizure may consist of nothing more crystallized than the "feeling" that something in the environment is not quite right.

One of my patients, a journalist, had an attack while in a restaurant. A usually pleasant and agreeable man of forty-five, he angrily ordered the waiter to return a steak to the kitchen because "it smells wretched and I have the feeling that it's not cooked enough." After the steak was brought out for the third time, to the embarrassment and consternation of his friends, my patient slumped to the floor in a convulsion and lost consciousness. Later, during his hospitalization, we discovered that the patient's feelings about the steak were due to a brain tumor located within a part of the limbic system.

In the last twenty-five years, brain scientists have refined their knowledge of the limbic system. Although it occupies only the lower one fifth of the brain, its influence on behavior is unbelievably extensive, with all its parts tied bidirectionally to the hypothalamus, a collection of cell groups with two connecting outposts: the septal areas in the front, and the amygdala along the sides. Think of the limbic system as a wheel, with the cortex marking the tire around it, and the hypothalamus, with its two outposts, marking the hub and its spokes. Brain and behavior experts have linked this interconnecting wheel arrangement to hormones, drives, temperature control, reward and punishment centers, and one part of it, the hippocampus, to memory formation. Its role in emotional reactions depended on the demonstration, in rabid dogs, that the limbic system was affected out of proportion to other brain areas. There is, in fact, a reverberating cycle by which messages can make full circle through five limbic system structures: hippocampus, mammilary body, anterior thalamus, cingulate cortex, and hypothalamus.

Experiments during the last twenty-five years in various animal species have demonstrated that profound alterations in behavior and emotional experiences can be brought about by altering the limbic system. If you like simple ways of thinking about complicated issues, think of the limbic system as the center for the four F's—feeding, fighting, fleeing, and sexual behavior.

Studies of the limbic system contributed two significant findings to brain evolution. In MacLean's words, "The evolutionary development in lower animals of a respectable cortex might be regarded as nature's attempt to provide the reptilian brain with a thinking cap and emancipate it from inappropriate stereotypes of behavior." Equally important, the limbic system introduced emotion as a unifying and, as we shall see, devisive factor in animal and human behavior.

Certainly, even the least introspective person recognizes the continuous interplay between what we feel and what we know. Usually these two mental processes work in harmony. When we are with friends who share our interests we have a feeling of security, of comfort, a sort of letting down of all barriers as we bask in the harmonious interplay of pleasurable thoughts coupled with comfortable emotions. In a courtroom, however, we may be in the presence of our enemies, and both our thoughts and our emotions function in a kill-or-be-killed law of the jungle. Even in this situation, however, both our thinking and our feeling selves are basically in tune. As an example of when they are not, consider the following:

You've accidentally bumped into an old lover at a weekend ski resort. While the wife and kids are up in the room readying for dinner, you step into the bar for a drink and—there she is. You smile and begin the usual pleasantries: "How long it's been," "Small world," "Hope things are well with you," etc. Your cerebral cortex can ramble on almost indefinitely with pleasant chitchat, all the while taking in countless observations about the person before you. At the same time, you begin to feel the rumbling of the limbic system, which doesn't deal in chitchat but remembers the old drives and the old feelings. It too speaks, but through its connections to the hypothalamus and down to the brainstem. Soon you may be embarrassingly and painfully aware of a racing heart, sweaty palms, a feeling of constriction around the neck, stomach churnings, and perhaps even the beginnings of an erection. Or perhaps the emotion felt is annoyance, the desire to get out of an uncomfortable, no longer "relevant" situation. In this case, too, there are limbic accompaniments as your face flushes, or your eyes dilate enough to make you uncomfortably aware of the glare of the light flashing in your face from over the bar.

While all this is going on, the R-complex, our old reptilian brain, is active with various forms of body language, spelling out for astute observers the contradictions between your pleasant verbal messages and the inner turmoil you are experiencing. Perhaps you are shaking your head a little too often, shifting position from one foot to another, engaging in expansive gestures that would be more appropriate to a large theater than a small cocktail lounge. But finally, after what seems an eternity, the former lover remembers she must be dressing for a dinner date, wishes you the best and is gone.

The relief you feel is almost immediate as your "three brains" come back into harmony. Your stomach is settled, the heart is languidly flip-flopping, and your hands are steady even before picking up "the double" you just ordered.

Such a hypothesis of brain function is, of course, difficult to prove. But the feelings are all there in each one of us, requiring only the appropriate trigger to release wide disharmony between our behavior and how we're feeling at the time. MacLean and others have demonstrated that sensory input—what we see and hear—can be transmitted not only to our cortex but to the limbic system as well. Therefore, such a cocktail lounge scenario is explicable, since all three brains would be experiencing the same scene, each in a different way.

From here, MacLean shifts to a computer analogy of how things are processed: "The reptilian and the old and new mammalian brains might be regarded as biological computers, each with its own functions."

Returning to the epileptic experiments again, we can think of the patient I described on page 65 as having a malfunctioning connection between thinking and feeling. In its most developed form a patient may possess such a strong feeling of disharmony that he may become depersonalized, doubt who he is, even talk about himself as if commenting on a stranger. Although not necessarily a sign of mental illness, this experience of "mental diplopia" (literally, mental double vision) is a common symptom of schizophrenia and one of the most commonly described effects from psychedelic drugs. The importance of such an experience for psychobiology is what it tells us about the relation of emotional feelings to external reality. In the examples we have looked at (a meeting with friends, a day defending oneself in court, a surprise meeting with an old lover) the emotions follow closely on the heels of some real experience (e.g., we *are* in court or engaged in an agonizing conversation with someone who once meant a lot to us). In the case of drug intoxication, mental illness, or epilepsy, however, the mental experience exists out of context, as with my journalist patient who attributed a disagreeable odor to his steak.

"It would thus seem that the raw stuff of emotion is built into the circuitry of the limbic brain," says MacLean. "Instead of explaining experience in terms of compulsion, as was implied in considering the reptilian brain (R-complex), or in terms of abstract thoughts, as was presumed in the case of the cerebral hemispheres, the mentation of the limbic system would appear to involve a process whereby information is encoded in terms of emotional feelings that influence its decisions and its course of action."

Such a view has implications for issues as far-ranging as mental illness and foreign policy. MacLean refers to the discrepancy between our feelings and our thinking as schizophysiology. In a sense, we can no longer trust to our innate instincts the feelings that govern many aspects of our lives. The study of epilepsy demonstrates conclusively that our *feelings* do not necessarily depend on anything going on in the real world around us, and that the strength of our feelings is not a measure of the authenticity of our experiences or the credibility of our beliefs. We can "feel" strongly about something and yet be dead wrong. In the instance where the housekeeper "felt a breeze" passing over her, she "had to look" even though her experience

68

and common sense told her the experience was imaginary. But what can we say of feelings so real that they begin to direct our behavior? MacLean suggests that this may be the basis for some forms of paranoid psychosis: a schizophysiology where believing is seeing rather than seeing is believing.

"Affective feelings provide the connecting bridge between our internal and external worlds and, perhaps more than any other form of psychic information, assures us of the reality of ourselves and the world around us," says MacLean. "The limbic system contributes to a sense of personal identity integrating internally and externally derived experiences."

As an example of how a "feeling" can be helpful at one time and destructive at another, MacLean recalls the experience of a young man returning from Europe on the S.S. *United States*. One morning while walking on the deck, the man had the distinct feeling that snow was in the air a short distance away over the fog-bound ocean. At the time, the ship was off the coast of Newfoundland. He spoke of this to several other passengers, who shrugged their shoulders and in some cases commented that they also felt that snow was "in the air." A few hours later, a news bulletin confirmed that it had indeed been snowing in Newfoundland at about the same time as the man's feeling. As MacLean comments, such an event, where "feeling" correctly mirrors a real event, is not unusual, given the circumstances. Presumably, the appearance of the sky and temperature and the moisture of the air affecting his skin and respiratory passages, added up to the snowy feeling. In other circumstances, however, similar feelings would be abnormal.

"Suppose the young man had experienced and communicated similar feelings while cruising in the subtropical waters of the Caribbean," asks MacLean. "And suppose, further, that the feeling persisted. Would he not perhaps become suspicious of fellow passengers who disagreed with his inappropriate feeling and challenged his sanity? We may imagine that the schizophrenic patient, with his persistent delusional symptoms, finds himself in a comparable situation."

From here it may be just a step to understanding how behavior based on intensely felt emotion can stir large numbers of people into irrational, even destructive, behavior. At times there can be widespread uneasy feelings among nations, as among individuals, and such feelings may result in the emergence of leaders who provide an "explanation." As an example of this, MacLean is fond of pointing to the Nazi experience. "The explanation that found appeal under the banner of the swastika was that the Fatherland was threatened by one of its minority groups and that the resulting widespread sense of uneasiness could be relieved by the torture and extermination of this group."

The Watergate experience provides an even more recent example of deranged feelings providing justification for irrational and, in the end, self-destructive behavior. "Paranoia for peace is not that bad," Richard Nixon told David Frost in the second nationally televised interview. And as Nixon so plainly revealed in the interview, it was the *feeling* of threat that motivated the plumbers and all of the other components of "dirty tricks."

Although never a popular President in the sense of a charismatic leader, Nixon's position was not realistically in jeopardy. Rather, his limbic "feelings" seemed to operate autonomously to direct behavior that should have been controlled by rational considerations. What is so disturbing about such schizophysiology is its tendency to spread like a virus that first convulses its victims and finally leaves them totally paralyzed. The "siege mentality," with its "stone-walling" and elaborate bugging system, were all based on pathological feelings that were never relieved even by the most repressive acts. In the Frost interview Nixon returns again and again to speak spontaneously of paranoia, a correct but belated appreciation of the role that distorted feelings can have on the behavior of otherwise rational people.

Psychobiologists such as Paul MacLean are providing us the beginnings of a new way of looking at national and international events, as well as a remarkable framework for evaluating our leaders. "How influential," MacLean asks us, "is the reptilian counterpart of man's brain in selecting and following a leader? Is it possible that this brain, in conjunction with the poorly discriminating limbic brain, mistakes the caricature of a leader for a genuine leader? With the increasing insights that are being obtained in the behavioral sciences, it is to be hoped that education through the mass media will make it possible for man to become increasingly thoughtful in his choice of leaders."

Limbic system research is not complete, by any means, and we must be careful about premature speculations. But what we know offers a more rational basis for informed judgment than anything suggested by the empty speculations of political scientists. Countries, nationalities, the names of leaders—all may change. But what remains is the common bond of the human brain. What is built into our brain over millions of years of evolution? What do we share with the reptiles and ancient mammals? How is it shaping our behavior and affecting our futures?

5
The Murder of a Child

The human brain contains the fossil memories of its past just as this stratified landscape contained earth's past in the shape of horned titanotheres and stalking dirk-toothed cats.

LOREN EISLEY

Edward O. Wilson is Professor of Zoology at the Museum of Comparative Zoology at Harvard University. A specialist in entomology—insects—Wilson is also the author of *Sociobiology: The New Synthesis,* published in 1975, and, in less than three years, already responsible for establishing Wilson as the most controversial biologist since Charles Darwin.

In his combination laboratory-office on the fourth floor of the museum, Wilson theorized to me about how our brain may have evolved:

"One of the problems in speaking about human brain development comes from our ideas about what it means for the human species to 'learn.' Learning is actually paradoxical. How can individual acts of learning be passed down from one generation to another? The idea that something like that can happen doesn't fit into our conception of evolution. You can learn, let's say, how to be a neurologist, but your children, if they also want to be neurologists, have to learn for themselves the principles of neurology. Now, if each brain is stamped afresh by experience, then, according to the traditional view, the role

73

of natural selection must be solely to keep the *tabula rasa* of the brain clean and malleable. But we don't believe this any more. Actually, only small parts of the brain resemble a *tabula rasa*. The remainder is more like an exposed negative, waiting to be slipped into developer fluid."

He gave another analogy. "Imagine a chess master sitting in his room playing over all the games of the great masters of the past. He is interested in technique and skill —how one game was played, how it could have been played better, etc. He is relatively unconcerned about questions that a young child might put to him: 'Where did the chess moves come from? Who gave you the rules?'

"A similiar situation exists when people discuss what we will call philosophical or ethical questions about the development of the mind and the brain. Bertrand Russell once remarked that philosophers know the results they want before they start reasoning. They are relying on unspoken premises. They almost never try to understand where the precepts come from, and this is understandable because they don't know evolution or biology. You have to know about biology, particularly about brain biology, before you can draw truly logical conclusions about why we think the way we do. You can't deduce ahead of time the principles of psychobiology. To put it differently, you can't learn psychobiology at the feet of a guru. In attempting to do so, contemporary philosophers have progressed no further than Sophocles' Antigone, who said of moral imperatives: 'They were not born today or yesterday: they die not, and none knoweth from whence they came.' "

The idea of learning about the brain and ethics from a man who has spent his life cataloguing the life patterns of insects seems, on the face of it, mildly ludicrous. This no doubt partially explains the vehement resistance Wilson continues to encounter whenever he launches into his favorite subject: the biological foundations of human values. To Wilson our deepest and, in some sense, our most human values are physically determined and, hence, explainable as "our overwhelming predisposition to register certain facial expressions when we are sad or angry or perhaps even sexually aroused." Human behavior, in this sense, is as appropriate for a scientist to study as the subject of Wilson's epic *The Insect Society*.

In *Sociobiology* he wrote the following statement, which has prompted a great deal of discussion: "The biologist who is concerned with questions of physiology and evolutionary history realizes that self-knowledge is contained and shaped by the emotional control centers in the hypothalamus and limbic systems of the brain. These centers flood our consciousness with all the emotions— hate, love, guilt, fear and others—that are consulted by ethical philosophers who wish to intuit the standards of good and evil. What, we are then compelled to ask, made the hypothalamus and limbic systems? They evolved by natural selection. That simple biologic statement must be pursued in order to explain ethics and ethical philosophy."

Despite Wilson's radically unorthodox orientation toward human values, he remains a traditionalist. To this day he writes on a huge mellow oak desk inherited from William Morton Wheeler, his predecessor as Curator at Harvard. On the wall behind him is a first edition of the September 3, 1871, *Punch* cartoon which characterizes a selection of "Men of the Day," including a bearded Charles Darwin astride a two-word caption: "Natural Selection." Wilson, like Darwin before him, is the target of scientists, one of whom alleged that Wilson's theories "operate as powerful forms of legitimization of past and present social institutions such as aggression, competition, domination of women by men, defense of national territory, individualism, and the appearance of a status and wealth hierarchy."

At its simplest, sociobiology seeks to discover the biologic basis for human behavior. Sociology, psychology, and even history are all right as far as they go, but, according to Wilson, they don't go far enough because they have never been integrated into the modern synthesis—the neo-Darwinian theory that seeks explanations for behavior in terms of its adaptive evolutionary value over millions of years.

Wilson has arrived at a position strikingly similar to Paul MacLean's. Whereas MacLean bases his theory on comparative studies of animal brains, particularly lizards,

Wilson, by his own admission, doesn't "know all that much about the brain." Instead, he has learned psychobiology in a context of painstakingly precise observations on the social patterns of thousands of animal species. This has led him, along with MacLean, to the realization that wherever mankind may be, and wherever we may be going, depends, to an as yet incalculable degree, on the workings of the limbic system.

"We have strong predispositions, strong emotional feelings about certain things. Xenophobia, for instance, has to do with treating aliens as hated objects and attacking them in some way. Here we have an emotion whose expression can vary from one culture to another. In a primitive jungle society, for instance, the expression may be to kill the feared and hated stranger. In a modern industrialized city it might take the form of shunning him, refusing him admission to a school or neighborhood. In each case the emotion is mediated by the limbic system. Whatever form it finally takes, however, depends on the conscious portions of the brain. This is where culture is important. This is a part of sociobiology that many people do not understand. I am not saying that culture is not important; I am focusing on the constraints that are biological—especially those widely believed to be centered in the limbic system—and responsible for the deep-seated and sometimes irrational responses we feel.

"It follows that value systems are probably influenced, again to an unknown extent, by emotional responses programed in the limbic system of the brain. The qualities that comprise human nature, from hunter-gatherer in the Kalahari Desert to skilled industrial workers in Pittsburgh, are surely due in part to the constraints within the unique human genotype. The challenge is to measure the degree of these constraints and infer their significance through the reconstruction of the evolutionary history of the mind."

As Wilson spoke, I suddenly remembered something mentioned to me by Arthur Koestler, an exchange of letters between Charles Darwin and A. R. Wallace, who coestablished with Darwin the theory of natural selection. Wallace wrote in 1869: "Natural selection could only have endowed the savage with a brain a little superior to that of the ape, whereas he possesses one very little inferior to that of an average member of our learned society." Darwin, apparently realizing the perilous implications of Wallace's statement, replied: "I hope you have not murdered completely your own and my child."

Behind Darwin's discomfiture was the dawning realization that the evolution of the brain vastly exceeded the needs of prehistoric man. This is, in fact, the only example in existence where a species was provided with an organ that it still has not learned how to use. "How can this be reconciled," I asked Wilson, "with evolution's most fundamental thesis: Natural selection proceeds in small steps, each of which must confer on its bearer a minimal, but nonetheless measurable, advantage?"

"It is entirely within the range of possibility," Wilson replied, "that the brain became so hypertrophied and complex that at some point its most important development was no longer related to the earlier genetic evolution. In other words, the influence of culture became predominant over strictly biological considerations. This cannot be ruled out. The opposite extreme is that we are 'hard-wired,' that we have inbred constraints, an actual structural basis for our emotions. The truth, I think, lies somewhere between the two extremes. You cannot of course explain the brain simply on the basis of evolution, and at the same time it is not totally dependent on the cultural environment either. The incest taboo, for instance, is present in virtually all animal societies. You can't know the basis for this, however, until you look at humans. Incest taboos are virtually universal in human cultures. One study in an Israeli kibbutz, for instance, disclosed that among 2,769 marriages recorded, none were between members of the same kibbutz peer group who had been together since birth. This suggests the influence of both genetics and culture. Somehow there is a predisposition to avoid sexuality with those people we have associated with up to the age of six years, regardless of whether or not we are genetically related to them. This is one of the constraints, in the form of a learning rule, that might be shaped by the emotional control centers in the hypothalamus and limbic systems of the brain."

77

The existence of "emotional control centers" within our heads stirs up visions of a race of robots created by "nature" to experience and act in certain ways. To an extent, our conscious self—interested in career advancement, personal happiness, or whatever—must negotiate a compromise within the brain's neural circuit between what we "know" and the ancient knowledge "hard-wired" within our limbic systems. Could this explain the ambivalences and paradoxes that have confounded man's attempts over thousands of years to formulate a "model" or theory of the human mind? In a sense, conflicts are built into the system; what we want for ourselves may not be the same thing that would favor the development of the species.

In *Sociobiology* Wilson suggested: "The hypothalamic-limbic complex taxes the conscious mind with ambivalence whenever the organism encounters stressful situations. Love joins hate; aggression, fear; expansiveness, withdrawal."

Looked at in this way, our hypothalamic-limbic brain structure seems an extreme liability. Indeed it is if, for instance, such things as the unlimited territorial expansion of our hunter-gatherer ancestors cannot be modified by the thoughtful realization that force doesn't justify wrenching land from our less powerful neighbors. Or maybe the difficulty is letting go of something we no longer need exclusively. Three months of bitter debate on the Panama Canal Treaty might be viewed as a fierce struggle between our limbic possessiveness and the cerebral cortex's insight into the unifying power of peaceful sharing.

"Although we can overcome the biological constraints imposed by our own limbic system," Wilson said to me, "we are forced to do so at great economic and social cost in terms of time, energy, and resources. Consider something like vegetarianism, for instance. Many people are now vegetarians, which represents a complete break with the meat-eating cultures of our ancestors. Co-operation during hunting provided in the past a new impetus for the evolution of intelligence, which in turn permitted still more sophistication in tool using. The shift to big game is believed to have accelerated the process of mental evolution. To this extent total vegetarianism is opposed to the limbic constraints. Now, just think about the efforts a vegetarian must continually expend in order to guarantee an adequate diet: nutritional planning based on a detailed knowledge of dietetics and nutrition, which the meat eater, who consumes an average diet, doesn't have to be concerned with. In other words, the meat-eating mentality, if you will, is built in and is relatively effortless. Vegetarianism, on the other hand, is a free choice that runs counter to our inborn limbic tendencies. In addition, that tendency would reassert itself under harsher conditions; people would fall back into the hunting and meat-eating style. Vegetarianism can be destructive if planned poorly, and, even if planned well, it exacts a certain cost.

"We are probably the most responsible generation in history in terms of our attempts to limit the possibility of global warfare. We want to avoid territorial and national conflict, which seems to be based on the natural predisposition of human beings to form hostile groups and expand at the expense of real or imagined enemies. But just look at the energy required to change this. The United Nations, for instance, is an extremely complicated and tremendously expensive diplomatic network spread throughout the world. Peaceful coexistence does not come 'naturally' but exacts a cost in time, effort, and money. I think we agree that peace is important enough to justify these costs, but I am less sure of the wisdom of some other goals."

One of the "other goals" involves current forms of social planning that ignore people's biological differences. Sexual roles are under heavy attack as women, quite justifiably, continue to mount an offense against discrimination and inequality of opportunity. But are these sex roles that cannot, or at least should not, be changed?

"We may choose to imitate the Israeli kibbutz, which is based on forcing similar role identities on both men and women. This flies in the face of thousands of years in which mammals demonstrated a strong tendency for sexual division of labor. Since this division of labor is persistent from hunter-gatherer through agricultural and industrial societies, it suggests a genetic origin. We do not know when this trait evolved in human evolution or how resistant it is to the continuing and justified pressures for human rights."

Such comments have not endeared Wilson to the more radical feminists who view sociobiology as yet another pseudo-scientific justification for a male-dominated society. But is such criticism really justified? Although such differences cannot preclude the sexes being forced into identical behavioral patterns, it does raise a question: Is the benefit worth the cost?

"Human beings are capable of living in an enormous variety of societies. A Maoist commune is already a reality for millions of people. Perhaps someday a self-styled genius might even seriously suggest we design our cities like termitariums, with people in strict caste systems like termites. The possibilities are unlimited. In such cases, however, the modifications required would be destructive. To this extent social control can rob us of our humanity."

It is too early to assess Wilson's contribution to psychobiology. One change he has already brought about, however, concerns our attitudes toward genetics and brain structures. Two thousand years of Western philosophy, unbuttressed by scientific observation, have emphasized the importance of individual identity: my thoughts, my feelings, my impressions, strung out in an endless continuum described as philosophical egoism or, more informally by the social critic Tom Wolfe, as "The Me Generation." But Wilson forces on us a more impersonal and disquieting vision where the "I" is less important than the "we."

"You cannot explain the brain fully by concentrating on how the individual mind works. You have to think in terms of gene pools, why some patterns of brain function were favored over others. We will need to know in the future how certain types of behavior are genetically linked to others. We must understand the mechanisms and the history of the human mind.

"Soon we may have to pick and choose among the emotional guides we have inherited and determine those that should be followed and those that should be sublimated or redirected, so that our behavioral patterns will both conform with biological principles and foster the growth of the human spirit."

It's a long way from an office at Harvard University to the open African savannahs of two million years ago. But we must try to make the mental journey if we are to understand how our brains evolved.

One of the first difficulties involves credibility. How can we say anything about what the brain was like millions of years ago? If left inside the skull after death, the human brain will soften and disintegrate. After only a few days it will be the consistency of a milkshake. How, then, can we determine what the brains of our prehistoric ancestors were like? The answer is, of course, that we can't do so by any direct means. Our knowledge of ancient brain structures rests on less direct, but nonetheless compelling, evidence.

The brains of primitive man share with us a bony skeletal covering, the human skull. The brain size of ancient man can thus be inferred from measurements made by pouring plaster into the fossil skulls which have been discovered by paleontologists. The endocasts that result from this procedure provide a general picture of the surface features of the brain as well as an estimate of brain volume: The larger the endocasts the larger the brain. By comparing the size of the endocast with the rest of the animal's skeletal remains, conclusions can be drawn about the size of the animal's brain compared with the rest of its body. From such studies several conclusions can be drawn.

For one thing, many of the lower vertebrates fail to show any greater enlargement of their brains than would be expected on the basis of their increased body size. In other words, the brain enlarged in the same proportion as the heart, lungs, or any other body part which could be expected to grow proportionately larger as the total size of the animal increased. Secondly, only about 80 per cent of the increase in brain size can be correlated with the increase in body size. But the remaining 20 per cent is responsible for the superiority of the human brain. Humans, along with dolphins, have a brain about six times larger than any other living mammal. The increase is due to the critically important 20-per cent increase in brain size which developed independently from the body-size enlargement.

81

Finally, the mammalian brain, which can be traced back at least one hundred and fifty million years, achieved its greatest development in the past three million years, culminating about two hundred and fifty thousand years ago in the evolution of the true human brain of *Homo sapiens*. In other words, the present level of brain enlargement in most mammals was reached early in evolution and maintained up until the present time. Our brain, in contrast, underwent its greatest growth spurt during the last two hundred and fifty thousand years. Even dolphins, who share with us an increased brain-to-body ratio, possess brains which have not developed much in the past twenty million years. In essence, a dolphin today is about as intelligent as one of its ancestors twenty million years ago.

Why did the human brain develop so rapidly in such a short period of time? What was the stimulus for this growth? How did it fit into the evolutionary scheme of things? The exponential growth of the human brain during the last two hundred and fifty thousand years is unique in the history of evolution. Even today we lack a satisfactory explanation how it came about.

Everything is not a total mystery, however. There are facts that we know with relative certainty about the brain's evolution. For one thing, the brain evolved in response to environmental changes, the most dramatic change occurring forty to fifty million years ago when weather patterns began shifting in many parts of the world toward temperate rather than tropical climates. This brought about great changes in the world's food supply. With vast jungles ultimately thinning out, nuts and fruits, usually present all year round, began to blossom only seasonally. In response to this change in the food supply, our ancestors emerged from the forest and began hunting on the savannahs, or grasslands, an event which occurred—judging from the remains of fossils—about fourteen million years ago.

The shift from the forest to the savannahs was not made by all the early apes, however. Chimpanzees, the most intelligent and socially sophisticated of all living non-human primates, remained forest dwellers and vegetarians. This is one reason why I think we must be cautious about drawing conclusions about human evolution from the study of chimpanzees. Their evolutionary history is different and intimately tied up with their continued residence in the forest, which includes tree-dwelling and meatless diets.

Out among the tall grass of the savannah, natural selection favored those primates who could stand up. For one thing, this made hunting easier, since prey could be spotted at a distance. In addition, biped locomotion enabled the early primates to spy hostile animals from a distance. Thus, those primates who could stand up survived long enough to produce offspring who shared the same physical characteristics. After many generations, the savannah dwellers evolved into a distinct class in which the forelimbs were freed from the task of locomotion.

An alternative and complementary theory holds that early man became biped in order to pick seeds. Support for such a theory comes from similarities in the skull and dental structures of modern man and contemporary seed-eating primates. In either case, the freeing of the hands of our early primate ancestors permitted the development of tools, with intelligence evolving as the consequence of early man's tool-using capacity.

At some point, in response to the increased demands of savannah living, these early primates began to eat meat. As a result of switching from a vegetarian to a carnivorous diet, those primates who could use weapons and tools began to prevail over those who remained largely scavengers and subsisted on spoils left behind by the large predators. Some scientists have gone so far as to postulate that *Homo habilis,* our direct ancestor, may have used his tool-and weapon-making abilities to kill off all the competing primate species. A South African anthropologist, Raymond Dart, concluded, after analyzing fifty-eight baboon skulls, that all had been killed by blows to the head, most likely resulting from the club-wielding capabilities of *Homo habilis* (*Homo habilis* means "handyman," a reference to the tools found in 1961 near the remains of the recently discovered *Homo habilis*). Although speculative, the possibility thus exists that our brains have evolved as organs for increasingly sophisticated killing!

Once early man became carnivorous, his brain developed neuronal networks that favored swiftness of foot and hand. Since the prey was often strong and highly dangerous, a measure of cooperation sprang up between several hunters, the first step toward culture and, some scientists feel, the beginning of language. From here cooperative efforts, using increasingly sophisticated tools and weapons against predators, stimulated further brain growth. Most anthropologists and psychobiologists are in agreement that meat eating and hunting are the primary stimulants of brain development and early culture.

During the last ten years we have witnessed a revolution in theories about man's origins. Since this is directly related to the subject of brain development, I am going to attempt a brief sketch.

According to traditional theory (a strange term, since the "tradition" dates back only a few years), man developed from a single unbranching line of descent: *Australopithecus* to *Homo erectus* to *Homo sapiens,* or modern man. Thanks to newly discovered fossil evidence, most anthropologists and paleontologists now favor the theory that modern man is only one branch in a life tree that began twenty million years ago with a primate called *Dryopithecus.* (See Figure 10.) Six million years later, the *Dryopithecus* line divided into three branches. The first evolved into the modern ape. The second, another line of apes known as *Gigantopithecus,* eventually became extinct. Only the third branch, *Ramapithecus,* continued to evolve. *Australopithecus africanus,* a meat eater, and *Australopithecus robustus,* a vegetarian, were two lines traceable from *Ramapithecus.* They both eventually died out one and a half million years ago. *Homo habilis,* the third line, eventually prevailed, thanks to enhanced brain development that resulted from tools and a formation of early hunter-gatherer societies.

The evidence of man's origin, although impressive, is far from complete. In the next decade we can probably expect to discover additional pieces of the jigsaw puzzle. Most scientists feel, however, that there will be no missing links. "We can no longer talk of a great chain of being in a nineteenth-century sense in which there is a missing link," according to Dr. Phillip Tobias, Professor of Anatomy at Johannesburg's Witwatersrand Medical School. "We should think rather of multiple strands forming a network of evolving populations diverging and converging, some strands disappearing and others giving rise to further evolutionary development."

Somewhere in this evolutionary development, however, we encounter a distinct discontinuity, the "murdered child" that troubled the sleep of Darwin's later years. How can evolution account for some of man's most impressive accomplishments? For example, was there a biological survival advantage that accrued to the inventor of the calculus? or to the artist who first discovered perspective? Such questions illustrate a fundamental misunderstanding. The intelligence that was favored by evolution is not the

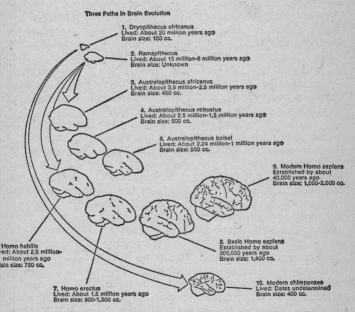

Three Paths in Brain Evolution

1. Dryopithecus africanus
Lived: About 20 million years ago
Brain size: 150 cc.

2. Ramapithecus
Lived: About 15 million–6 million years ago
Brain size: Unknown

3. Australopithecus africanus
Lived: About 3.5 million–2.5 million years ago
Brain size: 450 cc.

4. Australopithecus robustus
Lived: About 2.5 million–1.5 million years ago
Brain size: 500 cc.

5. Australopithecus boisei
Lived: About 2.24 million–1 million years ago
Brain size: 550 cc.

9. Modern Homo sapiens
Established by about 40,000 years ago
Brain size: 1,000–2,000 cc.

8. Basic Homo sapiens
Established by about 300,000 years ago
Brain size: 1,400 cc.

Homo habilis
Lived: About 2.5 million–million years ago
Brain size: 750 cc.

7. Homo erectus
Lived: About 1.5 million years ago
Brain size: 900–1,300 cc.

10. Modern chimpanzee
Lived: Dates undetermined
Brain size: 400 cc.

FIGURE 10. In Leakey's view, depicted here, the brain of *Homo* grew bigger and more complex than those of his primate and hominid cousins over a span of about twenty million years. The larger temporal lobe (the finger-shaped section), for instance, permitted language comprehension.

same intelligence we think about when we review our children's report cards. Proof of this comes from the study of brain size in different animals.

As a general rule, the larger the brain size across species, the smarter the animal. The qualifying phrase "across species" is the key to an important distinction missed by many theorists who should know better. According to evolutionary theory, intelligence resulted from natural selection for encephalization: the evolution of larger brains beyond the enlargement that would be expected from the evolution of larger bodies—the critical 20 per cent I referred to earlier.

Larger brains are correlated with a greater number of nerve cells in the cerebral cortex as well as a denser concentration of acetylcholine, an important chemical for nerve transmission. In addition, there is a greater ratio of neurons to glia, the general supporting cells, as well as an enlargement of the size of the subcortical brain structures. In other words, if we know the weight of an animal's brain, we can accurately measure the number of cortical neurons in the brain and arrive at a reasoned estimate of brain structure and chemical composition.

From here some people have jumped to the mistaken conclusion that brain size in a human can be correlated with intelligence. Goethe and Darwin are often quoted as examples of the enlarged brain—superior-intelligence concept. Scientific investigation of the matter, however, gives no support to this theory. Some of the most severe forms of mental retardation, for instance, demonstrate megalencephaly (enlarged brain), while recognized geniuses are often discovered after death to possess brains at the lower range of normal (usually less than three pounds). What this means is that the "intelligence" favored by natural selection is not necessarily the same "intelligence" we are speaking of when we refer to I.Q. No one has ever demonstrated, for instance, that genius, or even increased intelligence, can enhance survival value in the human species. Certainly contemporary demographic data do not support the view that human intelligence is correlated with fecundity. Rather, evolution is helpful only up to a point where new explanatory principles must be sought.

"Biological intelligence is a measure of the quality of the real world created by the brain of a particular species," says Dr. Harry J. Jerison, author of *Evolution of the Brain and Intelligence*. "The world as we know it ourselves represents the human grade of biological intelligence. Different worlds are presumably constructed by different species."

From here Dr. Jerison offers a series of philosophically intriguing speculations about the development of intelligence in different animal species. "The real world we know intuitively is a creation of the nervous system, a model of a possible world which enables the nervous system to handle the enormous amount of information it receives and processes. The 'true' or 'real' world is specific to a species and is dependent on how the brain of that species works. This is as true for our own world—the world as we know it—as it is for the world of any species. The work of the brain is to create the model of a possible world rather than record or transmit a world that is metaphysically true.

"A very simple construction of a world may be characteristic of the lower vertebrates. As a matter of fact, no transformation of neural information—no construction— at all may be required in the lower vertebrates. Their behavior is tightly bound to specific stimuli by fixed action patterns of response, in contrast to an intelligence system in which varied patterns of stimuli are transformed into invariant objects. Intelligence in biological perspective is clearly only one of several dimensions of behavior."

Man's rapid, one could almost say explosive, brain growth during the last 250,000 years, remains the most inexplicable aspect of evolution. "Natural selection could only have endowed the savage with a brain a little superior to that of the ape, whereas he possesses one very little inferior to that of an average member of our learned society," is as true today as when Wallace wrote it to Darwin.

One possible solution to this puzzle involves a reemphasis of the role of cultural rather than strictly biological evolution. Let me illustrate what I mean by a brief scenario.

Imagine a Stone Age hunting culture languishing and dying because nobody was smart enough to figure out the

87

solution to an important problem. It doesn't matter for our purposes what the problem was, but as an example let us imagine that the wheel had not yet been invented. Certainly any culture, particularly a nomadic one, operated at a great disadvantage before the invention of the wheel. After the backs of all available animals were loaded, there remained only the backs of those humans strong and willing enough to transport the possessions of a village.

There is nothing about hunting and killing that endows a hunter with an intuitive feeling that would lead necessarily to the invention of the wheel. Something else would be required—let's call it analytical thinking—and let's further assume that nobody had it. Now when we introduce even one person, the product of genetic mutation, who has the ability, hitherto unknown in the world, to sit quietly and think, the invention of the wheel becomes not only possible but probable. If such a person lived long enough and was encouraged to do so, he might well turn his analytical powers to the solution of the nomadic tribesman's transportation problem. And in the process, such a person might invent a primitive wheel composed of a circular stone fitted to two birch barks and pulled along by a tribesman.

Such inventiveness would no doubt be rewarded, since the concept of the wheel, once apprehended, works to the benefit of everybody in the society. One possible reward for our inventor might have involved a tribal dispensation from the rigors and dangers of the hunt. We might even speculate that the type of cerebral processing responsible for the invention of the wheel might actually conflict with the prowess and stealth required of the early hunters.

Naturally, dispensation from the hunt could be expected to confer some survival advantages on our hypothetical inventor. Free to remain home to solve other pressing problems, he would also be less likely to be killed and, thus, more likely to breed successfully. One or two generations later, such analytical thinkers could be expected to increase in numbers but, paradoxically, to their own disadvantage. At a critical point their analytical model for cerebral processing could be expected to be fairly common, perhaps even predominant and with a most predictable result.

At some point inventive thinking would reach a low enough value that hunting would once again be required. Predictably, the number of analytic thinkers would then decline, a natural consequence of their increased vulnerability to the dangers of the hunt. In this way a balance could be maintained between the need for hunting and the need for inventive thinking.

How could one fit such a scenario into an evolutionary framework? As previously mentioned, I would favor considering the whole process of brain evolution as a cultural rather than a strictly biological process. The society decides what is needed and favors it ("wheel inventors need not be engaged in hunting activities"), and at a later time limits its unchecked expansion. In this way certain types of brains are favored and can multiply for a time, but always under the constraints of culture rather than strictly biological forces. Stated simply, a culture recognizes and rewards those cerebral processes that accrue to the community's advantage, while ruling against the merely novel.

Such an interpretation is consistent with what we know about the brain's explosive development during the last 250,000 years. As Darwin himself realized, evolutionary theory alone cannot explain why the human brain developed far beyond what was needed to confer survival advantage. Cultural factors, however, do offer a partial explanation. We think and act the way we do because at some time or other in the past the kinds of brains we now possess conferred on our culture a survival advantage.

As early man shifted from the forest to the savannahs, close cooperation favored successful hunting and the formation of culture. Later, when hunting was replaced by farming, even stricter conformity was required. In each case certain capacities of the brain were favored over others and developed in response to cultural change. Was there any need for diplomacy before the formation of nations? Could there have been poets prior to the development of language?

Cultural evolution is best thought of as a dynamic

process in which survival considerations may encourage certain behaviors at one time and punish them severely at another. In some cases the behavioral changes may span millions of years, while in others, the whole process may take place in less than a century. The ability to quick-draw a six-shooter, for instance, conferred survival advantage in the early American West. Today such a talent is of dubious value to, let's say, a job applicant at a law firm. While one Albert Einstein in a century makes cultural evolutionary sense, how about a whole continent of physicists preoccupied with subatomic particles? No doubt our culture would recoil at a society so overbalanced in one area and would condemn its geniuses to the modern equivalent of the hunt.

Let's now look at some of the ways the brain is shaped by its experiences.

PART
3

6
Mirror of Reality

A young man said to me:
I am interested in the problem of Reality.

I said: Really!
Then I saw him turn to glance, surreptitiously,
in the big mirror, at his own fascinating shadow.

D. H. LAWRENCE,
Ultimate Reality, 1930

Philosophers have speculated for centuries on the nature of reality. Do our senses provide us with a trustworthy model of the "real world"? And what about the reality of animals? When a cat is placed in a room with its master, do both experience the same reality? Most are agreed they do not. But can studies of the perceptions of a cat tell us anything about our own perceptions? Only recently have we developed the tools to answer such questions, and in the process we are learning about the nature of our own reality.

Two Harvard psychologists, David Hubel and Torsten Wiesel, inserted tiny electrodes into single cells in a cat's visual area while simultaneously flashing simple visual shapes onto a screen in front of the cat. Usually these were bars of light, presented each time from a different visual angle. At a certain angle the cat's brain cells would fire, yielding a burst of impulses transmitted from the implanted electrode. As the bar continued to turn, the firing would stop, only to be taken up again by a neighboring cell which responded best to the new angle. The researchers discovered that different cells respond at different angles, with cells deep in the brain responding to more generalized characteristics, no matter where on the cat's retina the signal originated. Other cells responded to movement in rather more complex patterns. The important point is that the cat's brain is programed to select certain aspects of its environment determined by the selective responsiveness of the cells in the visual cortex.

Hubel and Wiesel's work still leaves us with the question of the possible usefulness of such an arrangement. A similar experiment by Jerome Lettvin, also of Harvard, provides the answer. Lettvin and his colleagues, in a paper entitled "What the Frog's Eye Tells the Frog's Brain," recorded impulses from the fibers in the frog's optic nerve. In the frog the programing was done in the optic nerve fibers rather than in the visual area of the brain. This doesn't negate any of Lettvin's conclusions, however, since the optic nerve is actually an outgrowth of the brain. When a doctor looks in the back of your eye with an opthalmoscope during your annual physical exam, he is actually looking at the optic nerve, a bundle of fibers that originate at the retina and convey impulses toward the brain's visual area. Lettvin discovered that a frog's optic nerve responds only to small objects that move with jerky motions, a quality that Lettvin aptly dubbed "bug perceivers." In other words, the frog's visual perception is geared to detect the quick, jerky movements of bugs as they leap about in the pond. If a frog is imprisoned in an aquarium, the frog will get along quite well if provided with living insects. But if someone does the frog a "favor" by killing the insects, the frog will starve to death. It is not so much insects as their movements which stimulate the frog's visual perception.

Other tricks can be played on an animal's visual perception. Colin Blakemore and Grant Cooper, of the Cambridge Psychological Laboratory, placed newborn kittens into special chambers lined with either vertical or horizontal stripes, never both. When the kittens were removed from this bizarre world their behavior was most puzzling. The "horizontal" cats had no problem jumping from the laboratory table down to the floor, but once there, they walked into the legs of the laboratory chairs. The "vertical" cats, in contrast, were able to avoid walking into the chairs, but were incapable of jumping across from one table to another of the same height. It was as if horizontal surfaces did not exist!

With electrodes placed inside the cats' heads, the "vertical" cats were discovered to be missing horizontal-feature detectors. As a result, these cats literally didn't see any object in the horizontal plane. The "horizontal" cats, in contrast, possessed no vertical detectors, hence their inability to perceive chair legs and the other vertical objects perfectly obvious to any cat raised in a normal environment. It seems that the cat's early visual-perceptual experiences programed the visual detectors to mirror these early experiences almost perfectly. If the environment is a normal one, visual perception will develop normally and the animal will be capable of normal processing. If the environment is an abnormal one (in this case a laboratory experiment), the visual processing will be programed according to the stimuli the cats encounter in their atypical environment.

Other researchers have carried the Blakemore experiment even further. A California team exposed test animals to an environment composed of spots rather than stripes. In this case the visual detectors failed to respond to lines of any orientation. Instead, their brain cells were sensitive to pinpoints of light in the environment. Although we can only guess how such an animal might perceive the world around it, it seems likely that the world conforms to a hazy formless perception, similar to what we see when we're caught in a fog. An alternative model might be a world similar to Georges Seurat's pointillistic style of painting.

Theoretically, at least, a laboratory animal's "reality" can be preset by its experiences during a critical period which occurs somewhere between three weeks and three months. In addition, there are vast differences in the perceptual world of different animals. While many are unable to distinguish one color from another, some have the capacity to detect wave lengths that are invisible to us. Whales are reported to be able to detect the sound of other whales across vast stretches of water, even a thousand miles away, while rabbits can respond to tiny variations in the sun's movement. In each instance the animals perceive only what their brains are organized to tell them.

The nature of visual perception, as with many of the questions we are discussing in this book, awaited experimental methods capable of putting different hypotheses to the test. A frog starving in the midst of a plentiful supply of dead bugs is an observation that could have been made, and perhaps was made, thousands of years ago. The exact explanation, however, that the frog's visual detectors are geared to certain patterns of movement, awaited the development of a technology capable of measuring impulses in a frog's visual system. I would suggest that such psychobiological explanations, all products of less than twenty years of research, have done more to help us explain "reality" than two thousand years of Western philosophy. But, as with D. H. Lawrence's young-man staring into his mirror at the beginning of this chapter, we cannot be faulted if we insist that our understanding of reality be based on something that reflects ourselves more directly than studies of cat brains and starving frogs. Let us move now to what psychobiologists can teach us about our own fascinating shadow.

In December 1958, a fifty-two-year-old man was operated on at the Royal Birmingham Eye Hospital in England in an effort to restore sight which he'd lost due to an eye infection contracted at ten months of age. Within a month of the operation, the patient (known forever after in the neurologic literature as S.B.) was the subject of a series of enthusiastic articles in the London *Daily Telegraph*. Widespread public interest followed in the wake of S.B.'s rapid visual recovery, which appears to have been almost miraculous. Within hours of leaving the operating room, he was recognizing things around him, reading, telling time at a glance from a wall clock, and generally engaging in all the activities normally carried out by a person with normal vision.

Among the readers of the *Daily Telegraph* article was a young Cambridge psychologist, Richard Gregory, who now directs the Brain and Perceptual Laboratory at the

University of Bristol. Gregory's special interest at the time was in the brain mechanisms responsible for human perception. The newspaper account reminded Gregory of a long-standing but unanswerable question about human perception.

"Suppose a man born blind and now an adult," wrote William Molyneux, a seventeenth-century philosopher, to the English philosopher John Locke in 1690, "and suppose that he has been taught by his touch to distinguish between a cube and a sphere of the same metal. Suppose then that the cube and sphere were placed on a table and the blind man were made to see: query, whether by his sight before he touched them, could he distinguish and tell which was the globe and which the cube? The acute and judicious proposer answers: not."

Locke's reply was intriguing but, in the absence of such a blind person, essentially untestable. In *An Essay Concerning Human Understanding* Locke commented: "I agree with this thinking gentleman whom I am proud to call my friend in answer to this his problem and am of the opinion that the blind man at first would not be able with certainty to say which was the globe and which was the cube."

For almost three hundred years the question concerning how a blind person would behave if he were granted sight remained unanswered. The publication of S.B.'s case by the *Daily Telegraph,* coincident with Gregory's reading of the article, laid the groundwork for one of the most intriguing investigations ever conducted into the workings of the human brain.

"The first examination of S.B. was carried out in a quiet private room in a hospital lit by winter daylight and lasted about three and a half hours," Gregory recalls. "At first impression he seemed like a normally sighted person, though differences soon became obvious."

Among the differences were S.B.'s failure to appreciate height. While looking down from his room, about forty feet above the street, S.B. mentioned casually that he could lower himself down with his hands. Later, when looking up at the window from outside on the street, he seemed amazed at his previous claim to be able to climb

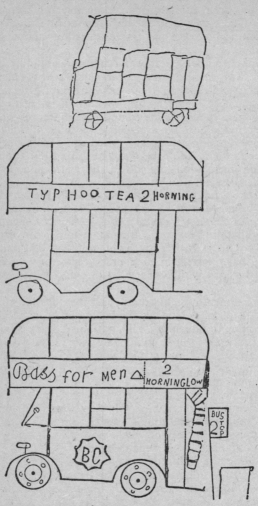

FIGURE 11. S.B.'s sequential drawings of a bus, beginning with the first drawing soon after recovery of his sight, and ending with the drawing made one year later.

down. On first seeing a bus, he commented that its height seemed excessive in comparison to its length. Regarding the recognition of faces, S.B. proved himself totally dependent on voice clues. His own description on first seeing his doctor's face demonstrates his problems in coming to terms with vision. "I saw a dark shape with a bump sticking out and heard a voice. I felt my nose and guessed that the bump was a nose, and I knew that if this was a nose I was seeing a face."

S.B.'s drawings provide a clue to what was going on. Requested to draw a bus, he sketched a bus with spokes on the wheels, despite the absence of spokes on London buses for the previous twenty years. In later drawings the spokes disappeared as he began to depend more on his new visual knowledge rather than his previous touch experiences with buses gained as a child. (He remembered being "shown," at about age eight, a bus which he was allowed to touch. The spokes were particularly memorable, since they corresponded roughly to his height at the time.)

In all the drawings the buses are shown in profile facing to the left, which corresponds to how S.B. would touch them as he got on and off over the years. (Recall that traffic is on the left side of the road in England.) In essence, S.B. was demonstrating what psychologists refer to as "cross-modal transfer": Objects and events perceived by touch are transferred into visual perceptions. Thus, spokes on a wheel, because they were once felt, are at a later point "seen" and put into a drawing, even though a photograph of a bus taken at the same time would not show spokes.

Despite improvements in his drawings, certain significant omissions continued. In all cases the radiator was missing, an unlikely object for a blind man to know by touch, since the front of a bus is a position of hazard to the unsighted. S.B.'s drawings explain both his perceptions and his subsequent tragic end. "These drawings," according to Gregory, "illustrated the general finding that although S.B. came to use vision, his ideas of the world arose from touch. His general life as a blind man remained with him until his death."

Along with these startling and unexpected perceptual

peculiarities, S.B. also underwent a gradual but unmistakable mental deterioration. "Before the operation he was regarded by everyone as a cheerful, dominant person," according to Gregory. Although severely handicapped, the blind S.B. had ridden on motorcycles, confidently stepped from the curb into heavy traffic, even earned a respectable wage as a shoe repairman. But several months after the operation which restored his sight, the self-confident blind man was reduced to a lonely recluse, who on most occasions never even bothered to turn on the house lights in the evening, but sat quietly alone in the darkness.

His comments to Gregory provide a touching commentary of a man losing his grip and sinking into an irreversible melancholy. "He complained to me that the world seemed a drab place," Gregory, a huge bear-like, friendly man, told me during our conversation on a brilliant summer afternoon on the quadrangle at St. Catherine's College in Oxford. "The one spontaneous comment S.B. made one evening was to describe the colors in the sky at sunset and to end sadly: 'Then we came down the hill and it all disappeared.'" On another occasion S.B. commented as if to himself: "I always felt that in my own way women were lovely, but now that I can see them I think they are ugly."

At the root of S.B.'s unhappiness was his inability to shift from a lifelong dependency on hearing and touch to his new "gift of sight." Never was this more clearly demonstrated than during a visit, arranged by Gregory, to the Science Museum in South Kensington.

"The most interesting episode was the reaction to a cutting lathe which is housed in a special glass case. It is a large and fairly simple tool. We chose this object because the lathe must be a tool that he often wished to see. We led him to the glass case, which was closed, and asked him to tell us what was in it. He was quite unable to say anything about it except that he thought the nearest part was a handle. He complained that he could not see the cutting edge, or the metal being worked, or anything else about it, and appeared rather agitated. We then asked a museum attendant for the case to be opened and S.B. was allowed to touch the lathe. The result was startling. He ran his hands deftly over the machine, with his eyes shut. He then

100

stood back a little, opened his eyes and said, 'Now that I have felt it, I can see it.' "

Of all the "sights" of London, only one gave him pleasure. "He found all buildings dull and of no interest. His only signs of appreciation were in moving objects, particularly the pigeons in Trafalgar Square. He took great interest in them and liked to touch them while he watched. He described how as a blind man he often felt isolated and sought such sounds of activity and movement."

S.B.'s attempt to come to terms with his new "reality" sunk him into an ever deeper melancholy. In a letter to Gregory, S.B.'s wife comments simply, "He is very disappointed about everything."

On August 2, 1960, a little over a year after the operation, the man whose sight had been miraculously restored died, in a deep depression, providing a tragic confirmation of Locke's dictum that the blind man would fail in the end.

"During all the years he suffered one of the greatest handicaps, he lived with energy and enthusiasm," says Gregory. "When his handicap was swept away as if by a miracle, he lost his peace and self-respect. We may feel disappointment that a private dream came true; S.B. found disappointment in what he took to be reality."

S.B.'s story, as related to me by Gregory, is deeply poignant and at the same time disturbing. At the root of my discomfiture, and perhaps Gregory's as well, although I did not ask him about it, is the harsh light that S.B.'s experience sheds on some of our most cherished illusions. Helping the blind to see provides stimulus for both medical research and religious belief. To me the most touching of Christ's miracles were the ones in which he restored sight to the blind. But, as S.B.'s sad experience makes compellingly clear, the real world doesn't fit our medical or religious preconceptions quite so neatly.

Just as no one could guess that a frog would starve to death in the midst of edible but unmoving insects, who could have predicted S.B.'s strange fate? After a lifetime of blindness, a fifty-two-year-old man's sight is restored through the work of a master surgeon. Yet the final result forces us to conclude that the man would have been better off living out the rest of his life in perpetual darkness. Why?

I pressed Gregory to speculate. "He learned to rely on his vision, once he had it, but this very reliance cost him his self-respect. As a blind man he had gotten on very well, but when he finally could see, his previously remarkable achievements seemed paltry and his position almost foolish. He felt that he had lost more than he had gained, by the recovery of sight."

From experiments with cats raised in an artificial environment, it appears that knowledge of "reality" can be determined by the brain's early learning experiences. S.B.'s tragedy suggests that, once established, our early perceptual patterns may be extremely resistant to change, even change toward what would seem to be improvement. This isn't to say that change can't occur, however, as we'll see in a later Chapter, Eighteen, where I'll describe some ongoing research on sensory substitution which may enable the blind to "see" through the use of their touch impulses. For the most part, however, our brain learns to perceive a "reality" and, after a certain critical point, is very resistant to change.

Although none of us were raised in rooms composed of either vertical or horizontal lines, the principle remains the same. Our brains have certain behavioral potentialities (feature detectors) that can be recruited to respond to whatever may be encountered in the environment. The way they respond, however, depends on the stimulation provided. But as the cat experiments make clear, once the patterns are set, they are likely to remain and determine how the animal or person will behave in the future.

Similar differences in perception, based on life experiences, may also exist for human beings. Dr. R. Annis and Dr. B. Frost tested visual acuity among the Cree Indians, a tribe living along the eastern coast of the James Bay in Quebec. The Crees alternate between living in summer cook tents and winter lodges. Both structures differ from our buildings inasmuch as they have fewer vertical and horizontal contours. When the visual acuity of the Crees was measured with an apparatus displaying lines in all orientations, their performance showed no preference for any particular orientation. Native-born Canadians, on the other hand, who lived in cities composed predominantly of vertical and horizontal structures, exhibited a strong preference for vertical and horizontal lines. The preferred orientation depends, it seems, on the visual angles most frequently encountered.

Such findings raise haunting questions about our own reality. In essence, do our perceptions provide us with a knowledge of what is really "out there"? Or, like the cats raised in a world of verticals and horizontals, do we respond only to those parts of reality determined for us by our earliest experiences?

A partial answer to these questions comes from studies of early brain and behavioral development in normal infants:

All knowledge bears the imprint of the mind's own structure

STEPHEN TOULMIN

In the Psychology Laboratory of the University of Edinburgh, a young mother and her six-day-old baby are seated before a camera. On cue, the mother sticks out her tongue, wiggles it briefly and then waits. Ten seconds later, the baby opens its mouth and performs a similar movement. Next, the mother flutters her eyelids, stops and opens her mouth, this time without extending her tongue. A few seconds later, the baby's eyelashes flutter, its mouth opens briefly and then closes.

The mother and baby are participants in a research project carried out by Dr. T. G. R. Bower, Professor of Psychology at the University of Edinburgh. Bower's goal is to discover those aspects of human development that are inborn and cannot be explained by learning or experience. The mother-and-baby experiment, Bower believes, proves that infants, even as young as six days, are endowed with complex perceptual abilities. Who taught the baby to recognize its tongue? How can it know that the tongue in its mouth corresponds to the red object sticking out between Mother's lips?

"Such a performance implies that the baby has a

103

perceptual image of parts of its own body which is elaborate enough to allow it to be able to identify parts of its own body with parts of other people's bodies," says Bower. "Remember that we are talking about newborn babies who have never examined themselves in mirrors or done any of the self-discovering things that adults can do."

Bower's observations have implications that go far beyond infant perception. The recognition of the mother's face and the baby's ability to correlate it with its own body suggest a primitive awareness of similarity. Is it possible that the baby, even this soon after birth, can recognize itself as human? Perhaps this is reading too much into a simple observation. Still, the fact that infants can carry out simple imitative gestures has laid to rest a two-thousand-year-old theory about brain development.

Until very recently, scientists could only guess about the mental world of infants. Based on the long period that infants remain dependent on their mothers, early pre-scientific observers suggested that babies are born with a largely nonfunctioning brain that requires development through experience. Modern psychobiology provides some small support for such a theory, since over five sixths of the brain's development takes place after birth. The theory is contradicted, however, by studies of other animal species. As a general rule, the more advanced an animal species, the quicker its newborns achieve independence. Why should humans be exceptions to this rule?

Bower and his associates base their research on the hunch that humans are not exceptions, but, on the contrary, possess at birth truly remarkable perceptual abilities. But to demonstrate these abilities, scientists must combine careful observation with more "natural" experiments, since, to no one's surprise, babies are notoriously bad research subjects. For one thing, their attention span lasts only a few seconds. For another, newborns seldom remain awake for longer than six minutes at a time. As an adaptation to these difficulties, most of Bower's experiments are carried out with a minimum of equipment and involve infants engaged in short, spontaneous interactions.

One experiment tests whether or not a baby will respond to danger by defending itself against a threatening object. When the baby is held facing an approaching box, it will thrust its head backward and raise its arms in front of its face, a clear demonstration that the infant is capable of detecting motion. Differences can also be demonstrated on the infant's response to soft objects compared to hard ones. Since "hard" depends on a sense of touch, such a finding is unexpected, according to the traditional theory that babies can combine their sensations only at a later stage of development. Bower's experiment suggests an alternate and exciting explanation: A baby is endowed from birth with the ability to perceive a unified perceptual world. Other experiments also tend to support the view that infants perceive a "whole world" rather than seeing one portion of it, hearing another portion, and, at a later point, combining their separate perceptions.

The key experiment proving this is concentrated on sound localization in infants only a few hours old. When a sound came from the baby's right side, the baby tended to turn its eyes to the right. When the sound came from the left, the opposite occurred. These results suggest an innate ability to localize sound. More important, it suggests the occurrence of intersensory co-ordination between seeing and hearing. Since the babies, in most cases, are only a few minutes old, they obviously haven't learned that a sound implies a strong possibility that there will be something to look at: the source of the sound. Turning the eyes toward the source of a sound is inborn and dependent on genetic and biological mechanisms. Learning and experience appear to play little, if any, part.

Bower's observations are important for the light they shed on an ancient dilemma. Which is more important for normal development, heredity or environment? For years, experts believed that man's capacities were strongly dependent on the environment. To Aristotle the brain was a *tabula rasa,* a blank slate, on which the senses wrote out the directives for "human nature." "There is nothing in the mind that was not first in the senses," was the intellectual pennant that was waved throughout two thousand years of Western philosophy. We now know that this is wrong. Even an infant born five minutes ago is capable of marvelous perceptual performances.

105

For the infant, perception is an exploratory act directed at specific events in the environment. It is also enriched by novelty. From birth, infants choose to look at novel rather than familiar things. The presentation of something new is enough to produce a state of attentiveness indicating that the infant is becoming interested in its environment. Moving objects are preferred to stationary ones, perhaps because the movement offers so much more information. They usually prefer noisy objects to silent ones for the same reason.

One researcher tested three-month-old infants by showing two different movies on adjacent screens. With one movie he played the sound track of that film; the other film was run without a sound track. The infants spent most of their time looking at the film with the sound track, proving that even at this early stage infants prefer stimulation that demands co-ordination of eye and ear. Infants expect to hear things that they have seen, and eventually to touch things that they have heard.

One of the most intriguing examples of inborn perceptual patterns comes from the work of Dr. R. Fanz. An infant lies comfortably on its back in Fanz's special test

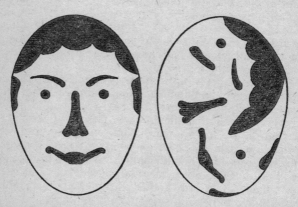

FIGURE 12. The test patterns used by Fanz to check for facial recognition. A baby will spend more time looking at the natural face patterns than at the randomized face-like design.

apparatus, while two test patterns are projected on the ceiling above. One of them is a human face; the other is composed of the elements of the face with its parts distorted. (See Figure 12.) Using a camera that traces the infant's eye movements, Dr. Fanz was able to determine which of the two figures commands the infant's attention. Since Fanz's experiments are performed on infants only a few days after birth, the contribution of learning or experience is minimal. By measuring the length of time the infant's eyes rest on each figure, Fanz found that infants prefer a face-like design over similar figures with the facial elements distorted or juxtaposed. Even at this early stage, the human face is a meaningful object!

In addition, infants are able to select against features in their environment that are potentially hazardous. Imagine an infant crawling along the edge of the Grand Canyon. Then imagine it looking over the edge into the vast abyss below. Would it reach out for something and, in the process, fall to its death on the rocks? Naturally, no one knows for sure what an infant would do in such a situation, and the experiment would obviously be unthinkable. Still, speculation on the behavior of a baby at the edge of the Grand Canyon stimulated Cornell University psychologist Eleanor Gibson to devise a close approximation to this tantalizing and terrifying scenario.

In Dr. Gibson's visual cliff experiment the baby is placed on a table designed with checkerboard patterns. At the edge of the table the wood suddenly merges into a transparent glass top through which, several feet below, the baby can see the floor. In essence, the design apparatus is the Grand Canyon experiment transferred to the laboratory, only the whole thing is completely safe, since the glass is strong enough to support the baby, and the "drop" only an illusion. In this experiment on visual depth, babies (or test animals, for that matter) will refuse to crawl out onto the glass. If placed on the glass the baby will immediately crawl back to the safe side. Anyone who has ever walked a small dog toward a subway grating has witnessed a similar reluctance as the dog skirts the edge of the grating through which it visualizes a sharp drop.

Gibson's experiment demonstrates that infants, even

at this early crawling stage, can visually appreciate depth. (Incidentally, the question posed by the "Grand Canyon Experiment" still isn't answered, since babies often "forget" the location of their legs and may inadvertently fall over the edge anyway.)

Based on experiments like those of Fanz and Gibson, scientists postulate that infants are born with certain inherent perceptual biases. Through the use of these biases they build up a behavioral repertoire. According to such a theory, perceptual-behavioral development is similar to something like the child's emerging capacity to play with blocks. At first, its activities are restricted to pushing blocks around on the floor (or, more likely, putting them in the mouth). Later, one block can be placed on top of another until, at a still later point, a tower of blocks results. At first glance, the infant seems to follow just such a pattern, with behavior becoming increasingly sophisticated.

Initially, the infant lies quietly on its back (the stage at which Dr. Fanz conducted his experiments). Later, at about six months, the infant can sit with only slight support. By twelve months it is able to walk while holding onto object (cruising), followed soon after by upright unsupported locomotion. From here, arm, hand, and leg coordination proceed at an irregular but progressive pace.

"The child is father to the man" speaks of deeply ingrained prejudices that the path from infancy to adulthood is a steadily progressive one. As with many other "common sense" notions of human development, however, the theory of a steadily progressive behavioral and perceptual maturation is wrong, based on insufficiently detailed observation. The trouble with everyday observation is that everyday observers aren't observant enough, or maybe they are unwilling to carry out the pertinent experiments. (Perhaps this is all to the good. Given the variations that exist in child-rearing, children are already involved in more than enough experiments!)

In any case, psychobiologists have recently arrived at some startling insights into the processes involved in normal infant perception and behavioral development.

Instead of a steady progression toward the attainment of behavioral skills, normal infants mature in starts and stops. A six-day-old infant, for instance, will walk with support, its legs carrying out a perfectly normal stepping movement. If the front of its ankles gently touch against the side of a table, the infant will try to step up onto the table. Both abilities disappear at about eight weeks of age, with walking returning in about a year, while the stepping response is never observed again.

An infant, during the first week after birth, will reach out for a ball placed several inches above its face, a remarkable exercise in eye-hand co-ordination that disappears at about a month of age, only to reappear once again at five months.

Similar early co-ordination can be seen in an infant's response to sound. From birth to about six months, an infant will turn its head and eyes toward the source of a bell. This response also fades at about six months of age.

"My observations seem to be at odds with the view that psychobiological growth is a continuous incremental process, like physical growth," says Bower. "Abilities seem to appear and then disappear, leaving the child worse off than he was when he was younger, perhaps no better off than when he was an infant."

Support for such a concept comes from Bower's experiments on how infants reach for objects. As we have already seen, infants lose this ability at about one month of age, recovering it about four months later. Bower's colleagues attempted to relate phase one (the early grasping seen in the first month) to phase two (its return four months later). Bower postulated that phase one must somehow be related to phase two and, as a result, practice during phase one might lead to an earlier and more efficient phase-two performance.

Bower's results confirmed the general hypothesis that those infants who had been practiced in the act of reaching for objects were more skilled at reaching during phase two, neatly snaring dangling toys presented to them from a variety of positions. More refined methods of testing were required, however, to come up with the reason why "practice makes perfect."

In one study Bower furnished the infants with a set of glasses consisting of wedged prisms instead of lenses.

We'll talk later about adult studies using prisms. For now it is enough to know that prisms displace an object's apparent location in space. Depending on the type of prism used, things can be made to appear upside down, horizontal instead of vertical, vertical instead of horizontal, or, as in Bower's experiments, shifted ten or fifteen degrees to the right or left.

Normal infants of five months, when confronted with an object, reach out and grasp it. With the prisms in place, they miss, withdraw their hand, try again, miss again, and finally give up. In contrast, older infants, as well as those with previous practice, are able to correct for the distortions by focusing on their hand and guiding it toward the displaced object.

As a further refinement in subtlety—the kind of "you haven't seen anything yet" experiment that Bower is famous for—the lights are turned out immediately after the infant spots the test object, usually a ball or a rattle, and then his behavior is observed via an infrared television camera. Those in phase one, the youngest, reach out unerringly in the darkness and grasp the stimulus. The older infants in phase two, in contrast, along with the infants who have outgrown phase one and are not yet into phase two but are practicing, fumble clumsily in the darkness. Only after several trials are they able to finally reach the object.

Under conditions of normal vision, it is difficult to tell what clues are used by the infants. Turn out the lights, however, and the infant must depend on his visual memory—the process almost certainly involved in phase-one reaching, since accurate grasps can be made in complete darkness. The phase-two infant, in contrast, does poorly and appears to depend on the observation of his own hands in front of him. Naturally, in the darkness such an infant cannot see his hands and therefore cannot correct for an error.

Bower's experiments demonstrate, in essence, that similar behavior (reaching out for an object) is based on different perceptual-behavioral mechanisms, depending on the infant's age. In the first phase, grasping is skillfully done but is less functional than in phase two, where the

110

infant can use clues from the visual position of his hand. In an ambiguous situation, a phase-two infant can literally follow his hand to an object, while a phase-one infant is dependent on vision, and thus is open to the effect of illusions and the tricks played on it by perspective. Hence, the discontinuity is explained: The infant "forgets" how to reach for an object, only to recover the capacity at a later date, based on more efficient mechanisms.

"My colleagues and I propose that there is a similar kind of process underlying perceptual development: as the child grows older, he progressively elaborates his internal description of events to make them more specific," says Bower. "Such a change in favor of more specific description acts to increase the likelihood of a smooth transfer from one skill to another."

Bower's experiments, along with the study of "vertical" and "horizontal" cats, suggest that perception is a dynamic rather than a purely passive or receptive process. We are literally created and transformed by our perception of the world around us. In addition, as in the case of S.B., perceptions are relatively permanent and resistant to change, even when the change seems to be an improvement (e.g., recovery of vision after a lifetime of blindness).

In a sense, each of us creates our own "reality" by the perceptual acts in which we are engaged. The truth of this is most clearly illustrated when we examine some of our perceptual acts.

Study the cube in Figure 13 for one minute. Keep your eyes fixed on the small horizontal crossbar. Within a

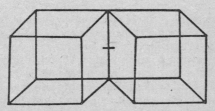

FIGURE 13. Double Necker cube. If you stare at the small horizontal crossbar in the center, the two cubes will shift in position —sometimes together, sometimes separately.

111

few seconds, the crossbar will suddenly shift from the back of the cube to the front. The inventor of this optical illusion was a Swiss geologist, Louis Necker, who described the experience as "a sudden and involuntary change." Without any effort on your part—in fact, despite your attempts to retain a constant perception—the figure shifts beneath your gaze. Rather than being dependent on your wishes, the phenomenon persists despite your deliberate efforts to stop it. It is an experience over which you have little or no control. In addition, you're not engaged in an intellectual exercise, since the cube is merely stared at without any attempt to examine any of its properties or deduce anything about it. Your perception in such an instance is best understood as an alternation between two equally possible interpretations of the cube. In fact, there is no "correct" choice to be made. Whether the crossbar is on the front or back of the figure isn't a question that can be settled by reference to any external "reality." The highest court of appeals in such a case is your own perception, which alternates between two equally plausible explanations. When it comes to perception, our brain can be thought of as a gambler who computes odds according to the statistical likelihood of one explanation or interpretation being the correct one. In most life experiences the bets are correct and we act in accordance with external "reality." But can our "reality" be manipulated? Can we be the victim of a sort of cosmic trickster?

In Figure 14, two people are standing against a wall; one of them appears a giant, the other shrunken to the size of a small child. But both people are, in "reality," approximately the same height; the distortion depends on our point of view. The room is specially constructed to give the perception of a normal rectangular room, despite a cleverly designed distortion. If we look at the people from the designated viewing point in Figure 15, our brain will make the wrong bet and decide that the problem must be with the people instead of the room. Giants and midgets are not outside our experience, although perhaps limited to the special circumstances of a circus; and those of us who are not experimental psychologists are unlikely ever to have encountered a room designed quite like this one. We are

112

vfewing point

FIGURES 14 & 15. A distorting room, where our perceptions are fooled and we see "a giant and a dwarf" instead of a cleverly constructed illusion.

fooled by our attempt to make sense out of a contradiction. Is there something wrong with the room or with the people? We bet on the people and lose.

We are able to correct our mistake by either looking at the room from the viewing point or (like S.B. touching the lathe in the British Museum) touching the room's walls with our hands or even a long stick. "Whatever I can touch I can see," S.B. explained, which corresponds in our case to, "Whenever my eyes and hands agree, I am less likely to be in error than when I depend on my eyes alone."

The perception of reality is best understood as a constructive process by which the brain builds useful models of the world. All of us possess useful internal models of what a room looks like, or how tall or small a person may actually be. We can spend a lifetime, and most of us do, without encountering a situation where our perceptual model of rooms and people are thrown into conflict. But such conflict situations can be constructed with rather haunting philosophical implications.

If our perceptions can be wrong, what does this tell us about the conclusions that may result from our logical thought processes? If we examine our beliefs, would they be just as parodoxical?

"Perceptions are guesses—hypotheses as to what object has produced the stimulation of the nerves," Richard Gregory explained when I put the question to him. "When we see things, what we see—what is present in the mind—is a hypothesis of what object or kind of object stimulated the eye. One might indeed think of perceptual hypothesis as the earliest form of belief. By developing elaborate internal descriptions of objects, the brain escapes the tyranny of reflexes: Decisions no longer depend directly on data available to the centers at that particular moment. The available sensory data can be used far more effectively, because the information can be expanded in terms of what has been learned in the past. The fact that there are perceptual illusions of many kinds, however, does seem to show that our perceptual hypotheses are only barely adequate. In unusual or specially misleading situations they can let us down badly."

114

Perhaps at this point you are thinking, "These experimental observations are very interesting, but life is not lived in the psychologist's laboratory. I would like to see something more lifelike, with a little more flesh and blood to it." In this you would be absolutely correct. Distrust of the results of laboratory experiments is a reasonable response to a long tradition of artificial experiments which have little relevance to everyday life. There is another but less well-known tradition about the brain's performance, however, that focuses on situations where the everyday world is bizarrely distorted.

In 1897, G. M. Stratton, a psychologist at the University of California, reported a series of experiments using distorted goggles. Using himself as the experimental subject, Stratton was the first man in history to see the world both upside down and reversed from side to side. Although the images of the world around him remained clear, things seemed strange and unreal, as if he were in a dream. "The memory images brought over from normal vision still continued to be the standard and criteria of reality. Things were seen in one way and thought of in a different way."

Stratton's contradictory impressions between what he "knew" and what he "saw" made even the simplest tasks extremely difficult and time-consuming. In addition, the experience created by this deliberately induced distortion was one of instability. "When I moved my head or body so that my sight swept over the scene, I did not feel as if I were visually ranging over a set of motionless objects, but the whole feel of things swept and swung before my eyes."

After less than a week, Stratton's ability to function in his strange world suddenly improved. He was able to walk around without banging into objects and, in addition, began reporting that his bizarre experiences seemed more natural. "When I looked at my legs and arms, what I saw seemed upright rather than inverted." At this point Stratton had become accustomed to an upside-down, right-to-left shifted world. Anxiously he brought his experiment to a close and removed his glasses only to find that "the reversal of everything from the order to which I had grown accustomed during the last week gave the scene a surprising, bewildering air which lasted several hours."

Stratton's experiments were the first to suggest that, while our perceptions are capable of profound modification, we may still be able to adjust to them. Even after turning the world upside down and shifting it from side to side, Stratton was able after a while to once again experience things normally. The reason he could do so depended on his active exploration of the new world by touch and movement.

A more modern goggle experiment, carried out by Irving Kohler, was even more revealing. In this experiment Kohler used glasses containing blue- and yellow-toned half lenses. When subjects looked to the left the world appeared blue, since they were seeing it through the blue half of the lenses. Looking to the right, things looked yellow as a result of viewing the world through the yellow half.

The subjects reported that after only a few days of wearing the goggles while going about their everyday activities, the colors returned to normal. The world no longer seemed blue when they looked to the left, or yellow when they looked to the right. If the subjects had passively worn the lenses and curtailed their activity, it's likely they would never have accommodated to the visual distortion imposed on them. By walking around, going to work, watching television—in short, living with the lenses—perception was changed back to normal. Proof of this came later when the lenses were removed. Instead of retaining the "normal" vision which the subjects had slowly acquired wearing the colored lenses, there were now complementary color changes which lasted for several days.

From experiments such as these, psychobiologists have modified our theories about behavior. For centuries scientists thought of behavior as a product of our senses. The goggle experiments, as well as the illusions described earlier, lead to just the opposite conclusion. Behavior is determined by what is perceived. If we sit passively with inverted goggles, the world will remain upside down. But by reaching out and interacting, our perceptions revert toward stability. If we are able to touch the walls of the distorted room, our perception of it is changed and what

116

was puzzling becomes clear. It's almost as if we're at a magic show and are suddenly able to view a magician's sleight-of-hand from behind the curtain!

Animal experiments confirm the belief that the accuracy of our perceptions depends on our ability to behaviorally explore the world. Dr. R. Held and A. Hein reared a litter of kittens in complete darkness until the time of their ingenious experiment. One kitten was then freed to explore the environment while dragging a basket containing a second kitten. (A pully arrangement enabled the second kitten's basket to move through the same territory covered by the active kitten.) These researchers discovered that the first kitten, the active explorer, developed perception, while the passive animal remained blind.

If a frog's eyes are rotated 180 degrees, it will move its tongue in the wrong direction for food and will literally starve to death as a result of the inability to compensate for the distortion. Goggles similar to Stratton's, when placed on a monkey, lead to immobilization rather than active exploration. In both of these instances, compensation for visual distortion seems not to take place. This capacity to compensate for drastic environmental alterations seems limited to human beings.

In summary, our conception of "reality" is best explained as an act of construction based on statistical probabilities. When these probabilities are skewed, as in the distorting rooms, our perceptions are wrong. But even then we can correct for errors, depending on our capacity to interact with the stimulus. In a sense, like S.B., we can also see the world best after we have "felt" it. From experiments on perception, psychobiologists are providing new and different answers to some of our most compelling philosophical questions. We are rapidly moving closer to the time when our inquiries about our place in the universe as well as the validity of our mental processes, our "reality," will be put to the test. This is an important first step in our attempt to understand the workings of the human brain. Any final assessment, however, must await our achieving a deeper understanding of how our brains are molded by our earliest experiences.

7
Children of the Moon

We have to regard the relation of mind to brain as not merely unsolved, but devoid of a basis for its very beginning.

SIR CHARLES SHERRINGTON

Imagine that you're a doctor in an intensive-care unit for newborns. Every day premature infants are brought to your unit, where they will either die or, after great efforts on your part, slowly begin to rally. Such work, while often challenging and exciting, can, if it ends in failure, be tragically disappointing.

For years, doctors working in neonatal intensive-care units have wondered about the intellectual development of premature infants under their care. What are the functional capacities of the premature brain? Are prematures in greater risk of brain damage? If not, will their brains develop normally? These are agonizing questions that many neonatologists—specialists in the care of prematures —are struggling with. They are also important questions, since psychobiological research indicates that in the premature baby of twenty-five weeks the most important aspect of brain growth has yet to take place. To make matters even more complicated, it's often not possible at birth to make any decisions about brain development, since infants, even those who will later grow up severely brain-damaged, often appear perfectly normal during the first few months of life.

Intensive-care units are providing the stimulus for psychobiologists to find out as much as possible about the stages of human brain development; and, as with many of the topics studied by psychobiology, the issues have ethical and social dimensions. If brain abnormalities are virtually certain at an early stage of prematurity, should infants be supported in an intensive-care unit, or should these scarce resources be channeled instead toward helping children more likely to develop normally? Most important, can future brain research discover ways to bring about normal brain development? To answer these questions, psychobiologists in recent years have increased their efforts to understand the early stages of brain growth.

The human brain approaches its adult size, weight, and number of cells by age two. It does so in a series of steps starting before birth. In the second trimester of pregnancy until about the age of six months, there is a "brain growth spurt" marked by rapid cell multiplication. This spurt can be broken down into two stages. The first extends throughout the second trimester and is largely restricted to nerve-cell multiplication. In fact, most researchers believe that the majority of our neurons are established during this period.

The second stage extends from the third trimester until the end of the normal period of breast feeding (usually six months). In this stage are formed the supporting cells of the human nervous system (the glia) along with the neuronal branches (dendrites) that extend from one neuron to another and establish the functional connections—synapses—between neurons.

These two stages of brain growth overlap, since nerve cells continue to multiply well into the second year of life. Of critical importance, however, is the process of dendritic extension, which results in the establishment of nerve cell connections. As we will discuss more fully in the next chapter, dendrites are numerous tiny extensions of the nerve cell, which are important in nerve impulse reception. Dendrites can be thought of as nerve-cell antennae which pick up impulses sent from other neurons. In many ways the number of dendritic connections is considered even more important than the total number of neurons in the

120

brain, since the density and complexity of dendritic connections probably has greater psychobiological consequences for brain development and human intelligence.

Animal studies indicate that there are also regional variations in brain growth, with some areas of the brain developing earlier than others. The visual cortex, for example, is established early, while some of the association cells linking widely separated areas of the brain don't develop completely until adulthood.

The earliest part of the human brain to begin functioning is the vestibular system, a center located largely in the brainstem and responsible for transmitting impulses important in co-ordination. According to psychobiologists,

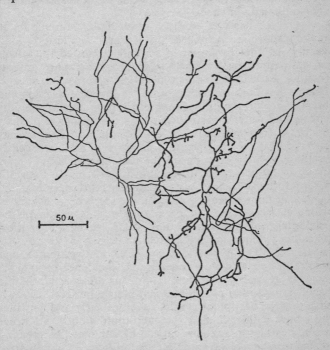

50 μ

FIGURE 16. Two neurons in the brain demonstrating multiple dendritic connections.

the vestibular system is operating as early as sixteen weeks. In fact, it is this early maturity that makes the vestibular system so important to early brain development. The fetus floating in its amniotic fluid registers its earliest perceptions via the activity of its vestibular system.

In recent years evidence has accumulated that the vestibular system is crucial for normal brain development. Infants who are given periodic vestibular stimulation by rocking, gain weight faster, develop vision and hearing earlier, and demonstrate distinct sleep cycles at a younger age. Vestibular dysfunction, on the other hand, is postulated as a possible cause of dyslexia and a contributing factor in childhood schizophrenia. In addition, there is also some evidence that the vestibular system may play a role in human violence.

On the basis of studies on vestibular stimulation in prematurity, Dr. Anneliese F. Korner, a Stanford University Adjunct Professor of Psychiatry, tested the effects of maintaining on waterbeds a group of premature infants ranging from twenty-seven to thirty-two weeks of age. Round-the-clock recordings were made of the sleeping and breathing patterns of the infants in order to discover the effects that waterbeds might have on the incidence of sleep apnea, a frequent cause of crib death. Dr. Korner discovered that merely placing premature infants in a gently oscillating waterbed reduced the incidence of sleep apnea by up to 60 per cent. But most intriguing of all, the most severe forms of apnea were reduced most dramatically. "One can only speculate about the underlying mechanism which reduces apnea in premature infants while on the oscillating waterbed," according to Dr. Korner. "The most plausible explanation appears to be that a continuous irregular oscillation provides input to the vestibular and respiratory centers, thus reducing the incidence of apnea."

Dr. Korner's waterbed therapy is an example of the benefits of applied psychobiology. Starting with the knowledge that the vestibular system develops early and provides the babies with their first form of stimulation, Dr. Korner speculates that infant survival can be enhanced by creating an environment that stimulates conditions in the uterus. Waterbeds are already standard in many neonatal intensive-care units, including the National Children's Medical Center in Washington, D.C.

Meanwhile, psychobiologists have recently developed a special tool to probe the functional state of the brain in premature infants. Known as Visual Evoked Response (VER), the method involves flashing a strobe light into the infant's eyes for the purpose of measuring its effect on recording electrodes placed on the back of the head in an area roughly corresponding to the visual cortex. It's now established that the VERs of premature infants undergo a series of measureable changes during development. Most important of all, however, is the timing of these changes.

In a young premature infant, the visual evoked response abruptly changes sometime between twenty-eight to thirty-two weeks. In essence, a baby of twenty-six weeks may show significant differences in tests of VER from a child only two weeks older. For several years, psychobiologists were puzzled about these changes. Did they mean anything, or were they only an artifact? With continued tests, the findings were repeatable and consistent, hence probably significant. But how do you correlate electrical findings recorded from the skull, with events transpiring deep within the brain?

Dr. Dominic P. Purpura, of the Department of Neurosciences at the Albert Einstein College of Medicine in New York, concentrated on one part of the developing brain. Using meticulously detailed microscopic studies, Purpura studied the visual cortex of prematures who died. He found that within the 28–32-week period the cerebral cortex establishes its basic pattern of nerve-cell connections. His findings suggest that there seems to be a critical period during those weeks corresponding to the VER changes. Rather than a smooth, continuous process extending throughout early development, neuronal connections are developed within a comparatively brief time during those four weeks. In fact, at thirty-two weeks the preterm infant has a visual cortex equivalent to that of a full-term infant.

Dr. Purpura and his colleagues are presently attempting to discover whether the enhanced synaptic organization seen at thirty-two weeks may be important in the evolution of newborn behavioral states. Body positions and sleep–wakefulness cycles, to mention only two, seem to follow patterns that are probably dependent on the establishment of synaptic connections laid down during that four-week period. Such speculations are part of a rapidly emerging new perspective on infant behavior which concentrates on the activities of fetuses in the womb.

With the aid of ultrasound and sonar—a kind of on-going television view of the interior of the womb—movies are already being made of fetal behavior. Later, these observations can be compressed via time-lapse photography and incorporated into a whole spectrum of prebirth behavioral activity.

Eventually, psychobiologists hope to be able to interrelate experiences within the womb, infant brain development, and the early mothering experiences. From this may emerge a new perspective on how our brains are formed, and even how we can change the brain when its development goes awry. The first step in this ambitious goal requires a revision in our thinking about the effects of the environment on brain development. In essence, can the environment bring about physical changes in the brain?

To find out, Dr. Mark Rosenzweig and his associates carried out a series of experiments evaluating the effects of three different environments on rat brains. In the first environment, a rat remained in the standard laboratory group cage with two others. The second environment consisted of a rat kept in solitary confinement with a minimum of stimulation and without the opportunity to interact with other rats. The third ("enriched") environment was a large cage fitted with a variety of toys. At the end of the experimental period, which ran from a few days to as long as several months, the animals' heads were cut off and their brains removed.

In all cases, the rats from the enriched environment showed an increase in the number of nerve cells and a greater weight of cerebral cortex, along with a thicker cortical covering. In addition, important brain enzymes showed significant elevations, which are thought to correlate with a greater number of connections between nerve cells and heightened cerebral arousal.

But the most consistent effect of experience on the brain turned out to be a boost in the ratio of the weight of cortex compared to the rest of the brain. Enrichment doesn't just act like an all-purpose "tonic," but has specific and localized effects on the development of the cerebral cortex.

Since the introduction of powerful scanning electron microscopes, it is now possible to subject the rat brain to more detailed analysis. Ordinarily, the number of connections between nerve cells can be roughly measured by counting tiny dendritic spines, or "thorns," along the length of the dendrite. A rough correlation ordinarily

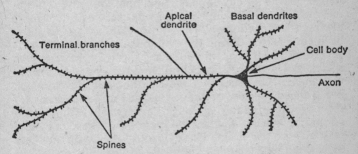

FIGURE 17. A nerve cell, illustrating multiple dendritic spines which serve as receivers for many of the synaptic contacts between neurons. The number of spines increases early in life when an organism is exposed to an enriched rather than an impoverished environment. A similar response is postulated in humans.

exists between the number of dendritic spines and the number of nerve-cell connections.

Albert Globus, of the University of California, discovered increased numbers of dendritic spines in rats from enriched environments, a possible cause for the enhanced learning capacity that other workers report is associated with environmental stimulation. Is it possible that learning, memory, and perhaps other brain functions depend on the quality of environmental stimulation? Psychobiol-

125

ogists are reluctant to go quite that far, since early enrichment improves some tasks while exerting little effect on others. All are agreed, however, that enrichment, even in small doses, exerts a measurable effect on brain development. As little as two hours a day of environmental enrichment over one month's time is sufficient to produce an "enriched" brain.

A particularly intriguing finding involves the rate of formation of certain critical brain enzymes in response to stimulation. Dr. J. Altman found that handling young rats increases the production of certain enzymes in parts of the rat brain concerned with movement. Even more provocative, however, was the discovery that in handled rats the rate of brain growth was temporarily slowed to allow for longer periods of development and, consequently, achievement of a better-developed brain with more cells and more complex connections. "This procedure seems to have been emphasized during the evolutionary process as well," according to Dr. Esmail Meisama, a psychobiologist who specializes in the relationship of brain enzymes to the environment. "One sees that the rate of brain development in advanced species is relatively slowed down, perhaps for the purpose of allowing longer interaction with the environment during the critical periods of developmental plasticity of the brain. Consequently, the brain in these species is more complex and well-developed."

In essence, the brain in experimental animals is shaped by the environment in a manner similar to the transformation of a block of stone through the skill of a sculptor. Naturally, when the sculptor's skill is deficient, the final product may be a grotesque distortion rather than a work of art. So, too, environmental variables may lead to disturbed brain organization and function. The brain does not so much *develop* as *respond* to tens of thousands of environmental variables. This revolutionary insight into human brain development is already bringing about profound changes in our attitudes toward the developing brain, and in particular, the care of premature infants.

"The moon is a cold planet, which has dominion over the child, and therefore doth bind it with its coldness, which is the cause of its death," was Aristotle's explanation for children born prematurely. Over succeeding centuries diverse theories have held sway. Some said labor began early because the fetuses were starving. Others claimed it was due to a change in the position of the infant's head within the mother's uterus. Whatever the cause, historical reports on prematurity stressed its frequent association with intellectual and even moral backwardness. "The immature infant becomes the backward schoolchild, and is a potential psychoneurotic or neuropathic patient and even a candidate for a home for imbeciles and idiots," according to a turn-of-the-century pediatrician.

Part of the difficulty in objectively evaluating prematurity came from the confusion that inevitably results when science is intermixed with moral self-righteousness. Even today, prematurity often occurs in a setting of poverty and illegitimacy. From here it wasn't difficult for some to prejudge prematures as defectives and social burdens. In a sense, illegitimacy, poverty, and immaturity were considered in the nineteenth century as moral rather than social problems.

Sporadically, however, a few reports began to appear in the 1930s claiming that premature birth unassociated with birth trauma had little effect on mental development. The developmental psychologist Arnold Gessell, writing in 1933, found no difference in the development of prematures and normal infants when the two groups were compared at later ages. Such reports, however, were contrary to the prevalent views of the time.

The truth about prematurity slowly emerged from studies first reported in the 1960s. A Scottish pediatrician, Cecil M. Drillien, did a follow-up study on premature infants showing that I.Q. changes between the ages of six months and four years depended principally on social class. In high social classes the I.Q.s for prematures increased by thirteen points, while prematures from the lower social grades showed a decrease of six points. When the children were restudied again at age seven, Drillien found very few retarded children among the upper classes except from among those with extremely low birth weights.

In poor homes, however, there was a "marked excess of retarded and very dull children" in all weight-category studies.

One significant difference between the upper- and lower-class homes described by Drillien was nutrition. The World Health Organization estimates that as many as three hundred million of the world's children are undernourished. In addition, undernourishment adversely affects some of the most important aspects of brain function. In an intensive study of the children of North American Indians, Dr. Ernesto Pollitt, of the Massachusetts Institute of Technology, demonstrated a 50 per cent decrease in behavioral performance in severely malnourished children. Memory, abstract reasoning, thinking, and verbal ability were most affected.

Studies on experimental animals subjected to malnutrition show many of the same findings associated with animals experimentally deprived of enriched environments: reduced brain weight, a decreased number of nerve cells, and a reduction in the thickness of the cerebral cortex—in essence, the picture of environmental isolation described by Rosenzweig. This has led some psychobiologists to postulate that malnutrition functionally isolates an animal from its environment, at least in terms of learning and memory. Both undernutrition and isolation result in lack of interest, apathy, a preference for sameness over novelty, and a reduction in exploratory behavior. In addition, irritability, fearfulness, and a heightened startle response appear.

Support for this malnutrition-isolation hypothesis comes from studies where malnourished animals are switched to enriched environments. Dr. Slovka Francova confirmed that fear in malnourished rats can be overcome by handling the animals on a regular basis as well as providing other environmental stimulation which can lead to brain enrichment.

"For optimum development of the brain, an adequate supply of both nutrients and external stimuli during decisive developmental periods is essential," says Dr. Francova. "A typical picture of malnutrition involves reduction both of the available nutrients and the environment stimulation. It may be assumed that environmental stimulation will potentiate adaptive processes which will result in more economic utilization of energy and nutrients. This is particularly important in animals with a low quantity of available food."

If Dr. Francova and other psychobiologists are correct, does environmental enrichment and stimulation affect all aspects of behavior, or only selected ones? Can the stimulation come at any time or only during critical periods where lack of environmental enrichment will inevitably result in irreversible brain damage? Finally, what are the effects of human malnutrition?

The brains of children who have died of malnutrition during the first year of life have fewer brain cells compared to normal children. They also have an overall decrease in whole brain size. In addition, it seems not to make much difference exactly when malnutrition occurs, as long as it's sometime in the first two years.

A study of Jamaican boys hospitalized with malnutrition divided the children into three groups: those hospitalized before eight months of age, those between eight and twelve months, and those between thirteen and twenty-four months. Dr. M. E. Hertzig and his associates found no significant difference in behavioral performance among the three groups when studied later between the ages of six and eleven years. This favors those scientists who claim that malnutrition at any time during the first two years has an equal impact on brain-behavioral development.

But answers are difficult to arrive at in this complicated area of brain-nutritional interaction. For one thing, it's almost impossible to develop an ethical experiment that can separate nutritional from environmental variables. Certainly scientists can't stand by while young children starve; thus, there can be no "control group." Despite these difficulties in designing an experiment, one study suggests that malnutrition can inflict permanent brain injuries that may never be reversed.

Adopted Korean children were tested several years after coming to the United States. In one group the children had been severely malnourished early in life. By age seven these children were normal in I.Q. testing. Another group of adopted Korean children had never been malnourished. Their performance on intelligence tests was not just normal but significantly above average. "These observations suggest that malnourished children, even though they are not retarded in later life, are never able to achieve their full intellectual potential," according to Dr. Merrill S. Read, Chief of the Growth and Development Branch of the National Institute of Child Health and Human Development.

The key question is, of course, whether subsequent good nutrition can entirely make up for deficiencies incurred during the critical first two years of life. This is a particularly agonizing question, since scientists are suggesting already that behavioral abnormalities of malnutrition in one generation may carry over two or even three generations, despite the eventual introduction of an adequate diet!

At first glance, the inheritance of behavioral characteristics across generations is difficult to explain. How could starvation in one generation affect the behavior and brain development two or, in some cases, even three generations later?

In an attempt to solve this riddle, Dr. Janina R. Galler, a child psychiatrist at Boston University Medical Center, has been working for the past several years with a colony of rats she has subjected to twenty generations of malnutrition. Dr. Galler explained her findings to me during a conversation at the International Conference on the Behavioral Effects of Energy and Protein Deficits.

"The rats, born after twenty generations of relative starvation, show stunting in physical growth and activity," says Dr. Galler. "Not surprisingly, nutritional rehabilitation reverses many of these deficits. What is surprising, however, is the presence of abnormal mother-infant interactions which persist beyond two generations of rehabilitation. These rats make poor mothers even though they are not nutritionally deprived. In fact, they do not act much differently than mothers of two generations earlier subjected to severe imposed nutritional deprivation."

Dr. Galler thinks the poor mothering behavior depends on disturbed patterns of infant-mother interaction repeated in each generation rather than inheritance across several generations of a genetic chance.

"I think the idea that malnutrition can cause a physical change in the genes of one animal which is then passed down over several generations is completely incorrect," Dr. Galler emphasized. "For one thing, such an idea doesn't conform to our conceptions about inheritance. For another, it really isn't necessary to be that complicated. Poor mother-infant interaction in one generation will produce females who make poor mothers in the next."

According to Dr. Galler, the initial nutritional insult resulted in behavioral consequences that interfere with mothering. It is this failure of mothering rather than genetic change that is passed on and can be measured several generations later.

International studies on brain development and human malnutrition seem to bear out Dr. Galler's hypothesis. Dr. Adolfo Chavez compared the effects of food supplements among the rural poor in a Mexican community. Beginning in the third month of pregnancy, an enriched diet was continued until the infant's birth, and was then resumed when the infant reached the age of three months (a point when the benefits of breast feeding begin to plateau). Chavez's interest was in the quality of mother-infant interaction. He focused attention on how the children with food supplements differed behaviorally from comparable children in other families not given the supplements.

The most dramatic change involved physical activity. The better-fed children were six to eight times more active. At age nine months, the children were crawling away from their mothers and seeking the company of older brothers and sisters. The poorly fed children, in contrast, began crawling later and stayed close to their mothers. They appeared timid, passive, easily frightened, and insecure.

So far, Chavez's findings seem predictable: better-fed babies have more energy and curiosity. What wasn't predictable, however, was the change in behavior noted between mothers with well-nourished infants and mothers with poorly nourished infants.

"The better-fed infants were more playful and required more attention and more protection against accidents, thus causing the mothers to feel a greater concern about them," says Chavez. "The mothers found that they had to talk to their children more often and had to begin what, under other circumstances, was rare in the community: establish closer emotional contact with their infants. Soon the participation of the father and the brothers and sisters became necessary.

"The better-fed infant—perhaps because of his greater caloric intake—is more active and consequently requires a greater response on the part of the mother. This sets up a feedback which raises the level of stimuli offered to or perceived by the child."

Chavez's findings suggest that the child rather than the mother sets the emotional balance between them and determines to a large extent how they will get along.

In addition, the mother-child interaction establishes patterns of brain development which will have later consequences. In the first year of life, for instance, the baby explores its environment and, in addition, begins to form secure ties with its mother. Malnutrition seems to interfere with both processes.

Dr. Pirkko Graves, of the Department of Pathobiology at Johns Hopkins University School of Hygiene and Public Health, examined the quality of mothering in both malnourished and adequately nourished infants in West Bengal, India. Along with Chavez, Dr. Graves found that "undernourished children showed lower levels of exploratory and play activity as well as increased attachment behavior to the mother." Malnourished babies are more likely to prefer the mother to toys and cling timidly to the breast, seeking comfort and pacification rather than food. In turn, the infant's passive behavior elicits poor mothering responses. "The Indian mothers showed less reciprocity with their undernourished infants when compared with the well-nourished mothers."

From here, Dr. Graves speculates about some of the possible behavioral consequences of malnutrition on mother-infant interaction. "Healthy psychological development depends on a proper 'distancing' between the infant and mother. If the child cannot separate itself from the mother or has no interest in the world around it, you can expect decreased intellectual and emotional development. We are really dealing with a case of isolation, which compounds the harmful effects of malnutrition. A clinging, timid infant evokes abnormal patterns of response in the mother. The child demands less, and as a result gets less in terms of new experiences. The mother, in turn, is less challenged by the infant and is likely to ignore it."

Others have noted dramatic examples of maternal rejection in malnourished children. "On several occasions mothers have left their malnourished children at our clinic and have given wrong addresses," according to Dr. Fernando Monckeberg, of the Institute of Nutrition and Food Technology at the University of Chile. "Later, when the children had been fed and partially restored to normal appearance, the mothers wanted the children back. They were even enthusiastic and quite contrite about previously abandoning the children."

Studies on infant malnutrition point up the subtleties involved in trying to separate the effects of the environment from induced physical and chemical brain changes. What started out as a search for the effects of malnutrition in human development has turned into a study of the complex interactions of malnutrition, infant responsiveness, and mothering. Although few conclusions can be drawn, authorities agree on a few basic points.

First, during the last period of pregnancy and the first two years of life, the brain is most vulnerable to permanent structural and dynamic brain consequences resulting from malnutrition.

Second, chronic prolonged malnutrition is more likely to bring about permanent brain damage than shorter periods of inadequate nutrition. Children born during the Dutch famine of World War II, for instance, show few effects today from their relatively brief period of severe food restriction.

Finally, the effects of malnutrition on the brain of developing infants has, as its most serious consequence, an alteration of healthy mother-child interactions. Here, psychobiology has placed on a firm experimental basis an

133

intuition held by many previous workers on malnutrition but, until recently, an intuition they were unable to prove: The infant, to a large extent, determines the quality of mothering.

So far, studies on malnutrition and brain development seem limited to infants and the very young. But recent findings from the Massachusetts Institute of Technology indicate that adult brains are also influenced to a significant extent by diet.

According to traditional medical teaching, the adult human brain enjoys a privileged status with respect to nutrition. While the other body organs must depend on the levels of nutrients available in the circulating blood, the brain has been thought capable of preferentially extracting whatever it needs from the blood. Thus, whatever you may have eaten for breakfast, the brain is able, like a persistent debtor, to take its share prior to the liver, the muscles, or any other bodily organ. Such thinking, long entrenched in authoritative medical textbooks as well as centuries of accumulated folk wisdom, is turning out to be a myth. It now appears that the brain's power of usurpation is overrated. Brain functioning depends very much on what you've eaten for breakfast.

Dr. Richard J. Wurtman, Professor of Endocrinology and Metabolism in the Department of Nutrition and Food Science at the Massachusetts Institute of Technology, found that the brain levels of two important neurotransmitters are dependent on diet. The first, serotonin, is formed from the amino acid tryptophan, which is an essential amino acid—i.e., the body is incapable of producing it. Rats fed diets rich in tryptophan immediately demonstrate elevations in blood and brain tryptophan levels. At the same time, brain serotonin levels rose, a consequence of increased serotonin formation from brain tryptophan. Even when rats were fed the tryptophan-high diets at unusual times, similar changes occurred, with the brain tryptophan and serotonin levels increasing.

A second neurotransmitter, acetylcholine, the most common neurotransmitter in the human brain, is also heavily dependent for its formation on a dietary precursor. Choline, contained principally in lecithin, a constituent of eggs, soybeans, and liver, varies directly with the dietary intake of these foods. Its conversion to acetylcholine can be correlated with the dietary choline intake.

Other neurotransmitters are also postulated to depend on the vagaries of diet. The catecholamine transmitters dopamine and norepinephrine—the neurotransmitters thought to be responsible, at least in part, for depression and mania—are produced from precursor compounds that the neurons cannot make for themselves. If the brain cannot produce a substance, then it must, like a heavily industrial nation, depend on other sources for its energy supplies. In this case, of course, the other source is the level of the needed substance within the blood stream or, ultimately, the dietary intake.

As a proof of the importance of dietary intake on brain neurotransmitter formation, psychobiologists have recently started treating some brain diseases with dietary supplements. Tardive dyskinesia, a movement disorder consisting of repetitive and often grotesque bodily movements, is thought to result from the release of inadequate amounts of acetylcholine in the brain. During 1977 and 1978, Dr. John Growdon, of the Department of Neurology at Tufts University Medical School, treated twenty patients suffering from tardive dyskinesia for two weeks with choline or a placebo. Nine patients showed "major improvement after choline: none showed any response from the placebo."

Growdon and his associates are now comparing choline with lecithin, the natural choline derivative, in an attempt to discover whether lecithin might act more effectively than choline itself. In addition, other studies are under way, at M.I.T. and elsewhere around the country, aimed at additional diseases thought to result from acetylcholine abnormalities. Memory loss, myasthenia gravis, even a study of the ability of acetylcholine to blockade the effects of morphine on pain sensitivity—these are only some of the ongoing studies. Although it's too early to be certain of the ultimate results, the possibility exists of modifying some forms of emotional or neurologic illness by altering the dietary precursors of some of the neurotransmitters. If successful, relatively simple dietary treatments may be available for presently refractory brain diseases.

What are the biosocial implications of the findings concerning malnutrition in human brain development? At first glance, the nutritional studies on children seem to suggest that resources should immediately be channeled into food programs that can provide a basic diet for children throughout the world. Although seemingly logical and compassionate, such an approach may be overly simplistic since it overlooks a terribly important finding: Behavioral changes brought about by improved nutrition cannot be sustained without extensive social changes. Put another way, malnutrition is usually only one aspect of more extensive and pernicious deprivations.

At a meeting of the Pan American Health Organization held in December 1977, Dr. Harrison McKay, of the Human Ecology Research Foundation of Cali, Colombia, presented a grim picture of the long-term benefits from food supplementation alone. McKay, a vigorous, indefatigable man, has worked in Colombia since the 1960s, spearheading pioneer efforts to combine food supplementation with instructions to mothers concerning forms of behavioral stimulation for the infants.

McKay's early findings were hopeful. He observed continued improvement in general activity and alertness among children receiving food supplementation. But when these children were re-examined, at the ages of six and seven, their performance lagged dramatically behind a comparison group of middle-class children. By the age of ten, their performance had either leveled off or, in some cases, had actually fallen back to performance levels observed at six years.

"In poverty-stricken environments there are obstacles to intellectual and behavioral development that become more oppressive as the child grows older," says McKay. "Intervention, such as food supplementation and educational efforts, can temporarily make up for some of this. But the improvements are difficult to sustain, and the environment of poverty continues to exert a negative and sometimes a destructive effect on human development."

McKay's experiences argue against total reliance on food-supplementation programs as sufficiently responsive to the problems of using better nutrition to build better brains. It's true that we now know that brain development can be adversely affected by early severe malnutrition. We also know it can be partially helped by improving the diet. Unfortunately, the benefits of nutritional supplementation cannot be sustained, because of the disturbing nature of the impoverished environments. In a sense, malnutrition is only the tip of the iceberg; below the surface lies a vicious cycle of malnutrition, an absence of environmental stimulation, impaired behavioral responses, and problems in the quality of mothering.

Here the findings of psychobiologists coincide with the pleas of humanitarians throughout history. A civilized society cannot tolerate a world situation where the majority are doomed never to reach their genetic potential because some refuse to surrender even a small part of a material legacy so huge they can never live long enough to deplete it. Our responsibility to each other does not stop at tossing bones and scraps to the hungry. (Recent international nutrition conferences on what to do with food surpluses, however, suggest that many countries still haven't arrived at the point of giving away what they don't need and cannot use!)

Studies on malnutrition and brain development lead, I believe, to more far-reaching conclusions than merely the distribution of food. The total environment, of which food is only a part, affects brain development and behavior. Intellectual and emotional stunting can result from either a poor diet or an oppressive and degrading social environment. Improvement of the diet, without changes in the conditions under which people live, is doomed to failure. The studies cited above prove that changing diets without changing lives does not make social or biological sense.

In addition, we cannot begin to measure the effects of starvation on a global scale. Many communities around the world are composed of people with significant psycho-behavioral deficits resulting from malnutrition, impoverished environment, and intermittent disease. For one thing, scientists have known for years that malnourishment lowers an individual's resistance to infectious diseases. This in turn leads to greater physical and mental limitations, which compound the harmful effects of malnutrition. "A hungry person lives only to satisfy his hunger. He does not have time for anything else," says Dr. David Levitsky, a Cornell University researcher specializing in the behavioral consequences of malnutrition.

The implications of malnutrition's effects on brain development are profound and disturbing. Infant malnutrition and subsequent brain development, it seems, cannot be reversed simply by the introduction of an adequate diet at a later point. If brain stunting is to be avoided, equal emphasis must be placed on training the mother to provide an interesting and stimulating environment for her baby. But this isn't easy either. Mothers of malnourished infants are usually malnourished themselves. They are relatively apathetic and show little interest in things around them, including their children. In turn, an inactive and undemanding infant will be less likely to elicit stimulating behavior from its mother. In effect, a combination of brain damage and social deprivation creates a destructive climate which erodes the mental capacities of three hundred million children around the world.

The research on malnutrition and brain development also points up the inadequacy of our current concepts. The question of nature versus nurture, or the environment versus genes, fails to do justice to the complexities of brain development. In addition, the studies of environmental enrichment and diet underscore the importance of applied psychobiology. Poverty, undernutrition, and social deprivation, although social phenomena, have psychobiological consequences. Our brains are literally shaped by our natural and man-made environments. We are not only what we eat but what we see and hear, the physical and emotional world in which we live.

Let's look now at one of the possible long-term consequences of faulty brain development: violence.

8

Cain's Curse

Is this a dagger which I see before me,
The handle toward my hand? Come, let me clutch thee:
I have thee not, and yet I see thee still.
Art thou not, fatal vision, sensible
To feeling as to sight? or art thou but
A dagger of the mind, a false creation,
Proceeding from the heat-oppressed brain?

MACBETH

We Americans are exquisitely conscious of violence. Our television shows and movies provide a steady fare of murders, muggings, and mayhem that mirrors our society to an increasingly distressful degree. Who in his right senses would walk alone at night down the streets of any of our major American cities? And violence permeates the workaday world as well: Airline hijackings compete for newspaper headlines with bombings by political terrorists.

Our response to violence meanwhile oscillates between ignoring the dimensions and seriousness of the problem to periodic "get tough" policies that meet violence with violence. If we are unsafe from muggers and rapists, our lives are no less endangered by periodic public "shootouts" between lawbreakers and law-enforcement officers. Throughout all this, little has been accomplished by way of understanding the basis for violence. Taken up with stop-gap measures, we've stopped asking ourselves the important questions. Why are some people violent? Is there any way to predict violence? Even better, Can violence be prevented?

In recent years psychobiologists have started assembling the pieces in a vast jigsaw puzzle whose solution may finally result in an understanding and reduction of violence in different cultures. Anthropologists are fond of pointing out that not every culture is violent. Some, in fact, are nonviolent. The key question is, of course, What accounts for the differences?

One intriguing association that emerges from cross-cultural studies relates early infant-care practices to the later development of violence. With few exceptions, societies which provide infants with a great deal of physical affection and bodily contact produce relatively nonviolent adults. In societies where infants are touched, held, and carried, the incidence of violence is much less than in societies where infant care is restricted to merely feeding and changing. Left at this, such findings tell us little about the mechanisms involved. A superficial interpretation might be that touching and carrying are two aspects of good mothering. Or put differently, infant stimulation and interaction with a mothering (or fathering) figure are important for normal development.

In the late 1950s and early '60s, Harry and Margaret Kuenne Harlow, at the Primate Laboratory of the University of Wisconsin, began a series of experimental observations on rhesus monkeys that provided the first experimental evidence to suggest that abnormal early brain development, brought on by poor mothering, might be the key to understanding some forms of violence. Rhesus monkeys make excellent experimental subjects for investigating such a relationship. Along with humans, rhesus monkeys undergo a long period of physical and emotional development which is heavily dependent on early physical contact and attachment between infant monkeys and their mothers. The Harlows first studied the effects of isolating the infant rhesus monkeys soon after birth. Alone in a cage, the isolated monkeys were able to view other monkeys in similar cages across the laboratory, but weren't able to make physical contact with them.

When later studied as adults, the isolated monkeys

were withdrawn, often sitting for long periods of time staring blankly into space. In addition, they exhibited self-mutilating behavior, often pinching a patch of skin between their fingers hundreds of times a day. On occasion, the animals bit parts of their own bodies, often in response to the approach of their human caretakers. At still a later point, self-aggression changed into aggression toward others, a "cringing fearfulness" merging into "prominent displays of hostility."

Although no one is completely certain why isolated monkeys behave as they do, certain observations provide a probable explanation. An isolated monkey will react in terror at the sight of its own hand if it appears suddenly in front of its face. Terror is instantly replaced by rage and the monkey will mutilate its hand, suggesting an inability to differentiate itself from its environment. Later, the monkey will attack others without warning or provocation. One isolated animal bit off a finger of a normal monkey who had been temporarily placed in the cage with it. Another killed a companion in a sudden unprovoked attack.

But the most intriguing observation of the isolated monkeys concerns their later behavior toward their young. Isolated females become brutal, indifferent mothers who, as a group, show less warmth and affection and, on occasion, kill their offspring. This is the first indication that the mother-infant interaction was a reciprocal one. Isolated, unstimulated monkeys grew up to become unstable, brutal parents. Deprivation of some aspect of the physical closeness usually bestowed on an infant resulted in a later inability to be a good parent. But what aspect of the mother-infant interaction is most important?

In an attempt to understand the mothering process, the Harlows placed a group of newborn monkeys in cages with two types of "surrogate" mothers—one made of cloth, the other made entirely of wires. In almost all instances the infant monkeys spent more time clinging to their cloth-covered mothers than to their wire mothers, even when the monkeys obtained milk from a bottle attached to the wire mothers! The Harlows interpreted these findings as showing the overwhelming importance of bod-

141

ily contact and the immediate comfort it supplies in forming the infant's attachment for its mother.

If the wire surrogate was replaced with a cloth "mother" capable of rocking, the monkeys, in most cases, preferred the rocking to the motionless cloth mother. When the two groups were later compared, their behavior contrasted dramatically. Those monkeys raised with motionless cloth mothers now showed repetitive rocking motions, while the monkeys provided with moving cloth surrogates failed to show abnormal movement patterns. From here, psychobiologists speculate that sensory stimulation might be an absolute requirement for normal brain development. If this is true, however, what kinds of stimulation are most important?

If an infant monkey is blinded at birth but reared with its mother, the abnormal rocking movements fail to develop, suggesting that visual stimulation may be less important than movement in establishing healthy behavior patterns and avoiding such things as rocking. Studies in blind human infants also fail to uncover abnormal movement patterns, provided the infants receive appropriate touch and movement stimulation. On the basis of these findings, psychobiologists began looking for other clues that might indicate the importance of movement and touch to normal brain development. One clue came from the pediatric wards.

Infants and children immobilized in bed for the treatment of bone fractures often develop emotional disturbances marked by hyperactivity and outbursts of rage and violence. Since only movement, and not sight or hearing, is restricted, a relationship is again suggested between early diminished movement and later abnormal, often violent, behavior.

A similar correlation can be demonstrated in adults. When a group of adult faculty members at a university took part in an experiment testing the effects of immobilizing their heads in a halter, 85 per cent reported the experience as "stressful." Vision and hearing were not interfered with and the subjects were free to converse with each other and engage in any activity they wished as long as their heads remained in the halter. Restriction of move-

142

ment alone resulted in "intellectual inefficiency, bizarre thoughts, exaggerated emotional reactions, and unusual bodily sensations." In adults, as well as infants, immobility alone, despite intact vision or hearing, can result in abnormal mental experiences, disturbed behavior and, in many cases, violence.

Observation of mother-infant interactions in the laboratory yields similar results. Physical holding and carrying of the infant turns out to be the most important factor responsible for the infant's normal mental and social development. Dr. Frank Pedersen, Chief of the Section on Parent-Child Interaction of the National Institute of Child Health and Development, compared the effects of visual, auditory, and movement stimulation on the infant's mental and psychomotor development. Out of six variables ranging from social responsiveness to object permanence, movement correlated positively with all six areas, whereas vision and hearing were important in only one.

So far, the results I've been describing are a matter of hard scientific fact. From here we move into issues that are more speculative. Not all psychobiologists are in agreement here, but the general trend is toward a revolution in our concept of the role of the environment in normal brain development and the subsequent tendency toward violence.

As we discussed in Chapter Seven, sensory enrichment leads to changes in the branching of neurons and the complexity of their dendritic interactions. The immature brain is dependent on sensory stimulation for normal growth, development, and function. Sensory stimulation has, in fact, been compared to a "nutrient." This dependence on sensory stimulation for normal brain development suggested a theory to Dr. James W. Prescott, a developmental neuropsychologist at the National Institute of Child Health and Human Development. The theory demands a major revision in our ideas about the role of movement in normal brain development.

Any type of body movement, even something as passive as being held or rocked, results in a train of impulses directed toward a specific part of the brain, the cerebellum. Traditionally, the cerebellum, a small three-lobed struc-

143

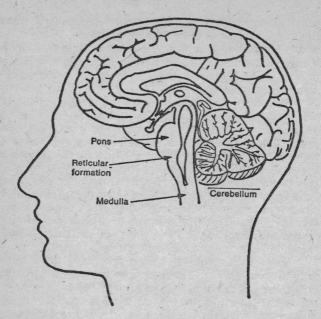

Pons

Reticular formation

Medulla

Cerebellum

ture behind the occipital lobe at the posterior part of the brain, is considered an important co-ordinator for movement. When you reach for a cup of coffee, for instance, the cerebellum is responsible for the smooth co-ordination necessary to bring the liquid to your lips. Patients with diseases of the cerebellum cannot lift a cup of coffee without spilling it. They also frequently sway when walking and, in some cases, cannot even sit without tilting to one side or the other. In a phrase, the cerebellum is responsible for smoothly co-ordinated muscular efforts. A ballet dancer represents perfect cerebellar functioning in which thousands of muscles are controlled with exquisite precision.

When an infant is rocked or cuddled, impulses are directed to the cerebellum that stimulate its development, a process that goes on until at least age two. In fact, the cerebellum is unique, since it is the only part of the brain where brain-cell multiplication continues long after birth.

"The cerebellum provides an explanatory model for the effects of social isolation," says Prescott. "The rocking behavior of isolation-reared monkeys and institutionalized children may result from insufficient body contact and movement. Consequently, both touch and movement receptors and their projections to other brain structures don't receive sufficient sensory stimulation for normal development and function."

To arrive at this conclusion, Prescott applied the well-known psychobiological principle that disuse or understimulation of a brain pathway may lead to a hyperactive response at a later point. If movement stimulation is deficient, for instance, the cerebellum and its connections with other brain pathways responsible for normal rocking or stereotyped behavior. Another suggestion that the cerebellum may be important in emotional behavior as well as movement came from direct electrical stimulation of a rat's cerebellum. Behavior similar to isolation-raised rats can be evoked in a normal rat if the cerebellum is electrically stimulated: Within seconds the rat may begin stereotyped circling behavior and compulsive biting.

"I think the cerebellum is involved in complex emotional behavior," says Prescott. "It may well serve as a master regulator mechanism for sensory-emotional and motor processes. It is intimately involved in many of the behavioral and emotional abnormalities of isolation-reared animals and infants and children deprived of movement stimulation."

Prescott's unorthodox theory emphasized the role of a brain structure, the cerebellum, which, according to traditional psychobiological thinking, has nothing to do with behavior. Not surprisingly, it was originally greeted with a good deal of disbelief and skepticism. For one thing, how could such a radical proposal be proven? What data exists to prove that emotional reactions, such as violence, are related to activities in the cerebellum?

Attempts to provide a psychobiological explanation of violence have been made for almost a hundred years. The first studies, carried out in 1892, pointed to the role of the cerebral cortex as an inhibitor of violence. Cats and dogs surgically deprived of their cerebral cortex lapsed into attacks of rage marked by fierce growling or hissing in response to even the mildest stimulation. Based on this, early psychobiologists speculated that the higher brain centers located in the cerebral cortex worked by inhibiting violence; and, with the cortex removed, the animals reverted to a violent state.

In 1928 a "violence center" was described in the hypothalamus, a tiny co-ordinating portion of the brain located above the reticular formation and responsible for the regulation of eating, drinking, temperature, and other activities. The posterior part of the hypothalamus turned out to be important in violence: When it was destroyed, animals became permanently savage. Soon other experimenters produced violent behavior in test animals by merely stimulating the posterior hypothalamus. This presented psychobiologists with a problem: The fact that destruction and stimulation of the same brain structure, the hypothalamus, produced similar results made it unlikely that violence can be explained simply on the basis of the release of higher centers.

The solution of this paradox awaited the development of thin needle-like electrodes that could be inserted into any part of the brain with a minimum of damage. With the electrodes permanently in place, the effects of stimulation and destruction of deep brain areas could be studied. Then, by hooking up the stimulation apparatus to a lever, the animals could learn to activate the stimulator themselves.

Using the self-stimulation apparatus, the late Dr. James Olds, of the California Institute of Technology, discovered that laboratory animals will self-stimulate for hours when electrodes are placed in the anterior portion of their hypothalamus. As the electrodes are moved to the posterior hypothalamus, however, the animals abruptly stop self-stimulation. From here, Olds and others postulated the existence of "pleasure" and "punishment" centers in the brain. The punishment centers roughly correspond to the areas of the hypothalamus responsible for rage reactions.

What emerges from these studies is a series of midline centers in the deeper portions of the brain that mediate pleasure and displeasure. With electrodes placed in the pleasure centers, a rat will bar-press up to five thousand times per hour. It will even prefer stimulation of the pleasure center to eating. The displeasure centers, in contrast, are associated with some extremely disagreeable sensations that often culminate in attack behavior. (The details regarding which brain centers are responsible for pleasure or displeasure are still being debated and need not concern us at this point. In addition, the physical and chemical environment of the stimulated region is often directly affected by neighboring regions in ways that can't always be anticipated.)

Despite these reservations, psychobiologists felt confident until recently that the pleasure and displeasure centers were largely confined to the brain's deep midline structures. If disagreeable mental states and violence were ever to be psychobiologically explained, they reasoned, the explanation would involve brain structures limited to these regions. It was commonly held that the midline structures, along with their connections to the limbic system, were responsible for the animal's complete emotional repertoire, including violence. Nowhere in this formulation was there a place for the cerebellum.

"The first requirement was to demonstrate that the cerebellum is connected with the emotional control centers of the limbic system," says Prescott. "There was really no reason to think that there were such connections other than the observation that lack of cerebellar stimulation, such as with isolation-reared monkeys or institutionalized children, leads to emotional disorders. The obvious place to look, it seemed, was in the brains of Harlow's isolation-reared monkeys."

In 1975, Prescott sent five isolation-reared monkeys from Harlow's laboratory to Robert G. Heath at Tulane University in New Orleans. Heath, a fifty-six-year-old neuropsychiatrist, is a fiercely controversial figure who, over the past twenty-five years, has implanted stimulating and recording electrodes deep within the brains of a wide variety of people. In the process Heath, a flamboyant but indefatigable researcher, has created for himself an international reputation for work which has ranged from

147

studies measuring the brain's electrical changes accompanying normal orgasm to the detection, by electron microscope, of the subtle alteration in the brains of monkeys trained to smoke marijuana. Heath's principal research interest, however, has always remained the psychobiology of psychosis and violence.

Psychosis, particularly schizophrenia, is marked by dramatic alterations in the patient's ability to experience pleasure. Anhedonia—literally the loss of the pleasure sense—is often accompanied by excessively painful emotional behaviors such as fear or rage. Although seemingly incapable of deriving pleasure, schizophrenics have heightened appreciations of pain, anguish, and loneliness. In 1954, Heath began a lifetime study looking for the brain mechanisms responsible for this pleasure-displeasure disturbance.

In animals, destruction of the septal region, a part of the limbic system, reduces awareness and profoundly lowers emotional expression. In contrast, electrical stimulation of the septum heightens alertness and induces what appears to be a pleasurable state. When Heath applied the same technique to humans, his subjects' reports confirmed the impression from the earlier animal studies that septal-region stimulation elicits pleasurable feelings. From here, Heath speculated that septal disturbances might be responsible for emotional abnormalities, particularly the inability to experience pleasure.

On the other side of the pleasure-displeasure story, Heath found activity in other parts of the brain's limbic system (hippocampus and amygdala) whenever patients experienced rage or fear. At other times, similar emotions could be artificially induced by electrical stimulation of the same brain sites. When implanted with electrodes in these sites, patients never self-stimulated and often inactivated the self-stimulating device in order to avoid inadvertent stimulation.

What emerged from Heath's rather unorthodox experiments (one patient retained an electric stimulator in his head for over three years) was a "map" of the brain's pleasure and displeasure centers. Patients implanted with electrodes into their septum, for instance, would selectively self-stimulate repeatedly. At other times during pleasure, such as an orgasm, activation could be detected in the septal region. With more intense sexual orgasm, faster frequency activity occurred which appeared similar to the brain waves seen during some forms of epileptic seizures. In some cases, even talking about sexual matters was enough to elicit similar septal activations.

From here, Heath formulated a theory about the brain's role in emotion. "Each system of the brain (pleasure and pain) is seemingly capable of overwhelming or inhibiting the other. Activation of the pleasure system by electrical stimulation or by administration of drugs eliminates signs and symptoms of emotional or physical pain, or both, and obliterates changes or recordings associated with the painful state. Similarly, activation of brain sites for adversive emotional responses replaces pleasurable feelings with painful feelings, and obliterates recording correlates of the pleasure state."

Pleasure and displeasure appear to operate like the opposite ends of a seesaw. Although it's possible to be angry and calm, or cheerful and inwardly unhappy, the experience of emotion is usually unitary. In most cases we are experiencing either pleasurable or unpleasurable emotions, not both. The balance can be tipped either way, however, sometimes to a pathological degree. Unhappiness can evolve into deep depression, or lightheartedness into mania. Later in the book we'll discuss the brain mechanisms which are thought to be responsible for these states, as well as some of the newer psychobiological methods of treatment. But for now I want to point out that Heath's experimental findings on pleasure-displeasure complemented Prescott's observations on the important role of early movement stimulation and violence. Certainly, according to traditional ideas about the cerebellum and the limbic system, there seems to be no way to correlate them.

"I mentioned to Dr. Heath that I would appreciate it if he would do some depth electrode studies on the cerebellum of several of Harlow's monkeys," Prescott recalls. "Heath told me, 'Why do you think we should do that?' He pointed out to me that no one had ever demonstrated direct connections between the cerebellum and the limbic system. After some encouragement he agreed, however, and five sensory isolation-reared monkeys were sent down to him."

At the base of Prescott's hunch that emotion and movement are interrelated, were some common everyday observations. Children like to be picked up, rocked, and taken on merry-go-rounds and roller coasters, as well as being whirled through the air. Most of us enjoy warm baths and soothing massages as calmers of tension and anxiety. Physical exertion such as sports or hard physical work can be exhilarating, even on occasion providing a "natural high." In all these instances, body movement seems to induce pleasurable emotions.

If these behavioral correlations exist, then, Prescott reasoned, there *must* be connections between the brain areas responsible for emotion (the limbic system) and those controlling movement (cerebellum). To look for these connections, Dr. Heath employed the method of evoked potentials. After implanting electrodes into several brain sites, a brief electrical stimulus was delivered to one of them while simultaneous recordings were made from the other sites. If activity could be detected, it provided an indication that the two sites were interconnected.

Heath's initial results with the Harlow monkey showed, not surprisingly, that the limbic areas responsible for emotion (hippocampus, amygdala, and septal areas) were directly linked. What was more surprising, however, was the discovery of connections between these emotional centers and the cerebellum. In addition, two-way communication could be shown between the centers in the cerebellum and the emotional areas of the brain for pleasure and displeasure.

"This was the first evidence that aberrant electrophysiological activity occurs in deep cerebellar nuclei, as well as other deep-brain structures—most pronounced in the limbic system—in association with severely disturbed behavior resulting from maternal-social deprivation," says Heath.

These results, Prescott believes, provide a preliminary basis for a psychobiological theory of the origin of human violence which may result from a permanent defect in the pleasure centers secondary to inadequate early mothering experiences. The infant who is deprived of movement and physical closeness will fail to develop the brain pathways that mediate pleasure. In essence, such people may be suffering at the neuronal level from stunted growth of their pleasure system.

In this model the expression of pleasure cannot be transmitted through the appropriate parts of the brain, because there are fewer cell connections. Since the infant is deprived of movement and touch stimuli, fewer impulses are relayed to the immature cerebellum; as a result, it develops abnormally. Fewer connections are then made between the cerebellum and the pleasure centers. Later, the person may have problems experiencing pleasure and, as a result, develop an insatiable need for it. In the absence of pleasure, the balance tips toward dysfunctional states such as violence.

"Think of these people as similar to diabetics," suggests Prescott. "As long as the diabetic gets his regular supply of insulin, everything is fine. Deprive him of insulin, and all kinds of physical and emotional disturbances may result. In a similar way, a child deprived of physical closeness will develop an extraordinary need for affection later in life which is unlikely to be fulfilled in the real adult world. This person often lapses into periodic violence."

Could sadomasochism be explained as a kind of short-circuiting phenomenon between the brain's pleasure and violence pathways? And could the love-hate feelings of schizophrenics, the ambivalence that forms the core of the disease, depend on similar disturbances? At this point psychobiologists can only speculate about such questions. But in the meantime research continues on ways to apply the new findings about the cerebellum's role in violence. For instance, Could a violent patient be treated by cerebellar stimulation?

In 1976, Robert Parker, a man considered the most severely ill mental patient in the Louisiana hospital system, was implanted with a cerebellar stimulator. Hospitalized since age thirteen, Parker had spent most of his time in a

151

ward for the severely disturbed. On numerous occasions he had savagely slashed his wrists, and at one point was barely prevented from killing his sister. Despite high doses of tranquilizers, Parker remained explosively violent and spent the major part of his time locked in his room under physical restraint. After the implantation of the cerebellar stimulator, Parker's violence ceased the first day after the operation. According to his doctor, he is now in "complete remission" and living outside the institution.

During 1976 and 1977 a total of ten more patients with different psychiatric diagnoses underwent procedures at the Tulane University Medical School for the implantation of cerebellar pacemakers. Among these were Robinson Smith, a forty-one-year-old physicist given to violent outbursts and an abiding compulsion to choke people; Ann Platt, a nineteen-year-old prostitute with a history of wrist-slashings, who on one occasion shot herself in the stomach with her husband's service revolver; and Julia Trone, a twenty-one-year-old librarian who had been committed to a state institution after attempting to kill her father with a knife. All three patients improved under the procedure—Robinson, to the point that he was well enough for discharge thirteen days after activation of the pacemaker.

"The favorable response in this patient group, representing a wide variety of psychopathology, substantiates our original working hypothesis—namely, the principal effect of stimulation is modulation of affect and, as a consequence, a gradual disappearance of the symptoms of the disorder. Symptoms of violence and uncontrollable aggression were most promptly eliminated," says Heath.

Obviously, the relationship of environmental stimulation, brain development, and subsequent violence is complex. So far, the theory that early physical contact and movement may play a later role in a tendency toward violence, remains an intriguing but unproven possibility which can only be settled by further psychobiological research. In the meantime, however, increasing violence on our streets and in our homes underscores the importance of carrying out whatever research will enable us to decide one way or another.

Child abuse is one promising area worthy of further research. Pediatricians have known for years that violent parents are often themselves the victims of child abuse. Psychiatric interviews with child abusers reveal a pattern of reduced pleasure in daily living. In addition, they typically report little enjoyment of sexuality, with only a few of the mothers reporting that they have ever experienced orgasm. Most appear so involved in catering to their own insatiable emotional needs that they have little time to respond to the basic needs of their children.

Open discussion with such parents often reveals an isolated, alienated person with a quick temper, low frustration tolerance, and a sense of personal deprivation. Since they cannot receive pleasure, they are also poor providers of pleasure to their infants. Child-abusing mothers rarely report having "fun" with their babies, and in a physician's office demonstrate little positive interaction, often spending the examination time concentrating on how to satisfy their own emotional needs. The abusive parent's behavior is, in essence, reminiscent of the mothering patterns of Harlow's isolated monkeys, who, in many cases, had to be physically restrained from abusing their infants. The most pernicious aspect of the child-abuse syndrome, of course, is its tendency to repeat itself from generation to generation.

Thanks to a greater awareness of child abuse on the part of physicians and the general public, child abusers are now coming to public attention sooner. In addition, major hospitals are setting up child-abuse clinics with established protocols designed to improve the quality of interaction between parent and child. Increased infant-care instruction, a lengthened period of maternal "lying in" arrangements, instructions on feeding patterns which emphasize physical closeness—these are only a few of the newer methods being tried, all of which tend to increase cerebellar stimulation through early and sustained physical contact between parent and child.

Child abuse, along with other forms of violence in

our society, is a multifaceted problem. To explain violence as entirely the result of a brain dysfunction is unrealistic and socially disastrous. Still, I consider the current attempts to relate pleasure-displeasure states, mental illness, and the early patterns of parenting as exciting and promising. If the technique of cerebellar stimulation works as well as early reports are suggesting, psychobiologists may soon be able to improve the lives of thousands of "hopeless" patients. Despite the success of medication in the control of mental illness, there remains a small core of untreated, frequently violent patients who are doomed to spend the rest of their lives in institutions. It is with these patients that the cerebellar-stimulation techniques seem to work best.

Along with basic brain research, however, we need parallel studies on the social aspects of violence and child abuse. Warm, demonstrative parents who are not afraid of physical closeness and intimacy, including touching and handling, should, according to the theory, be less likely than their cold, aloof counterparts to produce violently disturbed children. Is this true? At this point, no one can be certain. In fact, research couldn't begin to investigate the question until the recent suggestion that movement might be important for healthy brain and behavioral development. The acceptance of such a view, however, requires a paradigm change in our attitude toward the effects of the environment on brain development. It isn't intuitively obvious that the cerebellum, a regulator of movement, might also be involved in the modulation of pleasure or the later propensity to violence. Although we are still not sure—I emphasize again that the theory is provisional—a new and exciting approach is now available which may reward the collaboration of psychobiologists and social scientists. Who would have even imagined ten years ago that studies on early brain development and infant stimulation might someday suggest a provisional theory about the origin and treatment of human violence?

In the meantime, psychobiologists are now providing insights into the mechanisms responsible for normal and abnormal physical and emotional health. Some of these insights conform to traditional social and psychological opinions. Others are new, even unorthodox, such as the possible role of movement stimulation in setting the psychological balance between the capacity to experience pleasure and a tendency toward violence.

The important unifying theme in all of this seems to be the interdependence that exists between brain development and our cultural attitudes and patterns. If we give up on prematures, for instance, their physical and emotional growth will be stunted. But if we stimulate them through techniques that mimic the processes taking place within the uterus, their development can be normal. In essence, we can help to determine whether or not an immature brain will develop normally.

Similar effects of our cultural attitudes on brain development will be seen, during the next decade, in our approach to malnutrition. Will we consider the brain development of the next generation important enough to invest the resources needed to compensate for the nutritional deprivations that exist in many parts of the world? And will we encourage cross-disciplinary studies aimed at understanding and preventing violence? If our answer to either of these questions is "No," psychobiological research indicates that we may soon look forward to a world in which millions of people suffer from neuro-behavioral abnormalities.

Brain and behavior, it's turning out, are intimately interwoven in a tapestry of infinite complexity. When we learn about the brain, we learn about ourselves and, in the process, begin contributing to the tremendous task of defining the outer limits of our humanity. As with any worthwhile effort, however, our initial attempts are often groping and uncertain.

Let's look now at the exciting story of how we've arrived at our present knowledge of brain structure and its relationship to human behavior.

PART
4

9

Jumping Frogs
and Purple People

The story is told of a scientist who removed one leg of his test animal each time the animal hurdled over a barrier. After removing all four legs, the scientist concluded that the animal's failure to perform was evidence of a memory deficit.

On March 29, 1948, a wizened fifty-one-year-old mental patient in a chronic psychiatric ward in Melbourne, Australia, swallowed the first of four bitter-tasting tablets. Over the next five days the patient's doctor, John Cade, watched for behavioral changes.

An incessant talker and pacer of the wards, the patient, over the previous six years, had nearly destroyed himself in a state of sustained mental excitement. "He was amiably restless, dirty, destructive, mischievous, and interfering," Cade recalls.

As he swallowed the last of the four pills, the patient resumed his pacing and was not noticeably improved twenty-four hours later. After two more days of drug treatment, the experiment—for the pills had never before been employed as a treatment for mania—seemed doomed to end in failure. At this point, with over twenty years of mental-hospital confinement behind him, Cade's patient stood an almost certain chance of remaining in a locked ward for the rest of his life.

On reflection, one would have suspected from the start that the experiment was unlikely to succeed. For one thing, Cade was a practicing psychiatrist, not a researcher. In addition, the preliminary work, which we will describe in a moment, had been done on guinea pigs, not humans. Coupled with this, the justification for the new treatment rested on nothing stronger than a vague hypothesis that sounded more like the mutterings of a medieval alchemist than a reasoned research strategy. "Could it be," Cade reflected, "that mania may be due to a state of intoxication by a normal product of the body circulating in excess, whilst melancholy is the corresponding deprivation condition?"

As if this wasn't enough, the drug lithium, in 1948, was beginning to be associated with an alarmingly high number of deaths among heart patients who were ingesting it as a salt substitute in an attempt to combat the swelling that often accompanies heart failure. As a result of these deaths the drug would, within the next year, be taken off the market in many countries around the world.

At the basis of Cade's experiment was the belief that mania, if caused by an intoxicating substance produced by the patient, might result in the excretion of the toxin in the patient's urine. To test this hunch, Cade injected fresh urine into the abdominal cavity of a litter of guinea pigs. Alternating the urine of manic patients with hospital personnel and other non-manic hospitalized patients, Cade's original findings pointed to a "something" in the urine of manics that consistently killed the guinea pigs in a dose less than one third the concentration of the controls. After twenty to thirty minutes the guinea pigs would begin to shake, their legs became rubbery, and complete paralysis intervened within minutes. From here, generalized seizures occurred, leading rapidly to unconsciousness and death.

Next, Cade studied the separate components of urine in order to precisely define the toxic agent. When he found it—urea—he set out to discover the effects of increasing or decreasing the other urine components, one of which, uric acid, proved tremendously difficult to dissolve. To solve this problem, Cade looked up a soluble form of uric acid—lithium urate—and injected it along with urea into the guinea pigs.

Lithium urate, it turns out, exerts a powerful pro-

160

tective effect against the killing power of urea. To determine whether the lithium salts alone had any discernible effect, Cade next injected lithium alone, with startling results. Not only did the guinea pigs fail to convulse, but they became lethargic and docile, suffering themselves to be turned on their backs, where they gazed placidly into the face of the now intensely earnest investigator. If lithium could neutralize the action of the "toxin" and, in addition, exert a quieting effect on usually restless guinea pigs, what would be the effect of injecting lithium salts alone into manic patients? Could lithium quiet the manic patients by neutralizing the "toxin" that is presumably responsible for the disorder? In an attempt to find an answer, Cade broke off his search for the urinary toxin and turned his attention instead to discovering the effects of lithium on manic-depressive patients.

On the fifth day after administering lithium to the fifty-one-year-old man, things began to change dramatically. The patient walked less, talked less, was tidier, more settled, and less distractible. For the first time in anyone's memory, he was capable of holding a sustained conversation. Three weeks later he was enjoying the freedom of an unlocked ward. Two months after that came discharge, after the completion of a hastily-put-together cram course on how to manage the activities of daily living outside an institution.

Looking back now on Cade's discovery, one can enjoy the rare privilege of observing the strange and unpredictable ways that psychobiology can develop. Cade was right, but for the wrong reasons. To this day, no one has ever discovered a "toxin" in the body of manic-depressives; Cade was, in fact, searching for an antidote for something that didn't exist, or at least has yet to be discovered. But in the process he stumbled onto a principle even more valuable: The brain is a physical system that can be modified by physical agents.

If the previous sentence strikes you as rather obvious, let's take just a moment to consider the cultural attitudes toward mental illness that existed when Cade carried out his lithium experiment. In the 1940s and '50s psychoanalysis and other psychological theories dominated

American psychiatry. Over 95 per cent of practicing psychiatrists at the time identified themselves as psychotherapists, only a small proportion of whom placed any emphasis at all on the brain events accompanying a mental breakdown. In this setting, which persisted in some parts of the United States well into the mid-1960s, schizophrenics or manic-depressives, if they could afford it, entered a long treatment plan involving "intensive" and "dynamic" psychotherapy. The poorer patients' options were more restricted: the chronic wards of the nation's mental hospitals. In either case, a curious attitude existed regarding mental illness, which, in light of developments in the last twenty years, seems as crazy as some of the illnesses these "treatments" were designed to cure.

Imagine going to a doctor to seek help for a serious heart condition that you've had since birth. After telling the doctor about your symptoms and the treatments you've received in the past, you sit back and begin to ask a few consumer-oriented questions. To your surprise, you find that the doctor has never examined a human heart—in fact, has never seen one close up—and, in addition, doesn't believe that it is necessary to know anything about the heart in order to treat heart disease.

At this point I suspect you would make a quick exit and try to find a trained cardiologist. It is a matter of "common sense," you'd probably say, that a doctor must know about the heart in order to properly understand and treat heart disease. Despite the impeccable logic leading to your conclusion, psychiatry at the time of Cade's experiment was firmly committed to an approach toward mental illness that roughly corresponded to the attitude of the "heart disease specialist" who never examined the human heart. Few psychiatrists, with the exception of an occasional one who was also a neurologist, knew or cared about the workings of the human brain. The prevailing attitude was a psychological one: Mental events (psychic traumas, "unconscious" conflicts, etc.) are directly responsible for disturbed behavior.

Some support for such a theory came from detailed studies of the symptoms and signs of mental illness. In some instances the illnesses seemed obviously related to pychological conflicts. A bride who developed hysterical blindness or paralysis on her wedding night, for instance, would provide "proof" that hysteria could be adequately explained on psychological grounds alone—in this case, sexual repression. When, in addition, the symptoms disappeared after hypnosis or psychotherapy, it seemed obvious that psychiatrists were on the right track. Such was the prevailing general attitude toward mental illness.

Although not realized until several years later, the implications of Cade's successful experiment with lithium were profound. A life-long manic-depressive was returned to normal life, not through a sudden "insight" or any modification of thought or attitude wrought by years of psychotherapy, but through the power of a chemical: lithium. What did this imply about the origin of mental illness?

It is often insufficiently appreciated that ideas, like clothes, must conform to rigidly defined but nonetheless capricious fashions. Concepts we now take for granted—such as the existence of viruses that can cause everything from polio to influenza—were at one time ridiculed as the product of the overactive imaginations of certain nineteenth-century physicians. So, too, the theory of a physical causation of mental illness was, in the 1940s, a distinct minority view.

Today, psychochemical agents—tranquilizers and antidepressants—have become established as the most effective means of treating the major mental illnesses. This transition from purely psychological treatments to psychochemicals began with Cade and others who demonstrated that chemicals could exert a calming action on mental patients without necessarily lessening their alertness. At first these biochemical insights proceeded by somewhat hit-or-miss methods. One drug, for instance, originated from the chance observation that tuberculosis patients, when treated with a new anti-TB drug, also began to show improved spirits. On a hunch, the drug was then tried, with stunning results, on depressed patients who were not tuberculous. This soon led to the somewhat anomalous situation where both tuberculous and depressed patients were treated with the same drug. Things might have continued in this "rough and ready" vein, were it not for the timely convergence of several lines of research.

The first and most productive change concerned advances in our knowledge of the way nerve impulses are transmitted in the brain. For over a hundred years, two schools of thought vied with each other on this question. The first believed the brain was composed of a continuous net of cells, each nerve cell directly connected to its neighbor. According to this view, the brain consisted of layer on layer of interconnected cells with impulses traveling through the brain like water through an intricate network of pipes.

The other view was proposed by a Spanish artist-turned-physician, Ramon y Cajal, who developed special chemical stains that showed the nerve cells to be composed of a central cell body, a long process known as an axon, and finally innumerable smaller processes, the dendrites. Cajal shared the Nobel Prize in 1906 for his concept (published in his own medical journal in Madrid)

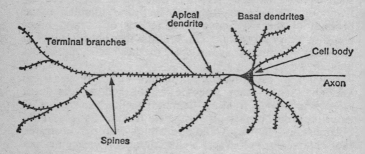

that nerve impulses are conducted along the axon of one neuron to the dendrite of a second neuron. If the stimulus passes to yet another neuron, the process is repeated. The impulse leaves via the axon of one neuron, and makes contact with the dendrite of another.

For years the debate raged as to whether the brain is composed of individual neurons completely separated from each other, or whether the true state of affairs is closer to an elaborate spider web. Some measure of the difficulty scientists had at the time in deciding between these two views can be deduced from the Nobel Prize Committee's choice, in 1906, for a person to share the prize in medicine with Cajal. Camillo Golgi, the foremost proponent of the net theory of brain function, along with his lifetime rival, Cajal, jointly shared the honor, since the scientific world was unable to decide between their competing theories.

Looking back now on the debate which, as I will discuss later, still has not been entirely settled, one can't help but suspect that occasionally Cajal must have felt that he was laboring on the short end of the odds. If the neurons are separate and not a part of a vast nerve net, then how does nerve-cell communication take place?

At one point Cajal's view seemed to suffer a fatal blow. A world-famous brain scientist announced that he had developed a stain showing tiny fibers connecting one nerve cell to another. Cajal, reacting in an inspired flash of brilliance, immediately substituted a milder solution of the same stain and demonstrated that the tiny fibers belonged to individual neurons rather than forming the interconnections between neurons. To prove his point, Cajal sketched his anatomical slides on the tablecloth of a nearby Madrid café where the earliest psychobiologists met daily to debate the microscopic architecture of the human brain.

Today we know that Cajal was correct. With increasingly sophisticated stains that would have appealed to Cajal the artist, brain scientists over the ensuing fifty years have demonstrated that neurons in reality are separate cells that communicate with each other but are never in direct physical contact. This rules out the view that nerve impulses pass through the brain like water through a system of pipes. But with the establishment of the "neuron doctrine," the even more vexing question remained: What is the nature of the nerve impulse?

The brain, we now know, is activated by tiny electrical circuits. From the time of the Greeks, scientists have observed that electricity can be produced by rubbing resin with cat's fur, or glass with silk. Today we often inadvertently produce the same "electricity" when, crossing a carpeted room, we shake hands with someone and each of us receives a mild shock of static electricity. By the nineteenth century, such observations had progressed to the point that electrostatic machines, consisting of a glob of sulphur on a spindle turned by hand, provided drawingroom entertainment as well as a focus for scientific speculation. As often happens, however, one of the key observations linking electricity with brain function came from an incident that occurred outside the science laboratory.

In 1780, the wife of a lecturer in anatomy at the University of Bologna was in bed with a minor illness for which her physician prescribed frog soup. The cook, while working with a batch of cleaned frogs legs, placed them on a table next to one of the now popular electrostatic machines. A few seconds later, the woman ran screaming from the kitchen, literally frightened out of her wits when she noted that the frogs legs had started jumping!

The frog-leg experiment is now routine in physiological psychology courses throughout the world. When an electric current is applied to a frog's leg it will twitch, even though the frog has been previously decapitated and can't in any way be experiencing an "electric shock." In addition, a twitch can be introduced in one frog when it's touched by the nerve of another frog, an observation dating back over 130 years. But most important of all in regard to our understanding of nerve-cell communication, the fluid surrounding a discharging nerve can be applied to an isolated muscle and it too will produce a contraction. In this last instance a chemical rather than an electrical force is at work. In some way electrical changes in one cell produce a chemical which is transferred to a second cell via a fluid medium. When the chemical reaches the second cell, events switch back again from chemical to electrical.

From experiments conducted over the last fifty years, scientists have developed a theory of brain action as both electrical and chemical. Generators of weak electrical currents lie on the surface membrane of nerve-cell bodies as well as along the dendrites and axons. With the aid of

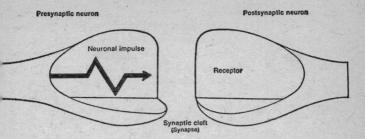

FIGURE 18. A diagramatic representation of a synapse.

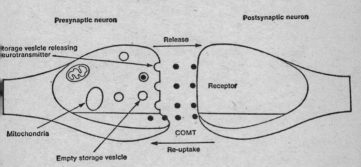

FIGURE 19. The normal process of release, re-uptake, and enzymatic metabolism of the neurotransmitter.

powerful electron microscopes, the surface structure of the neuron can be observed to possess many points where the axons of one cell approach very closely, but don't quite touch, the dendrites of other cells.

Communication between neurons is achieved when an electrical potential is generated in one cell and travels along its axon at a constant rate and coded sequence. The code can be compared to the letters of the alphabet—not quite the message, but the elements from which the message will be composed. You can also think of it as Morse code, with the signals transmitted unchanged from the sender to the receiver. Along the greatest distances, usually the length of the axon, the process is purely electrical. But when the signal reaches the end of the axon, it is

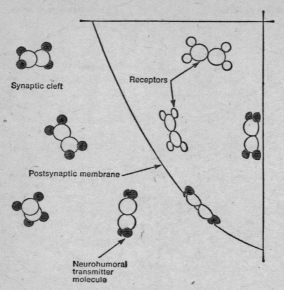

FIGURE 20. Neurohumoral transmission across a synapse. The transmitter molecule fits only specific receptors much as a key fits only a specific lock.

ready for transfer to the next neuron, and the most important aspect of nerve communication, a chemical process, is ready to take place.

High-power microphotographs of the brain depict discrete areas where brain cells almost touch, the synapse, as well as a narrow synaptic cleft, or space, between the neurons. At the moment an electric current reaches the end of the axon, special chemicals called neurotransmitters are released. These cross the synaptic cleft and selectively alter the membrane of the second cell by linking to a specific receptor. (See Figure 19.) Scientists refer to the match between a neurotransmitter released from the first cell (the presynaptic membrane) and its special receptor site in the second cell (postsynaptic membrane) as a "lock and key" arrangement. (See Figure 20.) Just as only a special key will open your apartment door, although there may exist many copies of the same key, so too only certain chemicals will fit the chemical "lock" on the post-synaptic membrane. Once there, the neurotransmitter alters the membrane in such a way as to generate an electric potential on the second cell. From here the electrical response is conducted as it was in the first cell, along the length of the cell's axon.

Within the last fifteen years psychobiologists have identified a family of chemicals with different molecular structures (Figure 20) which exist throughout the brain and are important in normal and abnormal mental functioning. Through the use of special techniques, these neurotransmitters can be made to stand out against the background of the nerve-cell membrane and their distribution mapped throughout the brain. At the same time, the neurotransmitter's chemical fate is revealed as special enzymes begin breaking it down into smaller inactive molecules or, alternately, it is reabsorbed by the membrane of the first cell.

Neurotransmitters are of two types—excitatory, which facilitate the development of an electric current in the neighboring cell; and inhibitory, rendering an electric charge less likely to develop on the second nerve cell. At any one moment the chance of a nerve cell's "firing" depends on the sum total of excitatory and inhibitory neurotransmitters at the synapses. For example, if five excitatory synapses exist, their influence can be counteracted by five inhibitory synapses. You might think of it as an exquisitely sensitive microbalance in which the development of an electric current depends on the ratio of excitatory to inhibitory neurotransmitters.

Our knowledge of the effects of different neurotransmitters developed hand in hand with drug treatments for mental illness. A drug derived from the plant rauwolfia, used in India for three hundred years as a sedative, was noted to decrease the concentration in certain brain areas of norepinephrine, the first neurotransmitter conclusively identified as involved in nerve-impulse transmission. Along with sedation, however, rauwolfia—or reserpine, as it came to be known—produced a slowly progressive depression that sometimes reached suicidal proportions. But when reserpine was preceded by the drug iproniazid, used in the treatment of tuberculosis, the depression could be halted.

Psychobiologists at the time speculated on the implications of this finding. Some suspected a link between tuberculosis and depression. One researcher even seriously suggested that the drug's antidepressant effects arose from its sharing a small part of its molecular structure with a contemporary rocket fuel! It turns out that the correct interpretation is that the antituberculosis drug prevents the action of an enzyme, monoamine oxidase, which ordinarily splits norepinephrine into smaller inactive chemicals.

The sequence goes like this: Reserpine causes sedation and depression by depleting the supply at the synapse of the excitatory neurotransmitter, norepinephrine. This can be counterbalanced by prior administration of the drug iproniazid, which blocks the action of the chemical monoamine oxidase before it inactivates norepinephrine. Thus, the first of a family of antidepressant drugs, the monoamine oxidase inhibitors, were introduced into clinical psychiatry.

Even more important than their usefulness in treating depression, the monoamine oxidase inhibitors also provided the first clue to a major biochemical hypothesis about depression. The brainchild of Harvard psychiatrist Joseph Schildkraut, the new catecholamine hypothesis—named after the chemical structure of norepinephrine and the other psychochemical, dopamine—postulated that depression is associated with an absolute or relative deficiency of the catecholamines at special receptor sites in the brain. Elation, or mania, on the other hand, was presumed to be associated with an excess of these neurotransmitters.

In the last fifteen years, a series of successful antidepressants have been developed based on the catecholamine hypothesis. Not all of them, however, work by altering the ability of monoamine oxidase to break down neurotransmitters. Another group increases the action of the neurotransmitters by a totally different means.

Along with the destruction of norepinephrine in the nerve cell, another way to terminate its action within the brain is by reaccumulating the neurotransmitter on the presynaptic membrane, where it resided prior to its release into the synaptic cleft. This is, in fact, the principal means by which the nerve impulse is ordinarily terminated. The presynaptic membrane might be thought of as a biological "Indian Giver" which can be depended upon to ultimately try to take back what it has previously given away.

This second class of antidepressants works by temporarily blocking the neurotransmitter's re-uptake mechanism. At a result, norepinephrine remains for a longer time in the synaptic cleft. Since, according to the catecholamine theory, increased epinephrine shifts the mood balance toward mania, the end result is a decrease in depression. Again, at the risk of oversimplification, think of mania and depression as exactly balanced on the opposite ends of a microscale.

Knowledge of the biochemistry of depression has advanced considerably since the catecholamine theory was first proposed in the 1960s. But, instead of only one or two neurotransmitters, there are now indications that the brain may possess hundreds. Workable drug treatments for depression can be devised, however, by concentrating on the role of the catecholamines, norepinephrine, and dopamine, along with another chemical, serotonin. The increase in the number of neurotransmitters has brought about an attitudinal change on the part of brain scientists somewhat similar to the attitudinal changes among nuclear physicists a few years ago in response to a volley of newly discovered elementary particles.

Psychobiologists, like their physicist colleagues, have stopped searching for *the* explanation—in this case, a single chemical that causes depression. Rather than a disordered chemical or the "toxin" that fired the imagination of John Cade, the true state of affairs in depression is likely to turn out to be an alteration in the balance between an as yet undetermined number of neurotransmitters. As we'll discuss in the last part of this book, the challenge now seems to be less technical (discoveries of more neurotransmitters) than conceptual: formulating the problems in ways that emphasize process rather than structure.

Our watches, cars, and television sets malfunction when some *thing* gets out of whack—a carburetor, a tube, whatever. From here, it's been the simplest of steps to assume that our brains, too, malfunction in a similar way: a missing or altered neurochemical. Although this clearly represents an advance over the voodoo mentality of the past, which associated mental illness with psychological factors alone, any future emphasis on finding a single neurochemical factor is unlikely to lead to a complete understanding of either mental illness or brain functioning.

While psychobiologists were making the observations that formed the basis for the catecholamine theory of depression, other workers, principally in France, were turning their attention to mankind's most perniciously destructive mental illness: schizophrenia. It's difficult today to picture the wards of mental hospitals prior to the introduction of the major tranquilizers. Although chains had been eliminated in the world's mental hospitals since the eighteenth century, no one had been able to outlaw straitjackets, seclusion rooms, or treatment by wet packs (swaddling agitated patients in iced sheets). For one thing, there simply were no other practical ways of controlling an acute psychotic. Although I'm too young to have seen the mental hospital wards of the pre-1950s, they would, in the words of an older colleague, "make the psych ward in *One Flew Over the Cuckoo's Nest* look like a country club." Given such a state of affairs, it's not surprising that innovative and sometimes bizarre treatments were being explored. One of these concerned the nature of artificial hibernation in animals.

Observing that some animals can remain for long periods of time in a semiconscious state, several investigators sought a drug that could reproduce in man the condition that exists in cold-blooded or hibernating animals. At the basis of the search was the quaint idea that somehow such a drug, when discovered, might also produce in humans the cold-blooded indifference, or "ataraxia," extolled by the Greeks.

A French anesthesiologist, H. Laborit, developed in 1948 a "lytic cocktail," a euphemism for a potent mixture composed of three drugs that quickly dispatched an agitated patient into a deep and prolonged sleep. Two of the components of the mixture were well known. The third, chlorpromazine, was a newly synthesized drug that was originally intended as an antihistamine. Although its antihistaminic properties were low, chlorpromazine exerted a calming action that intrigued the French psychiatrist Pierre Deniker enough to try it on ten acutely disturbed patients.

A measure of the drug's success is that between 1955 and 1967 the population of the public mental hospitals of the United States dropped about 20 per cent.

Although everyone was pleased with the drug's success, it wasn't until ten years later that its mechanism of action was understood. A Swedish psychobiologist observed that chlorpromazine and other effective antipsychotic drugs stimulated certain brain cells which specialized in the formation and release of the neurotransmitter dopamine. By an ingenious insight, the investigator postulated that the cells were producing more dopamine because chlorpromazine was occupying dopamine's receptor cells on the postsynaptic membrane. He suggested that a feedback system existed which carried a message back to the dopamine-producing cells on the presynaptic membrane: "Please send more dopamine, we're not receiving enough." The reason the receptor cells on the postsynaptic membrane were not receiving sufficient dopamine was that the chlorpromazine was occupying dopamine's usual special receptors.

Through the application of X-ray crystal analysis, psychobiologists discovered that part of the chlorpromazine molecule could be imposed on dopamine's molecular structure. As a further proof, the effectiveness of other antipsychotic drugs could be predicted beforehand according to how closely their molecular structure resembled dopamine's.

The humble model of a "lock and key" arrangement provided neuroscientists with a new way of thinking about psychosis. From here, newer and more powerful drugs emerged, which in some cases turned out to produce rather bizarre side effects. A faint purple discoloration of the skin developed in a few patients, stimulating psychiatrists around the world to look carefully within their patient population for the newly described "purple people." In other cases, movement abnormalities occurred where the patient engaged in incessant twitching or licking. Fortunately, in many of these instances the side effects could be eliminated either by a change in tranquilizer or the addition of another drug.

Psychosis, like depression, was yielding to the new emphasis on neurochemistry. And, in the process, psychobiologists were learning more about how nerve cells communicate. But, like a series of Chinese boxes, each discovery led to further challenges. One seemingly insoluble puzzle concerned the events which took place *after* the neurotransmitter had slipped into its receptor site on the postsynaptic membrane. Obviously, it doesn't just sit there. Something has to happen. But what?

One helpful way of thinking about the brain's action is to concentrate on the diverse functions the brain cells are called upon to carry out at different times. Withdrawing a hand from a hot stove must occur instantaneously if a burn is to be avoided. In this case the mediating neurochemical process must be rapid. The formation of a memory, in contrast, involves the establishment of a neuronal change which is relatively permanent. Whatever you can recall about your fourth-grade teacher (assuming you can recall anything at all!) must depend on a relatively resistant long-term change set down years ago somewhere in the neurochemical circuitry of your brain's fifteen billion neurons. Although no one knows for sure, the formation of such a memory probably takes a comparatively long period to establish, since it lasts longer. If you are building a house that you expect to last for several generations, you spend time and effort in its construction. Similarly, if the brain is to establish a relatively permanent memory trace, then the constructive process too is likely to take longer. In essence, there exists in the brain a continuous series of events that range from the very brief to the relatively permanent. Recently, psychobiologists have discovered a "second messenger" system which provides the basis for this diversity.

As often happens in psychobiology—or any aspect of science, for that matter—the researcher who searches for a mechanism of action is rewarded over his counterpart who retains a "building block" or jigsaw-puzzle orientation. As we have seen, depression and schizophrenia are turning out to be, in part, disturbances in the functional interrelationship of multiple neurochemicals. There isn't any "missing link" or "toxin," and such conceptual models inhibit rather than inspire promising research. But even the insight that one should search for a *process* rather than a *thing* leaves open the level of process that is likely to reward further study. In the case of nerve-cell transmission, psychobiologists were faced again with the Chinese puzzle situation: Open one box and it is discovered to contain an unknown number of additional boxes. In addition, at a certain point the issues become sufficiently complex that the relevant research must be carried out by removing tiny portions of the brain and studying them by splitting nerve cells into their component parts.

As we have seen, the cell membrane regulates what goes in and out of the nerve cell. After the neurotransmitter slides into place on the postsynaptic membrane, it changes the permeability of the membrane and modifies its action, sometimes by generating a nerve impulse which travels onto the next neuron. But at such a time, what is going on inside the individual nerve cell?

In the late 1950s a research group at Case Western Reserve University was studying how the sugar glucose is released from liver cells. Ordinarily, glucose is stored in the form of glycogen (animal starch), which breaks down, under the influence of the hormone adrenalin, into multiple molecules of glucose. When you are suddenly frightened or excited, adrenalin is released from your adrenal glands into your blood stream, where it is carried to the liver and stimulates the release of glucose, which you need under conditions of stress.

The researchers discovered that the cell membrane of liver cells, when exposed to adrenalin, releases a mysterious substance which, when placed inside the cell, does the job usually attributed to adrenalin: breaking glycogen into glucose. What had been considered a one-step process turned out to involve two steps. After adrenalin interacted with the liver cell membrane, a second (so far unidentified) substance took over and carried out the process of glycogen breakdown.

Several months and hundreds of trials later, the Case Western group hit upon the correct interpretation. Hormones, like adrenalin, activate a special enzyme within the cell membrane which generates a "second messenger" that relays the message for the first messenger (the hormone) into the cell's interior, where it can act on the biochemical machinery of the cell.

Turning next to nerve cells, psychobiologists found a similar enzyme within brain-cell membranes. Electrical stimulation of nerve-cell clusters removed from rabbits showed an increase in the level of this second messenger (Figure 21). The critical question was, of course, Can the same thing happen if a neurotransmitter is placed directly on the cell membrane?

The application of the neurotransmitter dopamine resulted in a large increase in the second messenger within the brain cell of a special area of the brain, the caudate, known to be rich in dopamine receptors. A short time later, researchers at the National Institute of Mental Health discovered enzymes specific for some of the other neurotransmitters. Dopamine and norepinephrine, for instance, were found to activate their own enzymes in cell membranes, which in turn led to the production of "second messengers" in the brain cells. But with this insight, the process of exploring nerve-cell action was just beginning. How does an increase in the level of the second messenger carry on the job begun by the neurotransmitter? And what possible advantage can such a two-message system provide to the brain?

Certainly, with respect to immediate brain interactions such as responding to pain, a double-messenger system would be of no advantage. In line with this reasoning, neurotransmitters are thought in such cases to act directly, altering the postsynaptic membrane and triggering either a nerve impulse or changing nerve membrane permeabil-

176

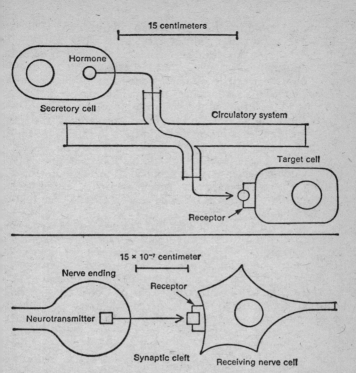

FIGURE 21. The action of neurotransmitters is thought to be similar to hormones. Both are chemicals that are released from one cell, travel outside the cell, eventually binding to specific receptor sites on the surface of the second cell. This activates the "second messenger system," which carries out a specific activity in the second cell.

ity. Events lasting a bit longer, however, probably involve the second messenger system. After a neurotransmitter reaches its receptor site (Step 1) it activates a specific enzyme (Step 2), which in turn triggers the production of a "second messenger" (Step 3), which does the actual work within the brain cell.

In order to appreciate the value of this three-step process, consider that much of the nerve-cell activity in the human brain is inhibitory rather than excitatory. Usually we imagine that a train of nerve cells, when activated, enables us to do something, whether reaching for a cup of coffee or composing a symphony. In reality, the most frequent chore performed by nerve cells is an inhibitory one. For example, the second messenger acts in the cerebellum, the part of the brain concerned with movement, by toning down the spontaneous firing rate of the cerebellar cells. This modulates movement and controls it so that unnecessary or clumsy motions are eliminated. In a similar way, a second messenger acts to regulate discharges in the deep-lying caudate nucleus. A disturbance in this smooth-functioning process results in other types of movement abnormalities: muscular rigidity and tremors, the hallmarks of Parkinson's Disease.

The most important value of the second messengers comes from their ability to provide a ready-made system for transfer of short-lived synaptic impulses into longer-lasting multi-component chains. For one thing, proteins in the synaptic membrane can be activated to enter the interior of the cell and chip off parts of DNA, the genetic material concentrated in the nerve-cell nucleus. These separated portions of DNA can then be picked up and ultimately translated into new, enduring proteins. Some psychobiologists are now speculating that such changes might represent the physical basis for long-term memory storage in the brain.

At this point it is too early to be sure about the importance of the brain's second-messenger system. If nothing else, however, its discovery provides another reminder that the brain doesn't always operate in ways that we can immediately intuit. "Entities are not to be multiplied unnecessarily" was the logical "razor" first articulated by the fourteenth-century philosopher William of Ockham. What may be true for philosophy, however, isn't necessarily so when it comes to psychobiology. Second messengers provide the mechanisms for variety, control, and long-term stability for the brain's neuronal network. Like a fine-tuner on a stereo, they enable the brain to achieve a purity of expression through exquisitely timed and controlled neuronal interactions.

So far we have talked about the crosstalk between two nerve cells as if this were all of the story—but, of course, it isn't. Of equal importance is the elaborate network of nerve circuits which exists within the brain, linking up the fifteen billion neurons into a meaningful pattern.

The turn-of-the-century physiologist Sir Charles Sherrington once referred to the pattern of brain-cell interaction as the workings of "an enchanted loom," which "weaves a dissolving pattern, always a meaningful pattern, though never an abiding one; a shifting harmony of subpatterns."

Our knowledge about the mechanisms responsible for the action of Sherrington's "loom" was only gathered during the past twenty-five years. From this research emerged a "classical theory" of nerve-cell communication. (It's ironic to speak of any theory of brain function as "classical" when the total period of scientific observation about the brain, even stretching the meaning of the word "scientific" to its absolute limits, spans at most 150 years, a pitifully short time in the history of ideas.)

As we have seen, nerve cells are believed to communicate via nerve impulses which travel down the axon of one cell toward the cell body or the dendrites of another. According to this theory, the dendrites exist as elaborate antennae extruded to pick up impulses sent along the axons of other cells. Brain activity thus could be depicted in an elaborate diagram such as one might draw after viewing a slow-motion movie of the lightning-fast interactions of billiard balls. As we briefly discussed in Chapter One, however, the potential interactions between single units in a complex system are, for all practical purposes, infinite. Even a typical chess position, calculated 25 moves ahead to cover all possible rejoinders, would take 10^{69} seconds to complete by a computer five times faster than any presently available. In order to put that figure into perspective, the total length of time since the Big Bang Theory concerning the origin of our solar system is put at a modest 10^{18} seconds!

From time to time, neuroscientists wondered about the function of the dendrites, the numerous branching

179

fibers extending in all directions from the nerve-cell body. Might they too be capable of propagating a nerve impulse? One group at the Mathematical Research Branch of the National Institute of Health demonstrated that tiny electric currents, created by synapses on the dendrites, flow through the whole dendritic tree. Since dendrites branch as they extend from the nerve cell, the spreading electric current can be visualized as a string of colored bulbs lighting up in sequence, culminating in a fully illuminated Christmas tree. To bring the analogy closer to the real situation, the balls should be considered not restricted to only an "on/off" state, but capable of dimming or brightening according to the number of excitatory or inhibitory impulses converging on them.

These "microcircuits," as they are called, add a totally new dimension to our already mind-boggling conception of how the brain works. It suggests that nerve cells can communicate with each other through their dendrites, independently of the axon or its nerve impulse. During the past five years, in laboratories throughout the world, microcircuits have been found in practically every part of the human nervous system. In the process, psychobiologists have had to construct a new paradigm of brain functioning to include direct dendrite-to-dendrite communication.

It is interesting to speculate what Golgi would make of the recently discovered microcircuits. In a way, they inch us closer to his "net theory" of brain function, which he continued to espouse until his death in 1926. Although dendrite-to-dendrite communication doesn't destroy the "single neuron theory" (the dendrites are clearly structural components of individual nerve cells), it does point up a rigidity in our thinking about brain functioning.

One of the oldest analogies about the brain compares it to a telephone switchboard, with all incoming sensory data relayed to the brain, where the appropriate "hookups" are made. As I trust I have shown by now, this model is hopelessly simplistic; but, for the moment, I would like to play with it a bit in order to give you a clear idea of the significance of the microcircuits.

A modern telephone system solves the problem of

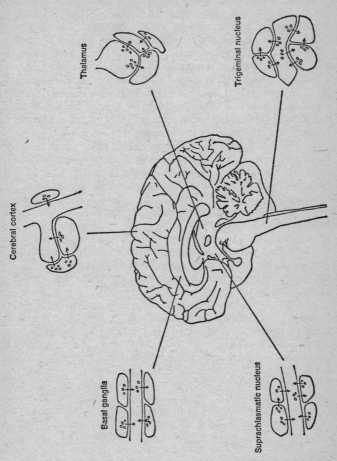

FIGURE 22. Typical microcircuits in different regions of the brain. Microcircuits are being discovered throughout the brain at all levels, starting at the cerebral cortex and extending all the way to the spinal cord.

information overload by microminiaturization and micro-circuitry, which perform integration functions at points quite distant from anything resembling a central switchboard. Microprocessors in television sets do the same thing. Instead of trouble shooting the whole system, a television repairman can now narrow his search, concentrating on different microcircuit units that enable him to repair the set by replacing a thin tray of microprocessors. In other words, it makes a good deal of sense for telephones, TVs, and human brains to possess small computational units that can provide localized as well as specialized information processing.

The microcircuit concept originated from the collaboration of a psychobiologist and a computer specialist at IBM. Dendrite-to-dendrite communication, in fact, was "discovered" by a computer! Instead of brain scientists telling the computer how brain cells interact, the computer simulation results suggested, for the first time, the highly unorthodox view that neurons communicate, at least partially, via the excitatory and inhibitory synapses existing between dendrites. Although initially skeptical of this computer conclusion, psychobiologists carried out the necessary experiments that eventually proved the computer correct.

Computers provide the latest in a long series of "models," all of which, at the time of their first introduction, purported to "explain" the brain. The construction of artificial models for real-life events is, of course, as old as the human species. One can envision, for instance, our early cave-man ancestors squatting down in the dust to trace with sticks the shortest and safest path leading back to their cave. Although eminently helpful on occasion, models often suffer from overselectivity. Only some parts of reality are modeled at any one time, and sometimes the most important parts are inadvertently left out.

When we try to understand how fifteen billion neurons in our brain interact, we naturally search for a model to help us. "It's like this," we begin, and then proceed to compare the brain to something we're all familiar with—a telephone exchange or, more recently, a computer. The modeling process comes apart, however, when we attempt to model something as complex as the human brain. As psychobiologists are discovering, the brain is not like anything else we've ever encountered.

For example, Dr. Kenneth Boulding, President of the American Association for the Advancement of Science, estimates that the capacity of the human brain is "literally inconceivably large." If each of the brain's fifteen billion neurons is capable of only two states (on or off), the capacity would be 2^{10} billionth power. To write out this number at the rate of one digit per second would take ninety years, or about the projected human lifespan in the twenty-first century. In comparison, the number of neutrinos (about the smallest particle in existence) that could be packed into the total astronomical universe ten billion light years across, could be written down in four minutes!

A contemporary Russian brain physiologist took a slightly more conservative view of the task of numerically writing out the number of possible neuron interactions. He estimates that the figure, when written out by hand on a flat surface, would span about 9½ kilometers. Although I don't think it would take ninety years to write out such a string of zeros, it's not something one would be likely to accomplish in less than several years, even when working at the task twelve hours a day. Granting that the true state of affairs probably lies somewhere between Boulding and the Russian brain physiologist's estimates, how do we go about constructing a "model" for such a process? Obviously, we do so at our peril.

Despite the dangers that are apt to ensnare the model builder, I think the brain can be usefully compared in some ways with a large city.

Take out a map of New York City and look at the vast web of intersecting communication channels that exist both above and below ground. In addition to the streets and subway routes, superimpose on it the flight patterns of all airplanes flying in and out of New York's major airports. Such a model, although incredibly complicated, is at least manageable. We could keep track, if we so desired, of all the traffic along a particular thoroughfare, or all of the subway passengers on Manhattan's West Side. These routes, along with their traffic, correspond to the brain "tracts" that psychobiologists have discovered over the past one hundred years.

Within the city lives a population variously estimated at between eight and ten million people. Imagine the people in the city as corresponding in our analogy to the neurons in the brain. Any person living in New York might, at any one time, be traveling on one of the streets or subways. So, too, a neuron may be involved in the traffic impulses along any one of the major tracts. This brings our analogy roughly to the point when microcircuits were discovered.

According to the traditional model, neurons are like lights that are turned either on or off. When a nerve impulse is generated, the relevant neuron blinks on and begins interacting with other neurons. Using such a model, it was hoped at one time that by tracing the connections between neurons, one eventually could come up with a map such as our map of New York where, say, starting at Christopher Street in Greenwich Village, we could eventually work our way to Fifth Avenue and Sixty-third Street.

But the introduction of microcircuits within the brain raised our conceptual model to a new order of complexity. Remember, in Chapter One, where we talked about the complications introduced by telephone systems that link people together who are geographically miles apart? With the introduction of microcircuits, a complete model would require us to superimpose the labyrinth of telephone cables serving New York City onto our previous diagram of street, subway, and airplane patterns. Only in this way could we keep track of all the interactions between New York City's millions of people. If this were done, our resulting network of lighted bulbs would consist of briefly illuminated flashes corresponding to short calls ("Just wanted to check whether you were home. I'll be right over") as well as illuminations that might correspond to lengthy business negotiations.

There are other interesting parallels between the brain and a city. As we have seen in Chapter Seven, malnourished brains may never reach their full potential, just as the residents of a city, cut off from their sources of food, will eventually starve to death. In addition, environmental stimulation is also needed, since the brain develops by interacting with its human and nonhuman environment. As with Rosenzweig's rats, new and more sophisticated neural networks are laid down in response to enriched environments. One can compare this to the loneliness and depression of a solitary person—the "isolated neuron" dwelling in the bowels of one of our nation's cities—who is cut off from interaction with other people. Here psychobiology underscores the truth of intuitions dating back thousands of years: Man is truly himself only when he's interacting with his fellow man. "An isolated brain is only a piece of biological nonsense as meaningless as an isolated individual," warned biologist Sir Julian Huxley.

Obviously, at some point our model of the brain as a city breaks down. There is certainly no proof that individual neurons possess even a rudimentary form of consciousness, as do the individuals living within a city. In addition, even if we had never been in a city we could reconstruct a rough but adequate idea of how a city "works" by interviewing one of the city's inhabitants. So far, no one has suggested that the working of the human brain can be understood by studying one neuron. In fact, as we've observed, the whole brain is qualitatively different from the sum of its constituent parts (the neurons).

As we'll discuss in the last part of this book, understanding the human brain demands new ways of organizing the vast array of emerging research findings in psychobiology. Since the brain is different and immeasurably more complicated than anything else in the known universe, we may have to change some of our most ardently held ideas before we're able to fathom the brain's mysterious structure. In the meantime we can only grope, like the proverbial blind man, attempting to "visualize" an elephant by touching now its trunk, then its legs, and finally its floppy ears. And, in the process, we'll develop models that we optimistically believe will eventually enable us to comprehend the infinite.

The latest of these models is, of course, the computer. Although computer models are more sophisticated than anything we've had previously, they are not likely to remain the last word on the subject of brain-cell communication. In addition, I'm uneasy that our "explanations" for brain functioning always conform so closely to the technological "toys" that happen to be popular at a particular time. Back in the days when a telephone switchboard was the ultimate in organization, people smugly asserted that the brain "worked" in a similar way. Now that our technology is greatly advanced, most of us know better. But thanks to increasingly sophisticated computers, some are now claiming that the brain is nothing but a computer. (Later in the book we'll say more about this conceptual virus, which, if we're not careful, can lead to a nasty case of mental malaise affecting chiefly the senses of curiosity and wonder.)

Will our future conceptions of the brain continue to depend so heavily on the prevailing future technological models? Almost certainly they will. Whether this means that God is a technologist, or that evolution is only the unfolding of some scientific imperative, I'll leave you to decide for yourself.

Another possibility—which I'm not pushing, but merely suggesting at this point for the sake of completeness—might be that the brain's attempt to understand itself involves an irreconcilable contradiction. No other living organism has ever demonstrated anything even remotely suggesting an understanding of its own functioning. Why do we think we are capable of such a feat?

Do our technological models—telephone switchboards, increasingly sophisticated computers, whatever else may be invented in the future—represent the way things really are in the brain? Or are they, rather, the only ways we are able to think about them?

10
The Jekyll and Hyde Solution

The only satisfactory way of existing in the modern, highly specialized world is to live with two personalities. A Dr. Jekyll that does the metaphysical and scientific thinking, that transacts business in the city, adds up figures, designs machines, and so forth. And a natural, spontaneous Mr. Hyde to do the physical, instinctive living in the intervals of work. The two personalities should lead their unconnected lives apart, without poaching on one another's preserves or inquiring too closely into one another's activities. Only by living discretely and inconsistently can we preserve both the man and the citizen, both the intellectual and the spontaneous animal being, alive within us. The solution may not be very satisfactory; but it is, I believe now (though once I thought differently), the best that, in the modern circumstances, can be devised.

ALDOUS HUXLEY,
Wordsworth in the Tropics

The most striking aspect of the human brain is its symmetry. Looked at from above, the brain presents two perfectly formed half shells that are completely indistinguishable to the naked eye. Since each cerebral hemisphere is the mirror of the other, there is nothing in their outward appearance that hints at the profound functional differences within. What we do know of these differences has come from studies of how the two hemispheres respond separately, sometimes even antagonistically, to the world around them.

As with any revolutionary scientific change, the discovery of hemisphere specialization came from a combination of luck and careful observation.

In 1844 an English physician, A. L. Wigan, published an account of the illness, death, and post mortem examination of a friend and patient he had known for many years. At the autopsy, Wigan made the startling discovery that his patient, who had been normal in every respect, was the possessor of only *one* cerebral hemisphere. Wigan realized the importance of his finding and published it in a book, *The Duality of the Mind,* which anticipated by a hundred years our explorations into the mysterious interrelationships of the two cerebral hemispheres. "If only one cerebrum is required to have a mind," he reasoned, "the possession of two hemispheres (the normal state) makes possible and perhaps even inevitable the possession of two minds."

Although ignored at the time, Wigan's speculations received some measure of support less than twenty years later when a French surgeon, Paul Broca, examined the brain of a patient who had been paralyzed and left almost speechless by a stroke affecting his left hemisphere. Broca suggested that the patient's language difficulties (aphasia) could be explained on the basis of a softening of one area of the left hemisphere. When other investigators subsequently demonstrated similar language difficulties in other patients with softening in the same location in the left hemisphere, it seemed reasonable to conclude that the left hemisphere possesses a language center.

Further support for this early scientific demonstration of hemispheric specialization came ten years later when a German neurologist, Carl Wernicke, discovered a second and rather different speech center. While "Broca's area" is responsible for the conversion of thoughts into smoothly articulated sounds, the speech area of Carl Wernicke is more involved in the process of conveying meaning. Instead of the stumbling, painfully slow sentences seen in patients with softening in Broca's area, a patient with destruction in Wernicke's area speaks with perfect articulation, but makes no sense. When asked about the weather, for instance, the two patients would answer very differently. The patient with a Broca's language center disturbance might say, "Sunny." If pushed, he might elaborate a bit more: "Sunny day." His speech is telegraphic, as if he is being charged by the word, but comprehensible.

In contrast, the person with Wernicke's aphasia might say, "I was over in the other one, and then after they had been in the department I was in this one." He may even respond with something as far out as, "Argentinian rifles." Such a patient lives in a linguistic tower of Babel, is unable to understand what is spoken to him, and is incapable of responding in a comprehensible manner to other people. Often such patients are misdiagnosed as psychotic, since to those unfamiliar with this type of language disturbance they speak in a "crazy manner."

Wernicke and others next made the logical assumption that these two language centers are connected to each other as well as to other regions of the brain. Wernicke's area is responsible for the comprehension and formulation of spoken and written language (with very few exceptions, aphasic patients write the same way they speak), while Broca's area provides the "muscle power" to produce speech.

A French neurologist, Joseph Jules DeJerine, at about the same time, suggested a role for the corpus callosum, the thick band of nerve fibers connecting the right and left hemispheres. In order to understand DeJerine's contribution, it's necessary to introduce a small dose of neuroanatomy at this point. It concerns the manner in which visual impressions are conveyed from the eyes to the brain.

A common misconception holds that the right eye scans objects on our right side and that the visual fibers from the right eye then travel to the right side of the brain. The left eye, according to this view, works just the opposite: It "sees" the left side of space and conveys its impulses to the left side of the brain. Unfortunately, things are a bit more complicated.

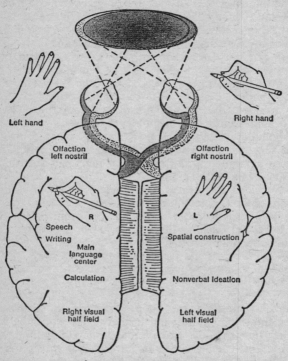

FIGURE 23. With the eyes fixated straight ahead, stimuli on the left of the fixation point go to the right cerebral hemisphere, and stimuli on the right go to the left hemisphere. The left hemisphere controls the right hand, and the right hemisphere controls the left hand. The left hemisphere is dominant for most people; it controls written and spoken language and mathematical abilities. The minor (right) hemisphere can understand only simple language. Its main ability seems to involve spatial construction and pattern sense.

Actually, the eyes can be thought of as divided vertically into two equal halves. The optic fibers from the outer sides of each eye do not cross but go directly to the same side of the brain, while the fibers from the inner sides (nasal) cross over just behind the eyeballs—the optic chasm—and proceed to the opposite side of the brain. Each eye thus contributes equally to the visual image in both eyes. Stereoscopic vision is one of the advantages of such an arrangement; it also precludes loss of sight to either "visual field" by destruction of one eye. A similar crossing-over exists for movement and sensation. When we move our left hand, we do so by activity originating in our right hemisphere; and when we pick up a hot cup of coffee with our left hand, the pain impulses cross over within the spinal cord and eventually end up within the sensory area of the right hemisphere. Once the stimulus reaches one hemisphere, it's immediately transferred to the other across the corpus callosum. If the two hemispheres are prevented from "talking" to each other across the corpus callosum, the hemispheres become functionally isolated, a phenomenon referred to as a "split brain."

DeJerine first established the importance of the corpus callosum by painstakingly analyzing one of his patients and deducing how the human brain *must* work in order to produce the patient's disability. DeJerine arrived at his conclusions by relying on a bit of medical detection. The patient was unable to see in his right visual field and couldn't read. He could copy words and write spontaneously as well as take dictation, but he could not understand what he had written. After his death a destructive lesion was found in the left visual cortex (explaining the right-sided loss of vision: Remember, the right visual field is linked to the left side of the brain) and in the corpus callosum. Although the patient's own written words could reach his right hemisphere via the intact left visual field, they could not be carried across the damaged corpus callosum to Wernicke's speech center, where visual images are decoded into language symbols, hence his inability to read his own writing.

This demonstration of the importance of the corpus

callosum was soon forgotten, and for the next sixty years brain scientists considered it little more than a fancy tethering system to hold the two hemispheres together. One reason for its decline in influence was the discovery of its total absence in some seemingly normal people. From here it is easy to jump to the erroneous conclusion that if "normal people" can do without it, it cannot be of much importance. Such a view ignores an elementary principle of scientific investigation: Statements about function must always be qualified according to the sophistication of the methods used for testing. We shall see later that the corpus callosum actually turns out to be the key to our understanding of how the two hemispheres of the brain work. To use a political analogy, it is as if the corpus callosum fired its lobbyists at about the turn of the century and, for over sixty years, survived without any representation. Naturally its influence on psychobiology dwindled as fierce lobbying efforts continued on behalf of less important brain areas.

As a result of these findings concerning the left hemisphere, brain scientists concluded that the left hemisphere was "dominant" with respect to language, reasoning, and symbolic thinking. The right hemisphere, in contrast, became known as a "minor hemisphere" because it seemed to lack distinct features. More sophisticated observations, however, suggested that things were not that simple. One particularly intriguing finding came from the study of patients with brain diseases. In those cases where disease, usually a brain tumor, occurred in the left hemisphere, one or another type of aphasia (either Broca's or Wernicke's) usually resulted, accompanied by paralysis on the right side of the body. In cases of right hemisphere disease, however, the findings were strikingly different. Let me describe the difference by reference to one of my own patients who suffered a stroke in his right hemisphere.

Lawrence Ross is sixty-five years old, a grandfather and a retired railway conductor. Although always in good health, he has been under treatment for the past year for high blood pressure. He is now in the hospital for the evaluation of a weakness of his left arm and left leg. His problem began about six months ago when his daughter noticed that her father was no longer shaving the left side of his face. When this was pointed out to him, he mumbled something about "being in a hurry" and returned to the bathroom to complete the job. About the same time he started wearing only one slipper, the right one, and seemed unconcerned that the left foot was uncovered, so that on one occasion he cut his left foot on some glass on the back porch. Eventually a weakness occurred on the left side involving both the arm and the leg.

At the time of the examination of Mr. Ross he was alert and in good spirits. During the first few seconds at his bedside I noticed a growth of stubble on his left cheek, a stark contrast to the recently shaved and smoothly pink right cheek. "Why are you in the hospital?" I asked.

"I have had a little blood-pressure problem in the past year," he replied.

"Anything else?" I prodded.

"No, just a checkup," he replied.

"Would you do me a favor? Let me see you pick up that glass with your left hand."

He extended the glass to me with his *right* hand.

"No. Please do it with your left hand," I persisted.

He repeated the performance with his right hand, only this time with a smile, as if vaguely aware that things were not quite right.

"Are you not able to use your left hand?"

"Of course, I just did. My left arm is a little stiff, though . . . some rheumatism in the shoulder . . . maybe that is why I'm having a problem."

"Then you admit that you are having a problem with your left arm?"

"What kind of problem?" he asked.

"Here, let me show you." I lifted Mr. Ross's left arm in front of his face. It fell like a rag doll to the bed.

"Can't you hold your left arm up?"

"Of course, I just did."

Mr. Ross is not crazy, nor is he being deliberately deceitful. Brain disease involving the right hemisphere is somehow disrupting Mr. Ross's perception of his own disability. In patients with weakness of the right side, resulting from a stroke in the left hemisphere, such strange

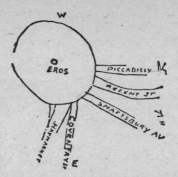

FIGURE 24. A sketch map of Piccadilly Circus drawn by a patient with a right-hemisphere brain lesion. Note the omission of streets entering from the left.

misperceptions are rarely observed. With right-hemisphere disease such as Mr. Ross's, in contrast, a whole spectrum of strange and inexplicable body distortions may result.

The earliest form of distortion involves a peculiar neglect of things off to the left. Lawrence Ross forgot to put a slipper on his left foot, and to shave the left side of his face. Figure 24 is a diagram of London's Piccadilly Circus drawn by a patient with right-hemisphere brain disease. Notice that all the avenues and streets on the left have been omitted. In other cases, the patient's behavior may become even more bizarre:

Mary Hastings suddenly collapsed at age forty-three with paralysis of her left arm and left leg. She tended to neglect her left side and minimized her paralysis. In addition, she began to develop unusual nocturnal fantasies. Here is a description by her doctor:

"She would wake in the night with a very intense feeling that someone was in the room—a person she knew; indeed, with whom she was very familiar. Sometimes, she was at a loss to decide who this could be, but on many occasions it would dawn on her that this person was none other than herself. She imagined that the 'person' or alter ego was somewhere on her left but just beyond her range of vision. The impression was so vivid that she would leave her bed and go from room to room on tiptoe trying to surprise this familiar interloper. At this point she would often get the idea that she was going crazy. She could not deny that the feelings were, to her, 'very very real' . . . 'Real enough to make me get up and look for this person whom I often knew quite well was myself all the time.' "

From the study of patients like Lawrence Ross and Mary Hastings, psychobiologists began to understand the role of the "minor" hemisphere. In fact, the right hemisphere is not "minor" at all. In essence, it is the right hemisphere that enables us to maintain a constant sense of who we are. When this is disturbed we may fail to incorporate important parts of the world and lose our sense of direction, ignore things off to the left, even have delusions that parts of our body don't belong to us. In addition, we may fail to grasp important distinctions. A wedding picture, for instance, may be interpreted as a funeral because "the woman is wearing a veil and several people in the picture are crying." Psychobiologists blame such failures on an inability to grasp a situation in its entirety. This capacity ordinarily depends on an intact and functioning right hemisphere. In summary, our perception of meaning, body scheme, and certain aspects of our sense of identity are disrupted in diseases affecting the right hemisphere.

The study of brain-damaged individuals is a useful but limited way to discover how the normal brain works. Critics have aptly suggested that conclusions about normal brain function can never come from the study of diseased brains. For one thing, we don't know the total effect on brain functioning created by disease in any one part. As mentioned earlier, the hemispheres are connected by the corpus callosum. Events in one hemisphere can be immediately telegraphed over this 200-million-fiber network to the opposite hemisphere. If we suppose that each fiber has an average firing frequency of twenty impulses a second, the corpus callosum is carrying something like four billion impulses a second right now as you are reading this sentence!

How can we be sure that Mr. Ross's neglect of his left side may not be due to some influence from parts of the brain other than the right hemisphere? How do we know that Mrs. Hastings' strange night visitor had anything to do with the damage her right hemisphere suffered from a stroke?

Psychobiologists wondered about questions like this for years and, until the 1960s, placed mental qualifications on any conclusions about right- and left-hemisphere functioning. Things were also complicated by an authoritative report in the 1940s which stated categorically that cutting the corpus callosum in man did not result in any changes in mental functioning. The breakthrough in our understanding of right- and left-hemisphere functioning awaited someone clever enough to devise tests capable of measuring the subtleties involved in hemisphere specialization.

In 1967, Dr. Roger Sperry, of the California Institute of Technology, reported his results in sixteen patients operated on for the control of life-threatening seizures. In each case all available medications had been tried, but to no avail. In a last, desperate effort to bring these severe epileptics under control, Sperry—or rather his neurosurgical colleague Joseph E. Bogen—cut the corpus callosum, in effect confining seizure discharges to only one hemisphere. Prior to the operation, Sperry and Bogen obtained proof that seizures were starting in one hemisphere and being conducted across the corpus callosum to the opposite hemisphere, resulting in uncontrolled generalized convulsions. After the surgery, most of the seizures ceased and "hopeless last resort cases" were controlled by ordinary antiepileptic medication. With the control of seizures, Sperry's real work began.

"The most interesting and striking features of the syndrome of hemisphere disconnection may be summarized as an apparent doubling in most of the realms of conscious awareness. Instead of the normally unified single stream of consciousness, these patients behave in many ways as if they have two independent streams of conscious awareness, one in each hemisphere, each of which is cut off from and out of contact with the mental experiences of the other. In other words, each hemisphere seems to have

196

its own separate and private sensations, its own perceptions, its own concepts, its own impulses to act . . . Following the surgery, each hemisphere also has, therefore, its own separate chain of memories that are rendered inaccessible to the recall process of the other."

Now let's look at some of the ways "two minds in one body" can be demonstrated. Ordinarily, the transfer of visual input to only one hemisphere can be compensated by the corpus callosum telegraphing the information to the other side. In "split brain" patients, such information transference is impossible in the absence of an intact corpus callosum. Each hemisphere thus operates in comparative isolation, enabling Sperry and other psychobiologists to devise tests aimed at tapping the individual capacities of the hemispheres.

In a test situation, one of Sperry's subjects sits with one eye taped, while fixing his gaze with the other eye on a designated fixation point. Visual images are then rapidly projected at a speed of less than a tenth of a second— much too fast for the subject to shift his eye fixation and pick up the image in both visual fields. Everything to the left of central fixation is projected to the right hemisphere; everything to the right, to the left hemisphere.

"When the visual perception of the patients is tested under these conditions, the results indicate that these people have not one inner visual world any longer, but rather two separate inner visual worlds," explains Sperry.

If two different figures are flashed on the screen, one to the left visual field and the other to the right, the person reacts, in Sperry's words, "as if one hemisphere doesn't know what the other has been doing." (See Figure 25.) In one example, a dollar sign is flashed to the left and a question mark to the right, and the subject is requested to draw what he just saw, using his left hand. Without hesitation the dollar sign is drawn by his left hand, which is powered by the right hemisphere. (Recall that the right hemisphere is involved with motor-power sensation and vision on the left side of the body.) If questioned what he has just drawn, the subject invariably responds, "A question mark."

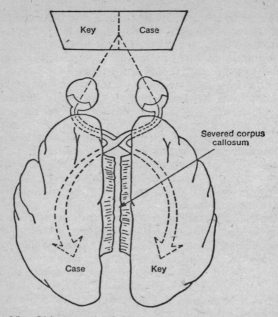

FIGURE 25. Object to the left (key) is projected to the right hemisphere, while object to the right (case) winds up in the left hemisphere. In instances where the fibers of the corpus callosum are severed, the two hemispheres are not able to cross-communicate.

Sperry's explanation for such puzzling performances draws on what we have already learned about the left hemisphere. We know, first of all, that speech is confined to the left hemisphere in most people. This is also true of the "split brain" subjects whose response is limited to what the left hemisphere has seen, since the corpus callosum is no longer present to convey the right hemisphere's perceptions. Since in this case the dollar sign was transmitted to the right hemisphere—and transmission to the left hemisphere is blocked by sectioning of the corpus callosum—the right hemisphere can only "program" the left hand to draw a dollar sign. When questioned, however, the subject responds with the stimulus projected to the speech-dominant left hemisphere. Thus, the patient draws one thing and verbally identifies another.

The existence of perceptions that cannot be verbalized —because the subject remains unaware of their existence —sounds strikingly similar to the Freudian concept of the unconscious. Sperry and his colleagues have photographed incidents that provide support for the contention that the right hemisphere is the repository for unconscious processes.

In one experiment a female patient was being tested with a series of geometrical shapes projected to either the right or left hemispheres. Suddenly a nude pinup was flashed to the right hemisphere. The girl began to chuckle, blushing slightly. "What did you see?" Sperry asked casually. She responded, "Nothing, just a flash of light," and chuckled again, covering her mouth with her hand. "Why are you laughing then?" asked Sperry. "Oh, Dr. Sperry, you have some machine," she replied, this time laughing out loud.

"Observations like the foregoing lead us to favor the view that in the minor hemisphere we deal with a second conscious entity that is characteristically human and runs in parallel with the more dominant stream of consciousness in the major hemisphere," says Sperry. Could this "second conscious entity" in the right hemisphere be none other than Freud's unconscious?

Dr. David Galin, of the Langley Porter Neuropsychiatric Institute in San Francisco, thinks the right hemisphere's performance is strikingly similar to the operation of "unconscious" processes. For one thing, both the right hemisphere and the unconscious deal with images which cannot ordinarily be verbalized. Second, their functioning depends less on logical analysis than on the perception of total pictures, which psychologists refer to as *Gestalt*.

"It is important to emphasize that what most characterizes the hemispheres is not that they are specialized to work with different types of materials" (the left with words, the right with spatial forms), says Dr. David Galin; "rather, each hemisphere is specialized for a different cognitive style—the left for an analytical, logical mode in which words are an excellent tool, and the right for a holistic, Gestalt mode, which happens to be particularly suitable for spatial relations."

If each hemisphere has its own way of "processing" the world, failures in "processing" might result from disturbances in the relations between the hemispheres, according to Dr. Galin. Isolated observations, such as the woman exposed to the nude pinup, suggest that the right hemisphere can sustain its own emotional responses. This provides a possible explanation for unconscious mental events, since the two hemispheres are capable of operating in their own way, each selecting different aspects in the same environment. As an example, Galin describes a young girl conversing with her mother.

In this hypothetical situation there is an unspoken conflict between the child and her parent that results in the mother expressing one message verbally while simultaneously conveying the opposite feeling through her tone of voice and body language. "I am doing this because I love you, dear," says the voice, while the face and tone says, "I hate you and will destroy you." In such an example each hemisphere would be exposed to the "same" experience, but, because of the differences in their cognitive styles, would extract a different meaning and emphasize a contradictory message. The left will respond to the mother's words and derive a sense of security from, "I love you, dear," while the right hemisphere becomes uneasy as it detects an ominous message from the mother's facial expression. In this example a different input is delivered to each hemisphere, just as in laboratory experiments where the two hemispheres are exposed to different pictures which they cannot share with each other.

In the laboratory experiments, of course, the commissures are cut. In the mother-daughter example of Galin, a similar result might be brought about by the inhibition of neuronal transmission across the corpus callosum. This hypothesis requires that aspects of the transmission from one hemisphere to the other be selectively and irreversibly blocked. Support for such a concept comes from a recent demonstration that the callosal fibers are capable of inhibiting as well as exciting neurons in the opposite hemisphere.

In addition, experimental subjects have been shown to strongly favor the right hemisphere for the memory storage of emotionally charged material. Dr. Max Suberi, of the Neuropsychiatric Center in Concord, New Hampshire, reported in late 1977 on his results with seventy-two subjects exposed to a spectrum of "stimulus faces," including neutral, happy, sad, and angry. Although the right hemisphere outperformed the left in facial memory storage across all the categories, the greatest differences occurred with the "emotional" rather than the "neutral" faces. Emotional facial expression is overwhelmingly processed in the right hemisphere and could, conceivably, in Galin's hypothetical example, be influencing the person to act in ways that may not be immediately accessible to consciousness.

Speculative models such as Galin's provide a physiological basis for certain forms of behavior. In the mother-daughter example, the daughter's two hemispheres might decide on totally different plans of action: the left to move closer to the loving mother, the right to flee from a hated enemy. If each hemisphere is equally strong, a state of ambivalence might result: the familiar love-hate relationship which, when fully developed, is one of the signs of schizophrenia. But in most cases the left hemisphere is able to gain control over the right and suppress it, largely as a result of the left hemisphere's superiority in language capability. "Be reasonable" or "Talk it out" are the commands in the left hemisphere which usually succeed in gaining control of the situation over the right hemisphere. If the left hemisphere is only partially successful, however, the right hemisphere may continue to function independently, like the girl who blushed at the pinup while her left hemisphere was incapable of explaining why. Such a sequence may also explain such things as slips of the tongue or the contents of certain dreams. In all of these instances the right hemisphere may be "speaking" in its own language of images and Gestalt patterns, which the left hemisphere has been only partially successful in suppressing.

At a later point in this book I'll say more about the split-brain research and its implications for our understanding of consciousness. But for now I want to emphasize its importance as one of the most philosophically provocative aspects of modern psychobiology. It attacks no less than our concept of the unity of the personality. This unity is a compelling subjective experience and, until the discovery of hemisphere disconnection, was rarely questioned. "The self is a unity—it regards itself as one, others treat it as one, it is addressed as one, by a name to which it answers. The Law and the State schedule it as one. It and they identify it with a body which is considered by it and them to belong to it integrally. In short, unchallenged and unargued conviction assumes it to be one. The logic of grammar endorses this by a pronoun in the singular. All its diversity is merged in oneness," according to the eminent philosopher-neurophysiologist Sir Charles Sherrington.

We now have good reason to believe that the experience of mental unity is an illusion. "We feel in one world; we think, we give names to things in another; between the two we can establish correspondence but we do not bridge the interval," wrote Marcel Proust. Modern psychobiology is proving what Proust and others knew intuitively. Here is Roger Sperry again on the duality of our consciousness: "These people do not complain spontaneously about a perceptual division. There is no indication that the dominant mental system of the left hemisphere is concerned about, or even aware of, the presence of the minor system under most ordinary conditions, except quite indirectly—for example, when an occasional response is triggered from the minor side. As one patient remarked immediately after seeing herself make a left-hand response of this kind: 'Now I know it was not me that did that!' "

The world was so recent that many things lacked names, and in order to indicate them it was necessary to point.

GABRIEL GARCIA MARQUEZ,
One Hundred Years of Solitude

From work with split-brain patients, psychobiologists have learned a great deal about the isolated capacities of each hemisphere. Although all this is helpful as far as it goes, it doesn't go far enough for our purposes. Of what relevance, for instance, has split-brain research in furthering our understanding of the cerebral function of the other 99.99 per cent of us with an intact corpus callosum whose two hemispheres must cohabit in a single skull?

And what are the normal relationships between the two occupants? Are they, for instance, marked by the harmonious and affectionate interplay of two newlyweds? Or is it more like the formalized, mutually respectful exchange of those couples who are long on marriage but short on love? Or are things even worse: a festering enmity between two roommates grimly determined to outlast each other in a rent-controlled apartment?

The study of cerebral-hemisphere specialization in normal people is the lifework of Dr. Marcel Kinsbourne, a forty-three-year-old Viennese neurologist trained in England and now at the Hospital for Sick Children in Toronto, Canada. Kinsbourne's interest in this aspect of psychobiology can be traced to an insatiable curiosity about how the mind "works" and, when it fails, the mechanisms for its decompensation.

Seated in the Neuropsychology Research Unit, pouring over slides of the brain with a hand lens, Kinsbourne at first reminded me of a well-dressed and charming savant skilled in the production of rare and expensive timepieces. I soon learned, however, that his interest in mechanics, the "nuts and bolts" aspect of cerebral function, is only a springboard for imaginative, challenging, and evocative speculations about human behavior.

Kinsbourne starts from a theory of communication and human language based on meticulous observation of infant behavior. The first and most cogent observations begin, not surprisingly, after birth.

Think of an infant lying quietly contented after finishing its bottle. In the few minutes before it lapses into sleep the infant gurgles, babbles, and emits primitive speech patterns while turning its head and eyes toward a sound or a quick movement. The infant starts by turning its head and eyes sideways and stretching its arms to the same side as the stimulus. At a later age it may point or

203

look at the object, and by age two it may even name it. In any case, naming is preceded by babbling and a primitive motor response, which in 80 per cent of the cases is to the right. From this, Marcel Kinsbourne speculates that the motor pattern responsible for early language (a rightward turning movement controlled by the left hemisphere) may be responsible for the better development of language in the left side of the brain. In other words, language is an expression of motor behavior, and the two of them can be expected to be closely linked.

How early in life can we detect evidence of cerebral lateralization? Studies on dichotic listening—the sending of separate messages to each ear for the purpose of measuring which ear is superior for different sounds—shows a right-ear superiority for sound in children five weeks of age. A complex interaction can be observed in which the organization of the newborn's movements is synchronized with ongoing adult speech behavior in the environment. This is similar to the adult pattern and is consistent with the theory that the right ear carries verbal messages predominantly to the speech-analyzing left hemisphere.

Brain damage in children interferes with language according to the age of the child. Children suffering left-hemisphere injury in infancy and early childhood (before the beginning of speech) may not show any language impairment. If the injury occurs later, at two or three years, there may be mild impairment, increasing in severity according to the age of the child when damage occurred. Not until around nine or ten years, however, is language impairment permanent and directly related to the left-hemisphere injury. From such studies psychobiologists formulated a theory that brain lateralization begins at the time language concentrates in a single specialized hemisphere. The trigger for this specialization was once thought to be the development of language.

Several observations cast doubt on this developmental model of cerebral localization. Anatomical studies carried out at Harvard and reported in February 1978 reveal structural differences in the two hemispheres, with the left hemisphere specialized, even in the fetus, for the production of language. In the vast majority of subjects, the left-hemisphere speech area was larger than the corresponding area in the right hemisphere. The tendency for speech to lateralize to the left hemisphere seems preset and genetically "programed."

At the same time, inborn behavioral tendencies were observed that complemented the Harvard studies. (It's fascinating how often the simplest observations are not made until someone provides a conceptual framework.) Dr. Gerald Turkewitz, a psychologist at the Albert Einstein School of Medicine, studied head positions in newborns ranging in age from several minutes to over a hundred hours. Turkewitz was curious to see whether the recently described anatomic differences in brains might have behavioral consequences. If the left hemisphere is highly developed at an earlier age, for instance, this might bias the infant toward head- and eye-turning to the right. Turkewitz's hunch was correct. When lying quietly on their backs, newborn infants have their heads turned to the right 88 per cent of the time. In fact, Turkewitz failed to discover even one infant with its head turned to the left more than the right. Even after holding the infant's head in the midline position, 75 per cent will still make their *initial* turn to the right. In addition, the infants respond more quickly to food stimulation to the right side of the mouth rather than the left. They will also turn toward sound originating from the right side quicker than sound from the left. In the case of heard speech, infants as young as twenty-four hours demonstrate evoked electrical responses that can be recorded from the speech-processing areas in the left hemisphere. With nonspeech sounds, the activity is recorded over the right hemisphere.

Turkewitz's research is a compelling demonstration of preset biases for attention, speech, and language which can be demonstrated as early as twelve hours after birth. Brain lateralization phenomena seem relatively independent of environmental stimuli. From here, some psychobiologists are cautiously speculating that many aspects of right/left hemisphere specialization begin before birth and are genetically programed. "It has become a truism in psychology that birth does not represent the point of origin for behavioral development," says Turkewitz.

Anatomic and behavioral studies in other primates fail to demonstrate similar hemisphere specialization. Although several nonhuman primates demonstrate asymmetry of the temporal lobes, most studies have failed to show evidence of higher specialization. Dr. Charles R. Hamilton, of the California Institute of Technology, carried out a series of split-brain studies on monkeys aimed at uncovering instances where one hemisphere is preferred over another. First the monkeys learned to perform a series of visual-discrimination tasks that closely paralleled the right/left hemisphere tests developed with humans (e.g., facial recognition in the right hemisphere, and the presence or absence of stimulus similarity in the left hemisphere). The tests were then switched so that the facial-recognition test was performed by the monkey's left hemisphere, while the similarities test went to the right hemisphere. "The task uniformly demonstrated hemisphere equivalents for solving all types of problems," according to Dr. Hamilton. "Review of the literature offers little support for the concept of hemisphere specialization in infrahuman animals."

Now let's return to Kinsbourne's studies on language and motor behavior. Naturally, each time we wish to speak it's not necessary to point or gesture, although many of us in fact do gesture almost constantly. If you think of pointing with your right hand while simultaneously calling out, "There it is," you have one end of a continuum, with silent speed reading at the other end.

At about age five, speech has liberated itself almost entirely from these motor mechanisms. The child of five is capable of silent thought, although, as many parents often learn to their embarrassment, children of five are notorious for the vocal expression of thoughts that would best be left unverbalized. Still, a casual observer would search in vain for anything similar to the gross head, hand, and trunk movements that accompany an infant's "speech." A more careful observer, however, will detect portions of this early reflex in the child's eye movements.

About 98 per cent of right-handed children—and

adults, for that matter—will respond to a question requiring verbal thought by eye movements to the right. Ask a question that is suited to a child's age (What is your favorite television program? Who is the Fonz?) and watch for a quick movement of both eyes to the right. According to Marcel Kinsbourne, this lateral eye movement is a remnant of the infant's gross turning movements of its whole body, which you recall was to the right about 80 per cent of the time. Kinsbourne suggests that language, which originates in the left hemisphere in 98 per cent of right-handers and about 60 per cent of left-handers, may be related to the bias which causes infants to turn to the right about 80 per cent of the time. (This also makes sense mathematically, since 80 per cent is just about the average of 98 per cent of right-handers and 60 per cent of left-handers.)

Raquel and Rubel Gur, a husband-and-wife team of psychologists from the University of Pennsylvania, studied adult personality traits associated with right and left eye-movers. They concluded that "gaze direction in response to questions is determined by problem type and by the individual's characteristic tendency to 'use' a certain hemisphere."

"If indeed left and right movers differ in a tendency to rely on one or the other hemisphere, one would also expect them to show corresponding differences in their characteristic modes of coping with problems and conflicts," reasoned Raquel Gur. Those with a tendency to look to the left might be expected to be more holistic, more nonverbal in their response to stress. Right-movers, in contrast, might be more analytic and intellectualized in their coping behavior.

To test this hypothesis the Gurs evaluated twenty-eight men with clear-cut eye-movement preferences. Those who usually looked to the left possessed personality characteristics such as a tendency to deny their problems, to hold in their hostility, to develop an appeasing "peace at any price" approach to personal conflict. Since these personality traits are also closely correlated with psychosomatic illnesses such as headaches, ulcers, and high blood pressure, the Gurs next measured the tendency of left-movers to develop psychosomatic illnesses. They found that left-movers reported nearly twice as many psychosomatic symptoms as right-movers. The results suggest that the preferential use of one hemisphere may provide a basis for predicting a person's tendency to develop certain diseases. It might also suggest a sensible approach to helping those left-movers who show a tendency to cope with stress by denying or escaping from it. Such people might be helped by therapy aimed at providing alternative coping methods which favor direct expressions of anxiety or even anger.

The importance of the eye-movement studies, according to the Gurs, stems from the light they shed on how we all differ in our tendency to preferentially use one or the other of our hemispheres. Their research is concerned with correlating eye movement with other measures of cerebral hemisphere specialization into a unitary description of personality and "cognitive style."

If successful, such a cognitive profile might enable us to understand aspects of everyday life which so far have eluded explanation. To mention just one of them, the Gurs recently completed a study of how classroom seating positions can be predicted on the basis of eye-movement preferences. They found that left eye-movers tended to select seats on the right side of the classroom, and right eye-movers preferred to sit on the left.

Even the course topic had an influence on the seating arrangements according to whether it activated the right or left hemispheres. For such left-hemisphere subjects as mathematics and the hard sciences, most students demonstrated a preference for sitting on the left so that their eye movements to the right correlated with the attentional bias to the right-sided space brought about by the activation of the left hemisphere. Art and music, in contrast, shifted the seating preference toward sitting on the right.

Studies on eye movements raised several intriguing questions. If eye-movement responses can tell us which hemisphere is involved in a specific mental task, what is the effect of multiple activities originating in the same hemisphere? Is there an interference effect? Or, to put it differently, was President Lyndon Johnson onto something when he said that Gerald Ford couldn't walk and chew gum at the same time?

To test the hypothesis that certain brain activities interfere with others, Marcel Kinsbourne taught his experimental subjects to balance a small metal rod on their index finger. After they became practiced at this simple task, Kinsbourne challenged them by requiring the repetition of a series of test phrases. He found that the performance of the subject's left hand was not affected by speaking, but the balancing performance of the right hand deteriorated dramatically. Since both speaking and right-hand performance use the same hemisphere (the left), Kinsbourne reasoned that the activity of one interfered with the other. He repeated the experiment, only this time with children as young as three years. Results were the same. From here, Kinsbourne and other investigators began to form the concept of cerebral geography.

The brain's performance capacity can be understood only in relation to its other ongoing activities. The effect is, in fact, an interactional one in which one activity will influence the brain's capacity to perform another activity, often in surprising and totally unpredictable combinations. Who could have predicted, for instance, that the balancing of a rod would be affected by simultaneous speech? After all, a woman can knit and converse at the same time. The key concept here is not knitting, talking, or rod balancing, but the *geography* of the brain, which co-ordinates these separate activities. What we wind up with are limitations on our activities that are imposed by the cerebral organization itself.

Kinsbourne is in the process of developing a model of what he calls "functional cerebral space." Stated at its simplest, we are capable sometimes of doing several different things at once, but not at other times. In some cases the first task—say, addressing Christmas cards—can be made easier and even more efficient if we are at the same time doing something relaxing, like humming Christmas songs to ourselves as we work. Other tasks, however, cannot be done together, and if we attempt to do so our efficiency in both of them is drastically reduced. An example of this is the student who grapples with a tough geometry problem while simultaneously watching a football game on television.

From such everyday experiences, Marcel Kinsbourne suggests a model of how the brain functions in carrying out several competing tasks simultaneously. Recall that the left hemisphere controls speech, rightward eye-turning, and the use of the opposite (right) hand. In the right hemisphere things are reversed: left eye-turning, the use of the left hand, and the capacity to carry out spatial and temporal rather than verbal tasks. In other words, each hemisphere contains the control centers for several activities that can occur simultaneously. According to Kinsbourne's hypothesis, speaking can interfere with the maximum effectiveness of the use of the right hand: The memorization of a test phrase disrupted the right hand's performance in balancing a rod.

In an even more sophisticated and impressive experiment, Kinsbourne trained two concert pianists to play two different tunes simultaneously on the piano. After they could do so without a single mistake, Kinsbourne requested the pianists to hum along first with the tune played by the right hand, then with the tune played by the left hand. He found that the performance accuracy was greatest when the pianists hummed the tune played by the right hand. When the humming accompanied the left hand's tune, the right hand's accuracy deteriorated with frequent errors. In other words, as Kinsbourne explains it, humming a tune and playing the same tune on the piano are related activities, with little or no interference effect. But when the speech center for humming (in the left hemisphere) is engaged in humming a completely different tune than the right hand (also controlled by the left hemisphere) is playing, there is an interference effect which results in frequent right-hand errors.

"From these kinds of experiments it is possible to begin to construct a functional space model based entirely on findings on the behavioral level," Kinsbourne pointed out to me while sifting through slides for an upcoming lecture (apparently finding no conflict between explaining a highly complicated theory of brain function while simultaneously operating his own internal slide projector).

Here Kinsbourne incorporates some elementary principles of brain organization. It's been known for years that in the brain similar body parts possess strong functional connections with each other. The right hand, for instance, is intimately related to the left hand via fibers traveling across the corpus callosum. It possesses much weaker connections to the right foot and the left foot. Proof of this comes from monitoring nerve and muscle electrical activity coming from all four limbs when only one limb is instructed to move. If the right hand moves, for instance, signals can be easily detected in the left hand, with fewer signals picked up from the right foot and almost none at all from the left foot. Put a different way, the interference effect is greatest when both hands are engaged in different tasks, and least of all with a task involving one hand and the opposite foot.

This research has already found application in a NASA-control panel. For years, space scientists had been trying to work out programs whereby astronauts could carry out several different spaceship control activities at the same time. Now a control panel is already in use based on a display in which the astronauts can do one task with their right hand and another with their left foot. Since these diagonally crossed limbs are farthest apart in Kinsbourne's functional space model, they exert the least interference effect on each other.

Kinsbourne's model is pulling together a large number of separate observations. One of these concerns the presence of mirror movements in hyperactive children. To demonstrate the movement, a child is instructed to open and close his hand as quickly as he can. At the same time, unknown to the child, the opposite hand is also engaged in less vigorous but nonetheless identical, or "mirror," movements. Mirror movements, as a rule, occur only in the hands. We can think of such a movement from one hand as overflowing across the corpus callosum to the hand area in the opposite hemisphere. This helps explain an intriguing finding by psychologist Doreen Kimura. From the analysis of videotape conversations, Dr. Kimura was able to demonstrate that gestures are more likely to be made by the right hand than the left when people are free to use either hand to illustrate or amplify a point. Since these gestures are not consciously intended (no one consciously thinks: "I will now raise my right hand slightly to explain that point"), it could be regarded, according to Kinsbourne, as a motor overflow phenomenon, with the greater use of the right hand predictable on the basis of a shorter functional distance between the vocal speech center and right-arm control centers, both located in the left hemisphere. As a corollary to Kimura's experiment, Kinsbourne measured the use of gestures when complex philosophical topics were discussed. In such situations, gestures tended to involve both hands rather than just the right, a finding that Kinsbourne thinks supports the view that both hemispheres are involved in imaginative and speculative thought.

In another experiment on functional cerebral space, Kinsbourne tested eighteen young adults who viewed a series of squares rapidly projected on a screen in front of them. The squares occupied the center of the subjects' visual field. The subjects were then requested to be on the alert for incompletion in the squares. In 50 per cent of the trials the squares were always complete ☐, 25 per cent were incomplete to the right ⌐, and 25 per cent were incomplete to the left ⌐. The subjects were watched carefully so that they kept their gaze at dead center of the square. The results showed no differences in recognition accuracy.

At this point, Kinsbourne imposed the additional task of reciting a memorized list of six words while the subject decided about gaps in the squares. His purpose was to compel the subject to subvocally rehearse the list while engaged in the perceptual task with the squares. The results this time indicated frequent errors in estimating left-sided gaps ⌐ with no loss of effectiveness in detecting right-sided gaps ⌐. In other words, using the left hemisphere for recalling the test words caused attention to shift to the right visual field. Objects in the left visual field were thus comparatively ignored and errors resulted.

Kinsbourne's findings indicate inherent attentional brain bias toward the right whenever we try thinking in words. Just how much we are biased is unclear, but the message seems to be that our perceptions are patterned by measurable biases that can determine what we are likely to see as well as those features of our world we are likely to ignore.

"If the features crucial to a task happen to be located on the right side," Kinsbourne predicts, "there will be a relatively high probability that will be detected. If it happens to be localized elsewhere, its detection has to await further attentional shift, in which time the briefly exposed stimulus might have faded into illegibility."

What are the implications of Kinsbourne's studies? Although we must be careful about premature conclusions, the results are philosophically intriguing. Speech, even involving something as automatic as reciting a memorized list, shifts our attention to the right side as a result of the activation of our left hemisphere. Our ability to experience, to perceive "reality," seems dependent on the mental "set" activated at a given time. If we think or speak out loud, it seems that we are likely to miss something in our left visual field. Merely telling someone, "I would like to ask you a question," preferentially activates their left hemisphere, biasing them toward the detection of objects in their right visual field. Again, psychobiology presents philosophy with questions more fundamental than "What is truth?" How can we know what truth is if the process of asking the question biases us to perceive some aspects of our environment to the exclusion of others?

In addition, there is evidence that merely telling another person, "I would like to show you a picture," sets in motion some highly predictable consequences.

Dr. Jerre Levy, of the Department of Psychology at the University of Chicago, presented test subjects with a series of mirror-image pictures. Starting with what we know about hemisphere specialization, Levy speculated that during the scanning of a picture the viewer's attention would be biased to the left side of the picture, giving rise to a perceptual asymmetry. Her reasoning was based on a series of studies of the right hemisphere in split-brain subjects which Levy carried out with her mentor, Roger Sperry, at the California Institute of Technology. When pictures are shown simultaneously to both hemispheres of a split-brain subject, the right hemisphere—the specialist in spatial visualization—will readily outperform the left when it comes to picture identification. Scanning a picture, therefore, can be expected to result in preferential activation of the right hemisphere, with increased attention to the object in the left visual field. (Recall that the visual fields cross to opposite cerebral hemispheres: The right cerebral hemisphere scans the left side of space, while the left hemisphere picks up information from the right visual field.) Based on this, visual processing, or even the expectation of it ("You'll now be shown a series of pictures") arouses the right hemisphere and biases attention and awareness toward the left half of space.

At first sight, such a theory contradicts one of the basic principles of esthetics: The most important contents of a work of art should be located in the right half of the picture. Throughout art history there are references to the lateral orientation of a painting in relation to its esthetic value. The art historian Heinrich Wölfflin suggested in 1928 that there is a strong tendency to scan a picture from left to right, with the eyes terminating on the right half of the picture where the most important content of the picture should be. Now psychobiological research on cerebral processing suggests the possible brain mechanisms responsible for esthetics: Esthetic balance will be created by a picture that corrects the brain's left spatial bias by depicting the most important parts of the picture on the right side.

"In viewing pictures, the right hemisphere is selectively activated, producing a bias of attention toward, and a psychological weighting of, the left side of space. Pictures which correct for this imbalance by having the more important content, or the greater heaviness, on the right, are considered, for this reason, to be more esthetically pleasing. The human esthetic sense is profoundly affected by the fact that the human brain is laterally differentiated," says Levy.

Returning now to the results of the Levy study, an overwhelming majority of right-handers preferred pictures in which the more important parts of the picture are on the right side. In essence, the activity of the right hemisphere shifts attention toward the left side of space. To compensate for this, most viewers select a more balanced picture with the most important element to the right.

As further proof of the right hemisphere's superiority in processing visual images, Levy designed an experiment involving split-brain subjects. In this situation a square is flashed onto the middle of a screen ☐. After the subjects correctly reported that they had seen the square, the square was then split in half and flashed so that one half was projected either to the right hemisphere ☐ or to the left hemisphere ☐. In either case, the subjects still reported a completed square. The split-brain subjects, in essence, hallucinated a completed square. As shown by Kinsbourne's earlier study on normals, a funtioning corpus callosum enables the observer to correctly identify a square as either whole or split.

"Each hemisphere of the commissurotomized patient, when presented with a stimulus, does not merely perceive that part of the stimulus which projects to the hemisphere, but instead effects a hallucinated completion," states Jerre Levy. "A square, for example, presented in midline to these patients is perceived as a complete square by each hemisphere, in spite of the fact that only the left half of the square stimulates the right hemisphere and the right half the left hemisphere. Neither hemisphere seems to be aware that only a half stimulus is responsible for its perception."

Next, Levy asked her split-brain subjects to stare at a fixation point and identify a series of human portraits flashed on the screen. (See Figure 26.) These were no ordinary portraits, however, but chimeric figures made from photographs that had been split down the middle and reassembled at random. The resulting chimeric figures were then projected. Nine times out of ten, the subjects reported that they had seen a whole person corresponding to the half face projected to the right hemisphere. After this, a series of split objects were projected, corresponding to various combinations of a rose, an eye, and a bee.

The results were clear. Irrespective of the stimulus and irrespective of the pointing hand, in the unbiased response situation the nonlanguage hemisphere was over-

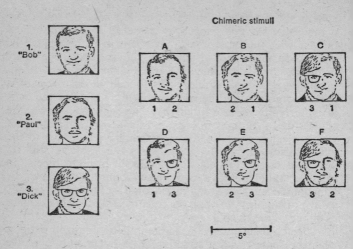

1. "Bob"

2. "Paul"

3. "Dick"

A   1   2

B   2   1

C   3   1

D   1   3

E   2   3

F   3   2

5°

FIGURE 26.   Stimuli and choices for face test.

whelmingly dominant in controlling the correct choice an average of 87 per cent of the time.

Levy's results point to an enhanced ability of the right hemisphere to process the identification of faces and common objects. Instantaneously, the right hemisphere outperforms the left and controls the subject's perception. In addition, such a performance does not require language. But what happens if we insert a language component into such an experiment?

"In order to look at internal language, we substituted for the original choice of pictures of a bee, an eye, and a rose, pictures of a key, a pie, and toes, and instructed the patients to point to the picture that rhymed with what they saw. The substitution of rhyming pictures for identical pictures produced a complete switch in dominance from left to right hemisphere."

Levy's studies illustrate the basic difference in right- and left-hemisphere operations. The right hemisphere's perception is holistic and doesn't depend on breaking things down into their component parts. We recognize a face instantaneously, for instance, and—with the excep-

216

tion of novelists—few people are skilled at verbally describing what someone "looks like." In line with this, patients with damaged right hemispheres have great difficulty recognizing faces, sometimes even their own, despite intact language powers. The left hemisphere, in contrast, is taken up with details and, when damaged, leaves the victim capable of only sketchy and incomplete verbal and written productions (e.g., the forced, telegraphic speech of Broca's aphasia).

Whether the right or left hemisphere is involved in a particular task seems to be determined also by a person's previous experience. Musically inexperienced listeners, for instance, recognize melodies better with their left ear (right hemisphere), while most concert-level musicians demonstrate a distinct right-ear (left-hemisphere) advantage. Such findings are leading psychobiologists to speculate that the right hemisphere processes relatively unfamiliar material, and, as the material becomes more familiar, it is taken over by the left hemisphere.

Each of the above conclusions holds true for right-handed people. But if the person is left-handed, all bets are off. Why? What is so different about left-handers? Although many explanations have been offered and countless theories exist about sinistrality (left-handedness), studies such as Dr. Levy's on cerebral lateralization seem to offer the best explanation. The studies also provide support for theories which suggest that right- and left-handers perceive the world in significantly different ways.

While over 99 per cent of right-handed people use their right hemisphere for spatial-temporal tasks and their left for language, the situation in left-handers is reversed about 44 per cent of the time. In practical terms, this means that almost half of any population of left-handers would give opposite responses to the pictures presented earlier. Since their spatial skills are integrated in the left hemisphere, they would preferentially activate their left hemispheres when engaged in such tasks as facial recognition or choosing from two mirror-image pictures. In their case, the right rather than the left half of space would likely be preferred, just the opposite choice of right-handers.

This may help to explain the difficulties encountered

in arriving at a general agreement as to the value of a work of art. Starting with the Greeks, attempts have been made to place aesthetics on a scientific basis. Why shouldn't people agree on the value of a work of art as easily as, say, the correctness of an algebraic equation? Part of the difficulty may stem from differences in cerebral lateralization. If almost half of the left-handed population has reversed cerebral lateralization and, in addition, the population of left-handers is on the increase, then even more differences of opinion regarding aesthetics can be expected in the future. Could the reluctance of many people to accept nonrepresentational art be based on differences of hemisphere lateralization?

Before answering such a question, it's necessary to find a reliable means of distinguishing the 56 per cent of left-handers who have the same lateralization as right-handers (left hemisphere for verbalizing, right hemisphere for spatial tasks) from the 44 per cent who have the reverse pattern. Do the two groups differ behaviorally?

"I was curious about this for several years," Levy

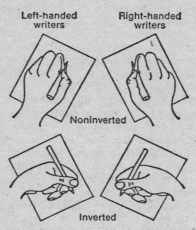

FIGURE 27. Typical writing postures of right-handed and left-handed people who write with inverted and noninverted hand positions.

recalls. "I had a feeling that there must be something that would enable me to distinguish the two types of left-handers, but I was not able to come up with any laboratory tests that were helpful."

Dr. Levy's eventual answer to this came from a quite unexpected source. One evening, while out on a date, she noticed that her companion signed a check with his right hand in an inverted position. "I did not know you were left-handed," Dr. Levy said.

"I'm not," her companion laughed as he held up his right hand, which was holding the pen. "Can't you see I'm using my right hand and not my left."

For the moment Dr. Levy was puzzled as to why she had become momentarily confused, thinking her companion was using his left hand when, plainly, he was right-handed. When she asked him to write something else, however, the reason for her confusion became immediately obvious. Her companion's hand position, with the greater part of the hand above the line of print and pointing toward the bottom of the paper, is extremely unusual in right-handers, occurring somewhere in the range of 1 per cent of the right-handed population. It is not unusual, however, and, in fact is quite common, in left-handers, occurring almost 60 per cent of the time.

The 60–40 ratio reminded Dr. Levy of the fact that close to 60 per cent of left-handed people have normal cerebral specialization, while in the other 40 per cent the situation is reversed (i.e., the left hemisphere is oriented for spatial tasks, the right hemisphere for language). She speculated that a hand position for writing might serve as an outward sign for brain lateralization. But how to test her speculation?

Over the next several months, Dr. Levy developed a test in which pictures could be rapidly flashed on a screen projecting to either the right or left hemisphere. While the subject stared at a central fixation point, a test stimulus was flashed testing either spatial abilities (the location of a dot briefly flashed onto a dark background) or verbal capacities (the recognition of a three-letter consonant/ vowel/consonant display).

Dr. Levy's hypothesis was simplicity itself. Since almost all right-handers use the normal (noninverted) hand position, and since their language hemisphere is on the opposite side (the left), the noninverted, or normal, hand position must indicate that the language-processing hemisphere is on the side *opposite* the writing hand. If this is true, right-handers using the normal, noninverted hand position should perform normally: recognize verbal test patterns better in the right visual field (left hemisphere) while doing better with spatial stimuli projected in the left visual field (right hemisphere). Dr. Levy's tests confirmed this.

Moving to the left-handers, one can reason the same way. Those left-handers who don't invert when writing are using the right hemisphere for language (remember, no inversion means the hemisphere opposite to the writing hand) and the left hemisphere for spatial tasks. They should do better on the tests when dots are flashed to their right visual field (processed by the left hemisphere, which in this case is a spatial processor), and perform best with words flashed into the left visual field (right hemisphere). This also tested out as true.

Stated at its simplest, the inverted hand position is a biological marker indicating that the hemisphere for language specialization is on the same side as the writing hand. Thus, left-handers who invert have the same pattern of hemisphere specialization as right-handers.

Levy's findings are scientifically exciting and at the same time socially provocative. Writing posture and handedness, along with eye movements, are turning out to be sensitive indicators of brain organization. From observation of handedness and hand position for writing, some very basic and important conclusions can be drawn about a person's cognitive profile. In essence, such studies are establishing sensitive and relatively unobtrusive ways of compiling information on how people think and perceive the world, what their reactions to hypothetical situations are likely to be, even what type of work they may perform best. Already there are indications that some of these attempts to define each person's "cognitive style" may reveal more about ourselves and the society in which we live than many of us are ready to accept.

Reasoning is of feminine nature: it can give only after it has received.

ARTHUR SCHOPENHAUER,
The World as Will and Representation

There is no female mind. The brain is not an organ of sex. As well speak of the female liver.

CHARLOTTE PERKINS GILMAN,
Women in Economics, 1898

Much of what we have discussed so far was learned from the study of adult male brains. Early psychobiologists had to rely on male brains, since most of their clinical material came from injured soldiers. From studies of the results of shrapnel wounds to the head, for instance, brain scientists established that the most posterior parts of the cerebral hemispheres are the final destination of those nerve fibers involved in vision. If a piece of shrapnel destroys that part of the brain, blindness results. If the wound is farther to the front, a more curious disability is produced: The soldier is able to see the world around him but cannot recognize objects when he encounters them.

From time to time, researchers wondered if the results they were getting from men would be duplicated in women. In other words, are there significant differences in the brain functioning of men and women? Before going into the data that provide the answer to this ticklish question, you might examine your own prejudices in this regard.

Certainly, anyone who has spent time with children in a playground or school setting is aware of differences in the way boys and girls respond to similar situations. Think of the last time you supervised a birthday party attended by five-year-olds. It's not usually the girls who pull hair, throw punches, or smear each other with food. Usually such differences are explained on a cultural basis. Boys are expected to be more aggressive and play tough

221

games, while girls are presumably encouraged to be gentle, nonassertive, and passive. After several years of exposure to such expectations, so the theory goes, men and women wind up with widely varying behavioral and intellectual repertoires. As a corollary to this, many people believe that if child-rearing practices could be equalized and sexual-role stereotypes eliminated, most of these differences would eventually disappear. As often happens, however, the true state of affairs is not that simple.

Recent psychobiological research indicates that many of the differences in brain function between the sexes are innate, biologically determined, and relatively resistant to change through the influences of culture. Before going into details, however, let me first say a word about the distinction that should be made between individual and statistical methods.

Mortality tables, along with population curves and Nielsen ratings, are based on samplings of large numbers of people for the purpose of discovering trends. They're useful in estimating how much longer a person might live, how many people inhabit a city, or what percentage of television viewers on a particular evening are watching a certain television program. They tell us nothing, however, about individuals.

A psychoanalytic case study or a well-written biography, on the other hand, is valuable for the details it presents about a single individual studied in depth. Few, if any, useful generalizations can be made from biographies or case studies that can be applied with any confidence to larger groups of people. Confusion frequently results when statistical and case-study approaches to a particular question are carelessly combined.

Research on sex differences in brain function are statistical, the result of testing large numbers of children throughout the world. The results, which we'll discuss in a moment, are useful for discovering *trends,* but they cannot and should not be applied to individual cases. I think that if we bear this in mind, the controversial area of brain-sex differences can be viewed with the necessary dispassion.

A review of hundreds of research projects carried

222

out over the last several years reveals the following "trends." From birth, female infants are more sensitive to sounds, particularly their mother's voice. In a laboratory, if the sound of the mother's voice is displaced by an experimental acoustic-deflecting apparatus, female babies will react, while males usually seem oblivious to the displacement. Female babies orient more to tones and are more startled by loud noises. In fact, their enhanced hearing performance persists throughout life, with females experiencing a falloff in hearing at a much later age than males.

Tests involving girls old enough to co-operate show increased skin sensitivity, particularly in the fingertips, which possess a lower threshold for touch identification. Females are also more proficient at fine motor performance. Rapid sequential movements are carried out quickly and more efficiently by girls than boys. In one test, measuring the ability of young children to tap their fingers to the beat of a metronome, girls consistently outperformed boys. The experimenters, Dr. Peter Wolff and Irving Horowitz, of Children's Hospital Medical Center in Boston, even attempted at one point to interfere with the children's performance by speeding up or slowing down the beat of the metronome. While this frequently disrupted the boys' rhythm, the girls were completely nonplussed and tapped on at the appropriate rate. The explanation for such findings depends on the association that exists between the left hemisphere and the performance of repetitive motor action. Just as the left hemisphere is more specialized in verbal functioning, such as speaking and reading, it also specializes in sequential analysis. Girls are at an advantage compared to boys in a wide range of abilities that require organizing data in sequence. For this reason, monitoring a steady rhythm despite the distraction created by the experimenter's deliberate speeding up or slowing down of the metronome was no problem for the girls, but resulted in clumsy start-and-stop disruptions in the boys' performance.

In addition, there are differences in what attracts a girl's attention. Generally, females are more attentive to social contexts: faces, speech patterns, and tones of voice.

By four months of age, a female infant can distinguish photographs of familiar people, a task rarely performed well by boys of the same age. Also at age four months, girls will babble to a mother's face, seemingly recognizing her as a person, while boys fail to distinguish between a face and a dangling toy and will babble equally to both.

Female infants also speak sooner, possess larger vocabularies, and rarely demonstrate speech defects. Stuttering, for instance, occurs almost exclusively among boys. Girls can also sing in tune at an earlier age. In fact, if we think of the muscles of the pharynx and larynx as muscles of fine control—those in which girls excel—then it should come as no surprise that girls exceed boys in language and linguistic abilities. This early linguistic bias often prevails throughout life. Girls read sooner, learn foreign languages easier, and, as a result, are more likely to enter occupations involving language mastery.

There is also research evidence that girls differ in their approaches to gaining knowledge about the world. They tend to favor a "communicative mode": asking others, taking advantage of other people's experiences, sparing themselves the need to personally encounter all the objects in their environment. For this reason, girls tend to conform by relying more on social cues. Since they are also better equipped in the auditory mode, they can pick up significant information from tones of voice and intensity of expression. Thus, interpersonal skills appear at an earlier age and form the basis for the "communicative mode" most women maintain throughout their lifetime.

Boys, in contrast, show an early superiority in visual acuity, which compensates to some extent for their lowered auditory capacities. Boys are also more clumsy, performing poorly in fine motor performance but doing better in gross total body movements, particularly those requiring fast reaction times. Their attentional mechanisms are also different. Nonsocial stimuli compete equally with social stimuli in eliciting a response. A male baby will often ignore the mother and babble to a toy or a blinking light, fixate on a geometric figure and, at a later point, manipulate it and attempt to take it apart.

A study of nursery preschool children carried out by Diane McGuinness, of the Department of Psychology at Stanford University, found boys more curious, especially in regard to exploring their environment. In addition, McGuinness' research confirmed earlier observations that males are better at manipulating three-dimensional space. When boys and girls are asked to mentally rotate or fold an object, the boys will overwhelmingly outperform girls. "I folded it in my mind" is a typical male response. Girls, when explaining how they perform the same task, are likely to produce elaborate verbal descriptions which, because they are less appropriate to the task, result in frequent errors.

In an attempt to understand the sex differences in spatial ability, EEG measurements have recently been made of the accompanying electrical events going on within the brain.

Ordinarily, the hemispheres show a synchronous rhythm, each side exactly the same as the other. When a person is involved in a mental task—say, subtracting 73 from 102—the hemisphere that is activated will demonstrate a change in its electrical background. It will become desynchronized. When boys are involved in tasks employing spatial concepts, such as figuring out mentally which of three folded shapes can be made from a flat, irregular piece of paper, the right hemisphere is activated consistently. Another way of finding out the same thing is to present the irregular piece of paper to the right and to the left visual fields. In boys, the fastest response always follows the presentation to the left visual field (right hemisphere). This is consistent with EEG findings of a desynchronization of the right hemisphere. Girls, in contrast, continue to do their best when the task is presented to the right visual field, indicating that they use their left hemisphere for both visual-spatial processing and verbal tasks.

The use of the left hemisphere by girls for both verbal and spatial processes results, according to psychologist Jerre Levy, in an interference phenomenon, a kind of "log jamming" in which the use of words to solve a spatial problem results in slowed, incorrect, or even absent responses. As we've discussed earlier, specialization of function in one hemisphere results in more efficient performance. "The male right-hemisphere specialization for visual functions contributes to their generally observed superiority on visual-spatial tasks," concluded one experimenter who tested differences between males and females in spatial localization.

When it comes to psychological measurements of brain functioning between the sexes, clear, unmistakable differences emerge. In eleven subtests of the WAIS (the most widely used test of general intelligence), only two subtests (digit span and picture arrangement) reveal similar mean scores for males and females. On six of the nine remaining subtests, males scored higher than females. The three tests where the females scored highest were similarities, vocabulary, and digit-symbol substitution.

Even more remarkable are the consistencies in sex differences displayed in the seven age groups into which Dr. D. Wechsler originally separated his test population. Higher mean scores for females were achieved in all seven age groups in vocabulary and digit-symbol substitution. In tests on similarities, females outperformed males in five of the seven age groups.

Wechsler's findings on sex differences, reported in *The Measurement and Appraisal of Adult Intelligence,* are rarely cited today. They have been substantiated across cultures, however, with the responses so consistent that the standard WAIS battery now contains a masculinity–femininity (M–F) index. By subtracting scores on the "feminine tests" (vocabulary, similarities, and digit-symbol substitution) from the WAIS scores on the masculine tests (information, arithmetic, and picture completion), an M–F score is obtained. A positive M–F score signifies a masculine trend; a negative M–F score, a feminine trend.

Further support for sex differences in brain functioning comes from Wechsler's experience with other subtests, which he eventually had to omit from the original WAIS test battery. A cube-analysis test, for example, was excluded because, after testing thousands of subjects, a large sex bias appeared to favor males. In all, over thirty tests eventually had to be eliminated because they discriminated in favor of one or the other sex. One test, involving mentally working oneself through a maze, favored boys so overwhelmingly that, for a while, some psychologists speculated that girls were totally lacking in a "spatial factor."

When it comes to personality characteristics, males and females tend to show some surprising differences. In four studies on curiosity, three found males to be more curious, one found no difference. In tests of field dependence (the ability to carry out a test procedure while ignoring irrelevant and extraneous components of the test), girls were found more field dependent than boys in a total of eight cultures.

A compilation of personality variables found boys less subject to their emotions in making decisions, while girls are less likely to take risks, more likely to have high grades in school, and do better under stress when engaged in problem solving. Most thought-provoking of all are a series of findings by Eleanor Maccoby and Carol Nagly Jacklin, of Stanford University, on personality traits and intellectual achievement. High activity, independence, competitiveness, and lack of fear or anxiety are correlated with intellectual achievement in girls, while in boys the correlation is with timidity, anxiety, lack of overt aggression, and lower activity level. In essence, Maccoby and Jacklin's findings suggest that intellectual performance is incompatible with our stereotype of femininity in girls or masculinity in boys.

Recent studies even suggest that high levels of intellectual achievement call for cross-sex typing: the ability to express traits and interests associated with the opposite sex. Dr. E. P. Torrance suggests that sexual stereotypes are a block to creativity, since creativity requires sensitivity—a female trait—as well as autonomy and independence—traits usually associated with males. A recent report by M. P. Honzik and J. W. McFarlane supports Torrance's speculation with a twenty-year follow-up on subjects who demonstrated significant I.Q. gains. Those with the greatest I.Q. gains displayed less dependency on traditional sex roles than those whose I.Q.'s remained substantially the same.

Throughout all this it's important to recall that we're not talking about one sex being generally superior or inferior to another. Rather, psychobiological research is turning up important functional differences between male and female brains. Our attitudes toward these findings should be no different than the psychobiologists who are presently engaged in this fascinating but controversial area of research: If there are sex differences in brain functioning, why do they exist? What purpose, if any, do they serve? Answers to these questions may come from studies of the effects of sex hormones on the developing brain.

It has been known for years that the injection of a male sex hormone, testosterone, at a critical time in embryologic development can change the sex of a developing animal from a female to a male. What effects can sex hormones have for later brain development? Most scientists now believe that sex hormones can "program" the brain to organize itself along either a male or female pattern. The most exciting and hopeful aspect of this whole thing, however, is that in some cases the ultimate sexual pattern can be significantly affected by the environment. In other words, in the case of hormone abnormalities, suitable environmental stimulation can tip the balance toward either a male or female brain pattern.

In the condition known as "testicular feminization" a male chromosome pattern coexists with a female body type. While genetically males, such individuals are females to all outward appearances. Although the debate is still raging about the "real sex" of people with testicular feminization, studies of brain functions support the view that they can be either sex, depending on how they are raised. Dr. D. N. Masica, of the Johns Hopkins Medical School, found that patients with testicular feminization raised as females had a predominantly feminine cognitive style (verbal performance exceeding spatial). Conversely, those raised as males developed a masculine pattern (spatial abilities far exceeding verbal).

There are also indications of sex differences in men's and women's response to environmental stresses. Dr. Lawrence S. Green, of the University of Massachusetts, is now completing a study of the effects of chronic low iodine

228

intake on the native population of the northern highlands in the Equadorian Andes. For the past five years, Green, who has been testing a group of iodine-deprived cretins, has shown that a dietary deficiency of iodine exerts a more destructive influence on female brains.

"Data indicate that under environmental conditions of extremely low iodine intake, there is a higher prevalence of neurological damage found among females than males. This is a function of sex-specific hormonal-dependent differences in adaptation to low iodine intake and is not a product of sex differences in role expectations," says Green.

In the event of brain damage, clear-cut sex differences occur. Among a series of right-handers who suffered left-hemisphere damage, three times more men than women showed language disturbances. According to the investigator, Dr. Jeannette McGlone, of the Department of Psychology, University Hospital, London, Ontario, the findings support the view that women show a more "heterogeneous pattern" for speech representation within the brain than is found in men.

From studies such as Lawrence Green's and Dr. McGlone's, Dr. Jerre Levy is now speculating that the male sex hormone may have a selective effect on right-brain functioning. Support for Dr. Levy's thesis comes from cross-cultural studies which show that females are inferior to males in tasks of spatial relations. To satisfy herself that these differences are indeed sex-related and right-hemisphere dependent, Levy recently repeated her studies measuring brain lateralization, sex, handedness, and hand position for writing. In all cases, females performed best in tasks performed by the left hemisphere, while males did best in right-hemisphere tasks.

As I mentioned in the introduction of this book, psychobiology provides insights into extremely diverse and sometimes unexpected and highly controversial areas. The discovery of brain differences between the sexes might possibly contribute to further resentments and divisions in our society. But must it? Why are sex differences in brain functioning disturbing to so many people? And why do

229

women react so vehemently to findings that, if anything, indicate enhanced bihemispheric capabilities in the female brain? I don't pretend to possess the answers to these questions.

But, on the basis of the information already available, it seems unrealistic to deny any longer the existence of male and female brain differences. Just as there are physical dissimilarities between males and females (size, body shape, skeleton, teeth, age of puberty, etc.), there are equally dramatic differences in brain functioning.

It seems to me that we can make two responses to the findings on brain-sex differences. First, we can use the findings to help bring about true social equity. One way of doing this might be to change such practices as nationwide competitive examinations. If boys, for instance, truly do excel in right-hemisphere tasks, then tests such as the National Merit Scholarship Examination should be radically redesigned to assure that both sexes have an equal chance. As things now stand, the tests are heavily weighted with items that virtually guarantee superior male performance.

Attitude changes are also needed in our approach to "hyperactive" or "learning disabled" children. The evidence for sex differences here is staggering: over 95 per cent of hyperactives are males. And why should this be surprising in light of the sex differences in brain function that we've just discussed?

The male brain learns by manipulating its environment, yet the typical student is forced to sit still for long hours in the classroom. The male brain is primarily visual, while classroom instruction demands attentive listening. Boys are clumsy in fine hand co-ordination, yet are forced at an early age to express themselves in writing. Finally, there is little opportunity in most schools, other than during recess periods, for gross motor movements or rapid muscular responses. In essence, the classrooms in most of our nation's primary grades are geared to skills that come naturally to girls but develop very slowly in boys. The results shouldn't be surprising: a "learning disabled" child who is also frequently "hyperactive."

"*He* can't sit still, can't write legibly, is always trying

230

to take things apart, won't follow instructions, is loud, and, oh yes, terribly clumsy," is a typical teacher description of male hyperactivity. We now have the opportunity, based on emerging evidence of sex differences in brain functioning, to restructure the elementary grades so that boys find their initial educational contacts less stressful.

At more advanced levels of instruction, efforts must be made to develop teaching methods that incorporate verbal and linguistic approaches to physics, engineering, and architecture (to mention only three fields where women are conspicuously underrepresented and, on competitive aptitude tests, score well below their male counterparts).

The second alternative is, of course, to do nothing about brain differences and perhaps even deny them altogether. Certainly there is something to be said for this approach too. In the recent past, enhanced social benefit has usually resulted from stressing the similarities between people rather than their differences. We ignore brain-sex differences, however, at the risk of confusing biology with sociology, and wishful thinking with scientific facts.

The question is not, "Are there brain-sex differences?" but rather, "What is going to be our response to these differences?" Psychobiological research is slowly but surely inching toward scientific proof of a premise first articulated by the psychologist David Wechsler over twenty years ago.

"The findings suggest that women seemingly call upon different resources or different degrees of like abilities in exercising whatever it is we call intelligence. For the moment, one need not be concerned as to which approach is better or 'superior.' But our findings *do* confirm what poets and novelists have often asserted, and the average layman long believed, namely, that men not only behave, but 'think' differently from women."

11

The Princess and the Philosopher

"The story is told of some peasants who were terrified at the sight of their first railway train. Their pastor therefore gave them a lecture explaining how a steam engine works. One of the peasants then said, 'Yes, pastor, we quite understand what you say about the steam engine. But there is really a horse inside, isn't there?' . . .

"The peasants examined the engine and peeped into every crevice of it. They then said, 'Certainly we cannot see, feel, or hear a horse there. We are foiled. But we know there is a horse there, so it must be a ghost-horse which, like the fairies, hides from mortal eyes.'

"The pastor objected, 'But after all, horses themselves are made of moving parts, just as the steam engine is made of moving parts. You know what their muscles, joints, and blood vessels do. So why is there a mystery in the self-propulsion of a steam engine, if there is none in that of a horse? What do you think makes the horse's hooves go to and fro?' After a pause a peasant replied, 'What makes the horse's hooves go is four extra little ghost-horses inside.'"

GILBERT RYLE,
The Concept of Mind

The proper way to exorcise the ghost in the machine is to determine the structure of the mind and its products.

NOAM CHOMSKY,
Reflections on Language

In May 1643, Princess Elizabeth of Bohemia wrote a letter to the philosopher René Descartes. The Princess, a shrewd and penetrating woman, posed a deceptively simple question to the forty-seven-year-old author of *The Passions of the Soul:* "I beg of you to tell me how the human soul can determine the movement of the body."

The Princess was probably the first philosopher in history—and an amateur at that—to point out the inadequacy of Descartes' much touted and recently published theory about the interactions between mind and brain.

"Let us then conceive," Descartes had written, "that the soul has its principal seat in the little gland which exists in the middle of the brain, from whence it radiates forth through all the remainder of the body by means of the animal spirits, nerves, and even the blood, which, participating in the impressions of the spirits, can carry them by the arteries into all the members . . . Let us here add that the small gland which is the main seat of the soul is so suspended between the cavities [of the brain] which contain the spirits that it can be moved by them in as many different ways as there are sensible diversities in the object, but that it may also be moved in diverse ways by the soul."

The little gland referred to by Descartes was the pineal, part of the brain which he postulated as the repository of the soul, or, as we would refer to it, the mind. The Princess was curious how an immaterial substance such as the soul could possibly interact with a material brain. At the basis of the problem, as the Princess so correctly pointed out, was Descartes' insistence that mind and brain, although intimately related, are nonetheless distinct.

According to Descartes, mind can only be apprehended subjectively ("I think, therefore I am") while the brain is mechanical and capable of being taken apart like a clock. In order to understand the correspondence between brain and mind, Descartes had hit upon the idea of the pineal as a kind of intermediary, making possible the "Incorporeal Soul in the Bodily Machine."

Three hundred years later, Descartes' position would be ridiculed by a professional philosopher, Gilbert Ryle, whose "Ghost in the Machine" pointed out the inadequacies of Descartes' position. By substituting "ghost" for "incorporeal soul," Ryle formulated in philosophical terms what the Princess had only dimly perceived, or perhaps had not dared to suggest: Why bother speculating about an "incorporeal soul"? Instead, consider the mind as, simply, the brain. In one fell swoop, Ryle disposed of Descartes' distinction and ushered in an era of confusion about mind and brain from which we are only now beginning to emerge.

At the basis of Descartes' separation of the mind and the brain was his conviction that the mind involved a qualitatively different process from the body. "The body is regarded as a machine, which, having been made by the hand of God, is incomparably better arranged and possesses in itself movements which are more admirable than any of those which can be invented by man," he wrote in 1637.

The body's responses, however, are not entirely typical of a machine, he pointed out, since they involve communication with a soul. "But the movements which are thus excited in the brain by the nerves affect in diverse ways the soul or mind which is intimately connected with the brain, according to the diversity of the motions themselves."

Since Descartes invoked two interacting, but nonetheless distinct, processes—mind and brain—his theory is known as dualism. Its value today, other than historical interest, derives from the emphasis Descartes placed on the mind as a process that defies localization in the brain and whose essential character remains mysterious. Princess Elizabeth detected the basic flaw in Descartes' proposal: If all material things, including the brain, can be influenced only in material ways, how can an immaterial mind bring about even so simple an act as raising one's hand?

After Ryle's devastating critique in 1947, dualism was replaced in the minds of many scientists by a cheeky confidence that the mind would ultimately turn out to be the burgeoning new field of psychobiology. Wasn't language already explainable by the discovery of language centers in the left hemisphere? And don't we know every-

235

thing that is important about vision as a result of studies on the effects of war injuries on the visual pathways?

Until recently, few neuroscientists realized that attempting to understand normal language and vision from studies of their disturbances is like trying to experience Beethoven's Ninth Symphony by reading a treatise on the construction and repair of violins. In this regard, at least, today's psychobiologists are far more humble. Most would agree that conclusions cannot safely be drawn about normal brain functions based solely on the study of what happens when things go awry.

In Chapter One I compared subatomic physics with psychobiology. In fact, at the beginning of the chapter I mentioned brain research as a probable area for future Nobel Prizes in physics. If we compare the development of physics with psychobiology, I think this initially puzzling contention will become more understandable. Our knowledge of the physical universe and the brain which attempts to comprehend it have both expanded at a dazzling rate. Unfortunately, however, the physicists hold about a fifty-year edge over their psychobiologist colleagues.

One of the most important events in the history of modern science was J. J. Thompson's discovery of the electron, which he and his colleague H. Lorenz described as a "splinter of the atom." Up until then, the atom had been considered the ultimate material unit in the universe. Over the ensuing years other particles and antiparticles have further lessened the physicist's enthusiasm for thinking of the universe as composed of static substances. A modern physicist admits that physical objects have an atomic structure. But the atom, too, has a structure, a structure that cannot be described as wholly material. Matter, it turns out, is highly packed energy communicating with other energy processes such as light, motion, and heat. As a result, physicists are no longer looking for an "ultimate" alternate physical unit. Instead, the emphasis is on structure, process, and interaction between a number of elementary particles and processes that may well turn out to be almost infinite.

Contrast this attitude with some of the things we still commonly believe about the human brain. On the one

236

hand, there are the latter-day Cartesians who believe we can never understand the relationship of mind and brain. This conveniently gets around the difficulty, but at the price of perpetual ignorance: We have both a brain and a mind, and the relationships between the two remain essentially unknowable. Obviously, if I believed this, I wouldn't have taken the trouble to write this book. Blind expressions of faith are appropriate only to religious conversions and grand passions.

On the other hand, there are those who think of themselves as hard-headed realists. According to them, we have only to come up with more data, a new neurochemical, or more sensitive electronic probes. "The brain is the mind," they tell us, and we have only to discover it neatly concealed like a crouching figure hidden somewhere in an oriental tapestry. Although initially appealing, this view is turning out to be far too simplistic. At this point our concepts of the brain and its relationship to the mind are undergoing subtle but important transformations.

In the 1940s and '50s a neurosurgeon at McGill University, Wilder Penfield, and his colleague Herbert Jasper, electrically stimulated portions of exposed brains of patients undergoing neurosurgery. They soon discovered that some of the past events in the patients' lives could be recreated by the power of a stimulating electrode applied to the temporal lobes—the "interpretive cortex," as Penfield referred to it. Crude bodily movements could also be produced, such as a sudden clutching action following stimulation of the hand area in the motor cortex. In addition, eerie feelings of familiarity were sometimes elicited in the waking patient. Although the patients remained alert and oriented, they reported that things and people in the operating room were somehow familiar. Some said they felt that everything around them had happened before or was part of a dream. Throughout all this, and despite the vividness of their experiences, the patients remained fully aware that their strange mental experience didn't correspond to any real events taking place in the operating theater, but were somehow caused by the surgeon's stimulating electrodes.

Here's Penfield's explanation for the memories: "A synaptic facilitation is established by each original experience. If so, that permanent facilitation could guide a subsequent stream of neuronal impulses activated by the current of the electrode even years later."

The familiarity-unfamiliarity experiences ("the experiential illusions") depended on the activation of circuits involving the temporal lobe and its connections with the limbic system. (In Chapter Four I described a patient who experienced similar flashes of unfamiliarity related to epileptic-seizure discharges originating in the area roughly corresponding to Penfield's interpretive cortex.)

Based on Penfield's findings, many psychobiologists began to move closer, in the 1950s and '60s, to the idea of a one-to-one correlation between mind and brain. If an electrode applied to the brain is able to retrieve a detailed memory, then the memory must somehow be "located" at that point in the brain. If a hand can be made to move by stimulation, then the brain area being stimulated must, under ordinary circumstances, be responsible for the motion of the hand, and so on. Taken to its logical conclusion, brain action leads to behavior through the activation of electrical circuits, similar to the motions of a puppet through the manipulation of its strings. (Such a model, of course, fails to account for the identity of the puppeteer.) But if we look back now to what Penfield and Jasper actually wrote at the time, we see that their interpretation of their experimental findings was quite different.

"We have to explain how it comes about that when an electrode is applied steadily to the cortex, it can cause a ganglionic complex to recreate a steadily unfolding phenomenon, a psychical phenomenon . . . It is obvious that there is, beneath the electrode, a recording mechanism for memories of events. But the mechanism seems to have recorded much more than the simple event. When activated, it may reproduce the emotions which attended the original experience . . . It seems obvious that such duplicating recording patterns can only be performed in the cerebral cortex after there has been complete co-ordination or integration of all the nerve impulses that passed through both hemispheres—that is to say, all the nerve impulses that are associated with or result from the experience. It seems to be the integrated whole that is recorded."

According to such a scheme, even the simple movement of a hand, following electrical stimulation of the motor cortex, cannot be explained as readily as many people were now claiming.

"When the electrode is applied to the hand area of the motor cortex, the delicate movements of the hand that the cortex makes possible are paralyzed, but the secondary station of gray matter in the spinal cord is activated again and crude movements, such as clutching, are carried out."

Ironically, Wilder Penfield is still erroneously cited as an exponent of the view that the mind can be reduced to some sort of clockwork mechanism operating within the brain. Such a misinterpretation of Penfield's research proves, if nothing else, that Descartes was well ahead of some contemporary psychobiologists who, after exorcising the Ghost, believe they are now ready to begin assembling and disassembling the Machine. Before committing ourselves to this "erector set" mentality, however, let's look at some subsequent research aimed at clarifying Penfield's electrical stimulation experiments.

Roger Shepard, at Berkeley, has measured the time it takes to rotate an asymmetrical object in his mind so as to confirm whether or not it is a mirror image of another object that is shown to him. The time is directly proportioned to the angle or rotation that is required and is therefore directly proportional to the time taken if he had actually held the object in his hand and moved it around.

AN INTERVIEW WITH J. BRONOWSKI,
The American Scholar, Volume 43 (1974), pp. 386–404.

Further experiments by Penfield and others involving electrical stimulation of the brain soon revealed unanticipated variability in the brain's response. Changes in the initial position of an arm prior to stimulation, for instance, led to widespread variations in the arm's excursions. This raised a key question: What is represented in the cortex—muscles or movements?

To find out, Stanford University's Dr. Karl Pribram began cutting out parts of the cerebral cortex in monkeys in order to study the effects of the operation on the monkeys' ability to open a complex latchbox containing a peanut reward. Pribram discovered that unilateral, or even bilateral, removal of the monkeys' motor cortex failed to destroy their learned ability to open the box. Instead, the time required to carry out the task doubled or even tripled. But the most illuminating aspect of the experiment came from studying slow-motion movies of the monkeys' postsurgical movements. In all cases, Pribram failed to observe an impairment of any movement or sequence of movements. In a phrase, the defect created by removing the motor cortex was *task specific,* not muscle or movement specific. "Neither muscles nor movements are present as such in the cortex," says Pribram. "Instead, actions with specific outcomes are represented."

This conceptual breakthrough—that acts, not movements or muscles, are represented in the brain—provides an explanation for a long-standing puzzle: How is it that a person can perform the same act in a variety of different ways? If the brain is organized around muscles, or even muscle movements, there should be only a limited number of ways to carry out a particular action. Instead, as we know, our actions can be carried out in an almost infinite variety of ways which draw on different muscle patterns. Let me illustrate what I mean by an example.

If you are right-handed, take a pencil or pen in your left hand now and write the word "syncytial," which means "net," another term that used to be applied to the net theory of brain organization. The action of writing "syncytial" with your left hand involves the skilled interaction of muscles that have never co-operated before in such an activity. It is totally new, and the cerebral process that makes it possible is unique. (I am assuming that, as with most right-handers, you have always written with your right hand and certainly prefer it when engaged in writing words that are uncommon.)

Soon other experiments confirmed Pribram's findings on the importance of actions. Dr. Edward Evarts, of the Laboratory of Neurophysiology at the National Institute of Mental Health, inserted microelectrodes into the motor cortex of monkeys trained to depress a lever to obtain a fruit juice reward. After the electrode entered a single cell in the motor cortex, the frequency of cell firing was plotted against the force necessary to move the lever. When the attached weight was slight and the lever easily moved, the cortical firing was infrequent. With heavier weights, requiring greater force to move the lever, the frequency of cell response increased. In other words, the action to which the muscles were put determined the activity of the cortical cells controlling them. Leaning down at the doorstep to pick up the morning paper requires a different set of muscles than leaning down to pick up a barbell. In addition, there are indications that the differences are established in the brain before the action is even started.

If you've been reading carefully, the last sentence in the previous paragraph should disturb you. How could the pattern of neuronal firing needed to carry out the act of picking up a paper or a barbell be preset before the picking-up process even starts? Isn't this a return to the "little man," an inglorious resurrection of the Ghost which psychobiologists' have so carefully exorcised?

Over the years, psychobiologists developed a theory of voluntary motion in which movements of body parts are entirely controlled by the opposite hemisphere. As we know, a stroke in the left hemisphere can paralyze movement of the right hand. From here it's easy to conclude that the act of moving the right hand is "caused by" neuronal firing in the left hemisphere. Recent experimental data reveal a much more exciting and philosophically intriguing series of events.

Edward Evarts first noted the effects on movement of damage in two brain areas, the basal ganglia and the cerebellum. The basal ganglia forms part of the R-complex, the ancient heirloom we've inherited from our primitive ancestors (see Chapter Four). The cerebellum is important in the co-ordination of movement and is speculated to play a role in normal emotional development. Damage to each of these structures results in separate alterations in movement without any accompanying paralysis.

In the case of damage to the basal-ganglia-slowed movement, muscle rigidity and loss of facial expression results: in essence, the symptoms observed in humans with

Parkinson's Disease, where the basal ganglia is depleted of its store of the neurotransmitter dopamine.

Damage to the cerebellum, on the other hand, produces an abnormality that is almost opposite to that found in the basal ganglia. Instead of slowed movement carried out with reasonable accuracy, cerebellar damage results in poorly co-ordinated ataxic movements that worsen as the action speeds up, and are least marked when the muscles are at rest.

"It seems clear that the three motor-control centers are functionally interdependent," Evarts reasoned. "But in what temporal order do they become active and what aspects of movement does each control?"

One of Evarts' microelectrode studies involved determining when the nerve cells in a monkey's motor cortex discharged in relation to a single hand movement. The monkeys were trained to press a telegraph key in response to the unpredictable appearance of a flashing light. If the monkeys reacted fast enough (350 milliseconds or less), they were rewarded with a few drops of fruit juice. Simultaneous recordings made from cells in the motor cortex and muscles moving the hand detected, to no one's surprise, that cells in the motor cortex fire *prior* to the muscular contractions. Shifting the electrodes to the sensory cortex located immediately behind the motor area, Evarts next discovered that the sensory cells fired *after* rather than before the muscle contraction—thus suggesting that the sensory cortex is not involved in the initial muscle movement, but later receives feedback signals from the muscles themselves that are important in repositioning the hand for the next movement.

When the electrodes were placed in the basal ganglia, cerebellum, and thalamus, Evarts and others discovered, to their amazement, that the firing of these structures corresponded to the motor rather than the sensory cortical cells. In other words, these brain areas, too, fired *prior* to the hand movement. To this extent, movement is carried out by the activation of all three brain regions.

Subsequent research by Evarts and other psychobiologists suggest that, as a first step, the entire cerebral cortex sends fibers to the basal ganglia and the cerebellum. Here, complex coding takes place which transforms the information received from the motor cortex into a complicated program, which is then returned to the cortex in the form of a new pattern. At this point, just prior to hand movement, activity can be recorded from all three brain areas rather than just the motor cells and the cerebral cortex.

Evarts' findings strike yet another blow to the narrow localization view that would explain behavior as the result of the activation of narrowly confined brain areas. "Indeed, a major change in our concept of central motor control involves a shift away from a tendency to think of motor control by individual components. More and more we are coming to think of motor control in terms of relations of different divisions of the central nervous system to one another and the parts of the body," says Evarts.

Turning now to work with humans, Dr. H. Kornhuber, a neurophysiologist at the University of Ulm, West Germany, measured brain potentials preceding voluntary movements—in this case a single movement of the right index finger. But in contrast to most similar experiments, Kornhuber's subjects were free to move their finger whenever they wished, so that the action truly represented an "act of the will." Kornhuber measured the cerebral potentials up to the instant the finger was observed to move. According to traditional thinking, the potentials should be recorded over the left motor cortex, the area where Penfield evoked right-finger motion by electrical stimulation. Instead, Kornhuber detected, at about eight tenths of a second before the finger movement, a bilateral potential (the readiness potential) which is widespread over both hemispheres! This is followed by yet another premotion positive potential (PMP), also present bilaterally at about ninety milliseconds prior to the finger movement. Finally, at only fifty milliseconds, a motor potential was recorded directly over the hand area of the opposite motor cortex.

Kornhuber's experiments present a partial answer to the question, "What is happening in my brain when I decide to carry out an action?" Psychobiologists are now convinced that during the readiness potential there develops a patterned impulse of neuronal discharges that spread widely throughout the brain. Later, this impulse activates

243

the cells in the motor areas responsible for programing the action—e.g., the motor cells in the left cortex activating the muscles in the right finger.

"The readiness potential can be regarded as the neuronal consequence of the voluntary command," says Sir John Eccles, the Nobel Prize-winning psychobiologist. "The surprising features of the readiness potential are its wide extent and its gradual buildup. Apparently, at the stage of willing movement, the influence of the voluntary command is widely distributed."

In order to appreciate the implications of Kornhuber's work, please reread the Bronowski quote at the beginning of this section. Willing an act—in this case rotating an object mentally in space—takes exactly the same time as physically manipulating it in one's hand! In the case of the physical manipulation, vision and touch are also involved, along with the train of connecting association fibers within both hemispheres. Cerebral potential recordings, just prior to beginning the rotation, would show activity throughout both hemispheres of the brain, not just the hemisphere opposite the hand doing the rotating. Later, this activity would weave a delicate pattern, like Sherrington's "enchanted loom," between the brain areas for touch, vision, and form.

"What is in charge of this shift is still partly a matter of speculation," according to Dr. Kornhuber. "One point seems clear: It is not a single superhomunculus in the brain, nor is it a single superdrive like Freud's libido."

Kornhuber's discovery of the readiness potential corresponds in importance, I believe, to Einstein's formulation of the relativity theory in physics. In 1905, physics was radically altered, and along with it came a revolution in our understanding of the physical world. So, too, the discovery that willed action is not localizable in any particular area of the brain represents a radically new way of conceptualizing the relationship between mind and brain. On the basis of these findings, other new and equally productive speculations have been put forth that are leading us closer to the solution of the mind-brain puzzle, aptly described by the philosopher Schopenhauer as "the world knot."

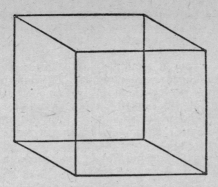

FIGURE 28. An example of deep structural ambiguity. One surface object, the Necker cube, but with two deep representations. Persons as both "mind" and "body" are similarly ambiguous.

A suggestion by the contemporary philosopher Walter B. Weimer relies on transformational rules. The Necker cube is structurally ambiguous, allowing two independent perceptions which alternate before our eyes. Weimer would have us think of the relation of brain and mind in terms of an equally ambiguous structure in which two distinct and separate representations can be discerned. As we discussed in Chapter Six, both perceptions are equally legitimate. It is not an instance of one being "correct" and the other "erroneous." There is in fact no way that one of the representations can be translated into the other. We see one *or* the other, not both.

"Transformationally related phenomena are distinct," Weimer reminds us. "The Necker cube is two objects manifested on one set of lines, not just one object seen from different perspectives. 'Persons' in this view are deep, structurally ambiguous objects like the Necker cube: They are simultaneously and indivisibly mind and body. The mental and physical are thus related and interact in the manner in which the alternative 'readings' of a common ambiguous structure are isolated and interact."

Weimer's model of structural ambiguity also gets around the most complicated aspect of the mind-brain puzzle. "How does the brain 'cause' the phenomena of mind?" Surface structures do not cause alternate surface structures in the transformational model. One perspective of the Necker cube does not *cause* the other.

Our attempts at equating mind with brain seem to encounter the same ambiguity that confronts our vision when gazing at a Necker cube or at one of Ames's distorted rooms, as we saw in Chapter Six. We experience difficulties whenever we attempt to translate one perception into another or, alternately, try to prove one aspect of ambiguity true and the other false. The mind and the brain are two aspects of a deeper, rather mysterious structure of human personality. When we apply psychological methods, we encounter "mind"; when we opt for measuring neuronal activity with microelectrodes, we deal with the "brain."

Although Weimer's analogy is powerful and compelling, it leaves open the question of how mind and brain are interrelated to yield two ambiguous objects. In the Necker cube, for example, both objects remain cubes, each seen from a different angle. The common feature, of course, is a set of lines strategically arranged. What is the common feature that we are thinking about when, at one moment, we view behavior as the product of the "brain," and then, a moment later, as a manifestation of the "mind"?

Such questions transport us into the ethereal realm of philosophy, where, I must admit, a neurologist may not be the best of all possible guides. The mind-brain relationship, however, like a country inn in a Victorian novel, provides a common meeting place for people of the most diverse persuasions. But if we are ever to understand the relationship of brain to mind, it may well require the genius of a psychobiologist who thinks like a philosopher.

One candidate capable of meeting such awesome intellectual requirements is Dr. Karl Pribram, Professor of Psychology at Stanford University. Pribram, at fifty-eight, is a board-certified neurosurgeon, an experimental psychologist of the first rank, one of the founders of neuropsychology, and a neurophysiologist with a penchant for brain paradoxes. Over the past twenty-five years, Pribram, the author of *Languages of the Brain* and hundreds of scientific papers, has slowly and painstakingly constructed a theory of the relationship of mind and brain based on what he calls "constructional realism." Pribram illustrates his philosophical position by a consideration of one of the most common substances in the world: water.

When we think about water we can consider it from several points of view. The wetness of water as a liquid can be changed into the solidity of ice, or boiled into steam. The physicist, on the other hand, knows that water is "really" H_2O, molecules of hydrogen and oxygen held together in a specific ratio by bonding forces. When the physicist enlarges his observation even further into the world of subatomic particles, water becomes a blur of interactive forces in which particles come into and go out of existence so rapidly that their presence, at any moment, can be expressed only in terms of probabilities.

These two views of water—the common-sense "Water is what you drink" and the physicist's "Water is a particular molecular configuration of hydrogen and oxygen"— can be reconciled, according to Pribram, by concentrating on structure. What are the differences in structure that result in ice one moment and liquid the next? Both are still H_2O in the sense of molecules of hydrogen and oxygen, but they differ according to some changes in the relationship of hydrogen and oxygen molecules to each other. According to Pribram, this shifting structural relationship provides the proper basis for reconciling the varying and, in some ways, irreconcilable views of what water "really is."

Another favorite example Pribram uses to illustrate his concept of structure is a Beethoven symphony. Originally the structure of the symphony existed only in the mind of Ludwig van Beethoven. Later, when he composed the symphony, the structure became transcribed onto paper. At that point, those who could read music were also capable of appreciating the symphony's structure. Still later, when the symphony was performed, the realization of structure was gained by those who listened to the performance.

"We know a Beethoven symphony by its structure," says Pribram, "but this structure must become realized in the notations on sheet music, the recorded imprint on a

247

plastic disc, the arrangement of magnetized minerals on a tape, or the orchestrations at a concert. The intrinsic properties of papermaking, printing, laboriously constructing 33⅓-rpm records and playback phonographs, the invention of wire recording and its gradual development into present-day tapes and cassettes, seem to have little to do with the structure of a symphony—yet they are essential to its realization."

The task of the psychobiologist, according to Pribram, is to pose questions about interaction in structural terms. What is the difference in the interaction between the H_2O molecules in a glass of water and those same molecules when water is frozen into a cube and dropped into a vodka gimlet? What structural similarities are there between the symphony as it existed in Beethoven's brain and the pattern the symphony creates in our brains? The question becomes less one of "Where is it? Where in the brain can we locate the symphony?" than a matter of "How does our brain extract the structure of Beethoven's symphony from sheet music, tape recording, whatever?"

"It is an understanding of structure and of the intrinsic organization in which structure becomes embedded that is elusive and that has to be worked toward by observation and analysis," according to Pribram.

In recent years, other areas of science have benefitted from the adoption of a structural approach. Molecular geneticists, for instance, are now busily probing the structural differences in the DNA molecule that account for the variations among cells in the human body. A single transformation of a DNA molecule may cause a cell to take on the specialized function of a liver cell. Another transformation of the DNA may result, instead, in a muscle or brain cell. Discovery of the mechanism responsible for these transformations—how they are turned on and off—is presently the aim of ongoing research in molecular genetics. A similar situation exists in regard to further breakthroughs in our understanding of the human brain. The key question seems to be, "How does the brain encode structure?"

At this moment your brain is involved in trying to

248

grasp the structure that existed in my brain at the time I was writing these words. To the extent that I am conveying what I wish to say, you and I are sharing the same structure. It doesn't matter that I am writing this in English, and you may be reading it in a Japanese translation. Both English and Japanese are merely the particular languages in which the structure of my thoughts happens to be embedded at this moment. Translating my thoughts into yet another language still leaves this structure essentially unchanged (assuming, of course, that the structure is correctly apprehended by the translator who conveys it into the new language).

In addition, structure can be conveyed in numerous ways. If you learn the main points of this book by listening to a radio or TV interview, you are also apprehending its structure, although probably less completely than by reading the book. The words are less important than the concepts expressed, which remain structural and, so far, essentially mysterious.

Strenuous and sometimes curious efforts were made to identify and trap the killer: the eyes of one of the murdered women were photographed in the belief (later discredited) that the assailant's image might be recorded on the retinas.
Encyclopaedia Britannica, Vol. V (1975), p. 491, *on Jack the Ripper*

Our knowledge of psychobiology has expanded so rapidly in the past century that even those of us who know little about the brain can now react with amusement at this curious Victorian effort to capture a killer by photographing the eyes of one of his victims. But to the Victorians, the human eye was only a kind of camera; therefore, why wasn't it possible to photograph a photograph? The question they addressed so naïvely was, of course, How are structures embedded in the brain?

We now know that brain function is unlikely ever to be explained in terms of a point-to-point correspondence between reality and changes in particular parts of our brains. Perception, for instance, demands the participation of all three functional units, none of which can be precisely localized.

After a rat is taught to run a simple maze, over 50 per cent of its cortex can be cut out and he can still run the maze with only insignificant errors. In more complex mazes, rats do less well; their performance, in fact, drops off in proportion to the amount of cortex removed. From this fact, the experimental physiologist Karl Lashley concluded that the more uncommitted cortex that is available, the more rapidly is the animal capable of learning. In other words, some of the functions formerly carried out by one part of the brain can be taken over by other parts which are, in Lashley's term, "equipotential." Some mechanism must therefore exist in the brain by which structure (in this case, the rat's maze-running performance) can be widely spread throughout the whole brain.

Imagine a jigsaw puzzle of which each piece contains the structure of the complete puzzle, certainly a fantastic concept unlike any jigsaw puzzle any of us has ever encountered. But in recent years Karl Pribram has suggested a model for brain function based on holography, a special type of photographic record in which parts of the picture are used, like the pieces of our hypothetical jigsaw puzzle, to reconstruct the whole photograph. It depends for its effectiveness on the use of light containing a single wave length.

Ordinary light—sunlight or the light from an incandescent bulb—is most concentrated at its source and dissipates as it travels through space (the flashbulb effect). It's made up of electromagnetic waves which oscillate at widely varying frequencies along the electromagnetic spectrum. (When white light is broken up by a prism, the resulting colors correspond to some of these different frequencies.) Light that oscillates at only one frequency (coherent light) behaves quite differently. Instead of a grab bag of widely varying frequencies, the coherent light of a laser beam remains concentrated and can travel great distances before dissipating its energy. A laser can be aimed like a gun, and the laser beam directed at distant targets for purposes that range from healing ulcers located deep within the eye to creating laser-induced fusion reactions with the power of miniature hydrogen bombs.

Scientists take advantage of a laser light's single frequency (its coherency) in the construction of holograms. An object whose hologram is to be obtained is first illuminated with the laser beam. The laser light strikes the object and impinges on a photographic plate at the same time as a reference beam—light from the same laser but reflected from a mirror—arrives at the film. Since the two waves have the same frequency but strike the plate at different angles, their interaction forms an interference pattern which is recorded on the plate. When looking at the resulting "photograph," the viewer sees only a smudge of rings and stripes corresponding to the interference pattern made by the two waves.

The word *hologram* is coined from the Greek word *holos,* meaning "whole," and it was chosen by its inventor, Dennis Gabor, to stress the fact that a hologram contains complete information about a wave. To this extent, holography is the science of light waves, their interference and defraction.

In order to "see" the hologram, the photographic plate is placed in its original position and the reference beam from the laser is switched on. At this moment, looking through the hologram is like looking through a window. The viewer can see the object at its previous position exactly as if the object were still present. In fact, the image seems so real that the viewer's eyes have to adjust their focus in moving from near to distant parts of the object. If he decides to take a picture of the holographic image, the camera lens would have to be adjusted the same way as it would if it were focusing on a real object. In fact, if the hologram is a good one, the viewer can't tell the projected image from its original!

Holograms offer several advantages over ordinary photographs. In conventional photography only the distribution of a wave's amplitude (actually the square of its amplitude) is recorded on the plane of the photograph via a two-dimensional projection of the object. For this reason, when we look at a photograph from various directions, we can't obtain new angles of approach—we can't see, for instance, what's happening behind people or objects in the foreground.

A hologram, in contrast, generates not a two-dimensional image of an object but the field of the wave front scattered by the object. By switching our point of view within the confines of the wave field, we can view the object from different angles, sensing it as a three-dimensional exact replica of the original object. In the movie *Star Wars,* Princess Leia is projected holographically to appear exactly as she did hundreds of years earlier.

The most intriguing aspect of holography, which Pribram suggests may be similar to brain function, concerns its method of information storage. The hologram stores visual information across an entire surface. Returning to our jigsaw analogy, the hologram corresponds to a jigsaw puzzle of which any piece, after removal from the completed puzzle, can be used to reconstruct the whole puzzle. A section of the hologram, when exposed to the reference beam of the laser, leads to the creation of the image of the entire object. Smaller pieces of the hologram lead to a loss of clarity in the projected image rather than loss of any part of the object itself. For this reason, it's always possible to reconstruct an adequate, though perhaps slightly blurred, image of an original object by using only a tiny section of the holographic plate.

An ordinary photograph, when cut in half or scissored into multiple sections, can be reconstructed only by placing the pieces in opposition to each other in accordance with their position in the original photograph. In holography, since the information is uniformly stored through the hologram, the original image can be reconstructed at any time merely by exposing a part of the hologram to light of the original frequency.

Pribram suggests that the brain, particularly the cerebral cortex, is the biological equivalent of a hologram. Certain neuronal assemblies, or perhaps even single neurons, are analogous to a section of the holographic plate. Information is then stored in the form of neuronal patterns, like images on a holographic plate.

Imagine a situation in which different "thoughts" correspond to the patterned interaction of many neurons.

Later, the same neurons may be involved in another "thought," only this time via different electrical patterns. Such a model would account for Lashley's findings, where multiple and extensive cortical removals had no effect on a rat's performance. The rat's maze-running abilities may have been stored holographically throughout the brain. A holographic model is also consistent with Pribram's findings that brain structure is built around "actions" which can be carried out by various means, using multiple muscles and muscle groups.

The decision to move the little finger in Kornhuber's experiment may correspond to a holistic pattern that spreads widely throughout the brain and is eventually structured in the form of a hologram. Although such a possibility remains speculative, the timing of events within the brain indicates that holographic structuring is certainly possible.

Transmission from one nerve cell to another takes no longer than one thousandth of a second, according to Sir John Eccles. The time required for the elaboration of a conscious experience—the act of "will" leading to finger movement, for example—is considered at least one fifth of a second. In Kornhuber's experiment the timing was even longer, eight tenths of a second, more than enough time for several hundred synaptic interactions to take place between cells. At least several thousand nerve cells spread throughout both hemispheres would eventually be activated. These, in turn, would, via synaptic relays, activate many others, leading to millions of neurons in eight tenths of a second! Thus, a hologram is not only possible but, at this moment, represents probably our best "model" for brain functioning.

Psychobiologists are hopeful that, as the hologram represents the dynamic patterning of the images of laser waves, some aspects of brain function will eventually be understood through the discovery of a biological hologram within the brain. So far, the work has proceeded only in terms of analogy. But even more important than its discovery, the search for a "brain hologram" has resulted in an emphasis on process rather than old preoccupation with localization. Perhaps up to this point our ideas about the brain have corresponded to the scrutiny by an unknown observer of the rings and lines of an unexposed hologram. What may be needed, to extend the analogy, may be a way of liberating the dynamic patterns entrapped within the brain's neuronal network. In at least one instance this has already been accomplished. Penfield's work with electrical stimulation resulted in the release of a holographic memory complete down to the finest details. This suggests that memories were somehow stored within the brain and were projected holographically coincident with the electrical stimulus to the patient's temporal lobes. But are there less intrusive ways than brain surgery to evoke holographic processes?

At this moment, psychobiologists are seeking new ways to study the dynamics of brain functioning. Before doing so, however, they must arrive at a unified conception of the meaning of human consciousness.

12
The Soul of a Peacock

I am persuaded that if a peacock could speak, he would boast of his soul, and would affirm that it inhabited his magnificent tail.

VOLTAIRE

Talk to someone for a quarter of an hour, then stop and ask him: "You've seen what's been happening. We are here, in a restaurant, eating. I talked about the sea holidays. Which would be the most realistic way of showing the scene we have just lived through during the last quarter of an hour? To show the two of us dining in a restaurant, or to show the beach and the waves we have been talking about? Or even to show them, not in the way we spoke of it, but in presenting the mental images in our heads, corresponding, interfering, even contradicting one another?"

A. RESNAIS AND A. ROBBE-GRILLET,
"Trying To Understand My Own Film,"
Film Makers on Film Making,
edited by H. M. Geduld, 1969

"Consciousness" is one of the most abused words in the English language. This partly explains the difficulty psychobiologists encounter in trying to define it. Part of the problem stems, no doubt, from our tendency to use the word in so many different and often contradictory ways:

A patient brought to a city emergency room in a coma may be described by the admitting physician as unconscious. Later, as the patient begins to awaken, the nurse will note on the chart "a rising level of consciousness" or "the patient is more conscious today than yesterday."

A three-month-old lying in a crib is less "conscious" than a Ph.D. candidate rehearsing a thesis for presentation to a panel of professors. Here, "consciousness" refers to a mental capacity more than it does to a state of alertness.

A Gay Rights activist, intent on obtaining our signature on a petition barring employment discrimination against homosexuals, might approach us with the intention of "raising our consciousness" to what he believes is a more enlightened attitude toward homosexuality.

At a party, someone may suggest we join them in smoking marijuana while they lecture us on "alternate states of consciousness." They may even refer to Eastern mysticism to bolster their arguments that our particular level of consciousness is unenlightened.

In each of the above examples, the word "consciousness" is used in a different way. This explains, I believe, why so much confusion exists about a definition of consciousness. Nobody can be really sure what the other person means.

If most of us remain unclear about the meaning of consciousness, we all refer casually to unconscious processes such as slips of the tongue. A friend who suddenly forgets our name during an introduction runs the risk of being the subject of our speculation about his "unconscious" hostility. One way to resolve much of the confusion, I believe, is to separate what we are sure of from what we are only guessing at.

The most striking aspect of consciousness is its discontinuity. Even as you are reading this page, widely varying mental images are popping up all the time. Perhaps suddenly you remember to cancel your newspaper delivery, since tomorrow you will be leaving on a two-week vacation. Or the imagery may be fantastic, even bizarre, with little or no relationship to the topic you're reading about. It's this kaleidoscopic and sometimes chaotic imagery which forms the basis for the stream-of-consciousness techniques in modern fiction, described in 1919 by Virginia Woolf:

"Examine for a moment an ordinary mind on an ordinary day. The mind receives a myriad of impressions—trivial, fantastic, evanescent, or engraved with the sharpness of steel. From all sides they come, an incessant shower of innumerable atoms; and as they fall, as they shape themselves into the life of Monday or Tuesday the accent falls differently from the old . . . life is not a series of gig lamps symmetrically arranged; but an illuminous halo, a semi-transparent envelope surrounding us from the beginning of consciousness to the end."

The actual extent of mental imagery varies from person to person. Experimental studies carried out by Dr. Jerome L. Singer, Professor of Psychology at Yale University, show that even during the most routine tasks, extraneous thoughts and images are occurring all the time. In one study, Singer's experimental subjects were engaged in a typical behavioral psychology experiment calling for rapid response to either sounds or movement. At fifteen-second intervals, Singer would interrupt the subject and ask him to report what was going on in his mind at the time. Even with subjects capable of 90 per cent accuracy in the performance tasks, random thoughts and images occurred which were fantastic, and occasionally wildly speculative. In another study, by Dr. M. Csikszentmihalyi, ongoing thoughts were recorded while his subjects went about their everyday activities. A group of surgeons, for instance, reported operating-room daydreams about such subjects as music, women, wine, and food. In most cases, the images occurred during the less demanding aspects of the operation, such as sewing up an incision or waiting for the scrub nurse to complete a sponge count.

In another phase of the study, the surgeons were encouraged to inhibit their imagery. Most reported the experiences as negative and discomforting, leading to moodiness, impatience and, on occasion, feelings of depression. Overall, the results tend to suggest that, as with nighttime dreaming, daytime imagery and random thoughts may play a critical role in determining our moods and perhaps even the maintenance of our personality. Just how much imagery can be considered "normal," however, can't be determined for certain, since, until the recent work of Singer and others, psychological research tended to downplay people's reports of their own inner experiences.

Despite the uncertainty, however, some aspects of consciousness are fairly well agreed upon. First, consciousness evolves in tandem with the development of our brain. To prove that, observe a group of children playing in a sandbox. At two years of age they are "loners," each one locked into a particular activity that absorbs him at the moment. One may be digging a hole, another reveling in the delights of tossing sand high in the air for the pleasure of feeling it drift downward on his head. Consciousness here is quite narrowly focused. At a slightly older age, say about four or five, a child's pleasure may depend on tossing the sand at another child and enjoying the spectacle of that child's turning quickly away in fright or anger. In this case, consciousness has evolved to recognizing the reaction of another person to a mildly sadistic act. The consciousness of the parent who witnesses this sand-throwing incident, meanwhile, represents a further elaboration of consciousness, expressed in a polite apology to the parent of the child who was the target of the sand-throwing incident. In each instance we've witnessed a "raising of consciousness." Starting with a solitary activity which gives pleasure in itself, we progressed to behavior that depends on a widening of consciousness to include another child's surprise response. Further up the scale of consciousness, we witness the adult's perception that a child's right to throw sand stops short of throwing it into someone else's face.

More formal observations on the evolution of consciousness in the developing child have been the lifetime study of Jean Piaget, a Swiss psychologist originally trained in zoology but with interests that are essentially philosophical. Since 1927, Piaget and his associates have accumulated thousands of factual and theoretical observations on children's mental development. Piaget postulates four developmental periods, each representing a different and gradually expanding level of consciousness.

In the sensory motor period, which extends from birth to about two years of age, the child demonstrates a series of reflexes: turning to light or sound, grasping an object dangling in space, sucking when the lips are touched, crying or waving the arms when startled. This is the only time in human development when stimulus-response considerations can adequately "explain" behavior. In fact, a pediatrician takes advantage of this when examining a newborn. If the doctor pretends to drop the baby, by suddenly lowering its head back toward the floor, the baby's arms open widely and it lets out an angry-sounding cry. This response, known as the Moro Reflex, is repeatable and, if absent at this early stage of human development, is a dependable sign of nervous-system disease. Infant behavior at this first stage of development is so programed that some children, later shown to totally lack cerebral hemispheres, have nonetheless still demonstrated normal reflexes during their newborn examination.

At about two months, a normal infant begins to coordinate certain acts: looking with hearing; seeing with grasping and, later, sucking. The infant starts to demonstrate a tendency to look at familiar objects such as mother or the family dog. At this period, according to Piaget, the child's reality consists only of those events occurring in the immediate environment. When the objects disappear they cease to exist. Consciousness under such circumstances is restricted to the immediate perception of objects and events succeeding each other without connection or relationship.

At about eight months, most children can reach out for something and even push away obstacles that may be in their path. At eight months, most children also demonstrate an elementary understanding of symbols and signs. A child may cry when the mother puts on her coat, since at this age such behavior is recognized as a preamble to departure. But the most intriguing of Piaget's observations concern the child's emerging capacity to realize that objects can exist quite apart from his own observations of them. In an ingeniously simple experiment, Piaget hid a watch under a cover and observed a nine-month-old boy lift the cover and pick up the watch. After three trials, Piaget hid the watch under a wool garment on the opposite side of the boy. Almost immediately, the child turned back to the original cover and lifted it only to find it empty. The process was repeated over and over again. Each time the child turned to the original cover, lifted it and seemed puzzled when he couldn't find the watch.

Such experiments, conducted in natural settings with very young children, support our earlier interpretation that reality is not so much recorded as constructed. The infant searching for the watch was betting on the likelihood that the hidden object actually remained where observation indicated that it should be. The child was handicapped, however, by its egocentric tendency to link the location of an object with his own previous perceptions. "I found the object under the cover; because *I* found it, it must always be there." Only later does the child learn that objects exist, and can change location independent of his own perceptions.

The final change in this earliest sensory motor period involves the child's manipulation of his world for the purpose of enjoying the spectacle of what happens: A cup of water will be held and then dashed to the floor as the child studies the spattering of fluid on the tiles.

The child's understanding of causality is also initially demonstrated at this period. To illustrate this, Piaget stealthily hid himself at the foot of a bed and swung a cane up to a child of eighteen months who was sitting on the edge of the bed with her mother. "Jacqueline is very much interested: she says 'cane, cane' and examines the swinging most attentively. At a certain moment she stops looking at the end of the cane and obviously tries to understand. She tries to perceive the other end of the cane and to do so leans in front of her mother and then behind her until she has seen me. She expresses no surprise, as though she knew I was the cause."

At about two years of age, the child is ready to enter the preoperational period, the second of Piaget's developmental levels. Here the child is capable for the first time of manipulating symbols according to his own private world. This is done, however, with certain limitations that are only gradually overcome. The concept of quantity, for instance, is missing from the child's mental operations.

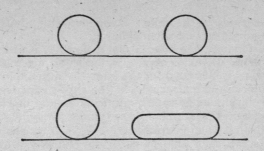

When presented with two similar plastic balls, the child will respond correctly that both balls are the same size. But when one is elongated, under his own observation, the child will usually pick the longer one as the "bigger." Such response indicates that the child's concept of "quantity" or "mass" depends on the size and shape of the container. If the container appears smaller, it must contain less; if larger, it must contain more. Similar errors in the estimation of size and quantity can be seen in adults, as we noted in Chapter Six when test subjects encountered illusions. Could the observers of Ames's cleverly constructed rooms be reverting to a preoperational mode of mental functioning?

I am introducing Piaget at this point because I think his theories of psychobiological development and intelligence fit best with recent psychobiological discoveries about consciousness. Although Piaget does not refer to brain processes as explanatory concepts (even now such correlations are only beginning to be drawn), his experiments are rooted in the soundest possible psychological methods: systematic study of how the brain evolves from a reflex machine (sensory motor period) to the level where consciousness first appears.

By age seven, the child is capable of what Piaget calls concrete operations. Here consciousness, in terms of an awareness of a self interacting with a real, objectively verifiable world, appears for the first time. "Only by means of friction against other minds," says Piaget, "by means of exchange and opposition, does thought come to be conscious of its own aims and tendencies." This shift in em-

261

phasis depends on the child's experience with adults and, more importantly, with other children.

In summary, the infant's reality consists at first of its own actions. All other objects in the world are relegated to a kind of background for the infant's motor expression. Later, objects are connected with the child's experience of them, as illustrated by the experimental failures of Piaget's young subjects to search in new places for objects they originally encountered elsewhere. Only later, at about age seven, is the child able to carry out the processes that underlie consciousness: to think in ways that are independent and may even contradict immediate perceptions.

To illustrate, consider one of Piaget's experiments carried out using the simplest of materials yet designed to delve deeply into the structural underpinnings of human thought:

A child is presented with a plastic bottle filled with colored water and another empty bottle of the same size. With the empty bottle tipped on its side, the child is requested to draw the water line when the colored water from the first bottle is poured into bottle number two.

The earliest response shows the water line drawn in reference to the shape of the bottle. Here the child's per-

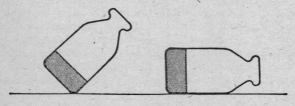

ceptions are limited to his assumption that the world will remain just as he experiences it. There is no law of gravity, no tendency for water to spread out along the length of the bottle.

At a later point, the child has a vague awareness of vertical and horizontal co-ordinates and indicates an altered water line. Only much later, at about nine years, is the child able to form a hypothesis of what might happen

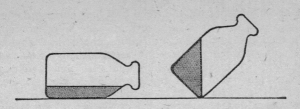

independent of his immediate perceptions. In this case he may indicate the correct solution.

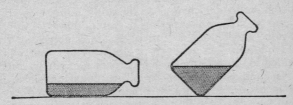

Although Piaget's theories are usually applied to the evolution of intelligence, I think they are equally valid in determining at what point consciousness begins. Consciousness involves comparisons between our perceptions and whatever internalized images we have built up over the years regarding the nature of the world around us. Think of the human brain as an information processor with a continually changing internal structure that is slowly being constructed over the span of a lifetime. The exact meaning of any stimulus will thus depend on the way the brain is programed to deal with it at that moment.

In another experiment along these lines, a small child is shown two glasses which have been filled with fruit juice to exactly the same level. After the child has indicated that each glass contains the same quantity of fruit juice, one glass is inverted and its contents poured into a short wide container, while the contents of the other glass are poured into a tall thin tube. "Which container has the most fruit juice?" the child is asked. Piaget has found that the correct response, "They both contain the same," involves four steps that evolve over time.

The youngest child is apt to select container 2, a clue

263

that he is focusing on the height. When corrected, he is likely to shift to width as a determiner and select container 1. In response to this second failure, the child then shifts back and forth for a short time between height and width until *suddenly* insight provides him with the correct approach: Height and width together provide the measure of quantity. In addition, he recognizes that since no fruit juice was added or subtracted from either of the original glasses, the contents of the two containers must remain the same. By a process of organization and reorganization of thought, the child is eventually able to decide something that is immediately obvious to the adult mind. The child's learning in such an experiment has nothing to do with fruit juice and bottles, but involves the establishment of a stable quantity—in essence, a construction that can be extended to all the situations he is ever likely to encounter. (If the child grows up to be a theoretical physicist, he'll have to experience another organization and reorganization of concepts before he'll grasp the interrelationship of mass and energy. In a world of subatomic particles, his decision about the conservation of mass—the quantity of fruit juice—would have to be altered by the knowledge that the whole experiment was being carried out on a spaceship plummeting away from the earth at the speed of light. In such circumstances, $E=mc^2$ will be the key to a workable solution to the fruit juice problem!)

Piaget's studies on children provide no less than a developmental theory of the mind. Although different children may enter each of the stages at different times, the sequence, Piaget claims, remains the same. Consciousness evolves in tandem with the maturing brain, and probably does not exist in any meaningful sense prior to the child's capacity to become aware of the separation between himself and the world around him (sometime probably in the late preoperational stage).

One of the implications of Piaget's work, I believe, concerns the current emphasis on establishing criteria for death. It has been suggested that "awareness" or "consciousness" can supply us with a means of deciding when life-support systems can be turned off. Alternately, "quality of life" decisions can thus be made on the basis of estimations of the patient's capacity to regain "consciousness." I think such proposals rest on overly simplified conceptions of mental development. As we discussed in Chapter One, alertness depends principally on the integrity of the reticular activating system and has nothing to do with the action of the cerebral cortex, where the distinctions between self and object are made that form the basis of consciousness.

Patients in a state known as Coma Vigil, for instance, remain alert with their eyes wide open, and they may even move their eyes and make following movements with their eyes as attendants enter and leave the room. Despite the impression of awareness, such patients are operating on a brainstem level, the cerebral cortex in some cases having been reduced to the consistency of mush. Consciousness is clearly not present here, despite the eerie appearance of alertness.

In less severe cases of brain damage, patients may be deeply comatose with no evidence of alertness; yet, later, such patients may return to their previous mental state. Thus, consciousness is best understood in terms of the developmental considerations suggested by Piaget. It's not an "all or none" or "once and forever" situation, but varies even within the same individual at different times and under different conditions (sleep is obviously a profound state of unconsciousness). There are strong indications, also, that societies as well as individuals undergo, or fail to undergo, similar stages in the development of consciousness.

It is not the consciousness of men which determines their existence; it is, on the contrary, their social existence which determines their consciousness.

KARL MARX in
Contribution to the Critique of Political Economy

The pioneer in the study of cultural differences in consciousness was the late Alexander Luria, who, until his death in late 1977, was Professor of Psychology at Moscow University. Luria's most famous work, which was described in Chapter Three, was in the field of neuropsychology. Of equal importance, however, is the study he undertook in 1931 in the steppes of central Asia. "It seems surprising that the science of psychology has avoided the idea that many mental processes are social and historical in origin or that important manifestations of human consciousness have been directly shaped by the basic practices of human activity in the actual forms of culture," Luria wrote. "My research takes the view that higher cognitive activities remain sociohistorical in nature and that the structure of mental activities changes in the course of historical development."

Luria's conception of consciousness, revolutionary at that time, anticipated by several years Piaget's findings on individual mental functions. "Consciousness—the highest form of the reflection of reality—is not given in advance, unchanging and passive, but shaped by activity and used by human beings to orient themselves to their environment, not only in adapting to conditions but in restructuring them," said Luria.

To test his theory of consciousness, Luria and his associates journeyed to the remote Russian village of Uzbekistan. Prior to the Russian Revolution, the people of Uzbekistan were peasants who spent their lives working cotton farms, did not attend school, and remained largely illiterate. At the time of Luria's study, an extensive network of schools was just beginning to be set up along with literacy programs aimed at introducing the elements of modern technology. At the time, the village was a microcosm with the most primitive and ancient lifestyles coexisting with the first hints of the emerging new society. Think of the situation as a twelfth-century medieval village suddenly set down somewhere in Long Island, and you have some idea of the unique opportunity Uzebekistan provided Luria for testing his thesis about consciousness.

In such a setting, Luria realized, the usual techniques of psychological research would be useless. Newcomers could be counted on to be greeted with suspicion and reserve; distrust of the researchers, in turn, might generate false data. To avoid both of these pitfalls, Luria engaged in long conversations with the villagers in the local teahouses, where many of them spent their free time. Rather than create a one-to-one situation, which the villagers distinctly distrusted, the interview was carried out in informal groups, often in the form of games, puzzles, or riddles. At all times, the experimenter encouraged the subjects to engage in a "clinical conversation": a response stimulating further questions and comments. Soon the psychologists and the peasants were volleying back and forth with questions, clarifications, further questions, and, finally, responses. "We used no standard psychometric test and we worked only with specially developed tests that the subjects found meaningful and open to several solutions," wrote Luria.

Luria hypothesized that among these primitive people a different level of consciousness would exist than would be found among those Russians who had already adapted to the technological changes introduced since the Revolution. In addition, proof of this would be available from studying the subjects' language and thought processes. "We had reason to suppose that word meanings would differ markedly (since words are the basic tools of thought), and experiments in the discovery of word meanings also reveal large differences in the content of consciousness."

Luria at first studied perception. As we discussed in Chapter Three, perception is an active constructive process rather than a passive recording of "what's out there": the eye is not a camera. But what about something as objectively verifiable as color? Can the conditions of daily life affect how a person sees something as basic as color?

Luria discovered that color naming depended on similarities between the test color and things in the peasants' immediate environment: peach, liver, fruit drop, calf's dung, cow dung were common responses.

When studying the "higher" thought processes, Luria discovered a similar dependency on personal experience rather than what we would call logical processes. One man was asked to pick the three objects that were alike from drawings of a hammer, a saw, a log, and a hatchet.

"Which of these things could you call by one word?"

"How's that? If you call all three of them a 'hammer,' that won't be right either."

"But one fellow picked three things—the hammer, saw, and hatchet—and said they were alike."

267

"A saw, a hammer, and a hatchet all have to work together. But the log has to be here too!"

"Why do you think he picked these three things and not the log?"

"Probably he's got a lot of firewood, but if we are left without firewood, we won't be able to do anything."

"True, but a hammer, a saw, and a hatchet are all tools."

"Yes, but even if we have tools, we still need wood, otherwise we can't build anything."

Such responses are similar to Piaget's preoperational child. "In this mode of thought," according to Luria, "the primary function of language is not to formulate abstractions and generalizations but to revive suitable graphic practical situations." A typical example of a graphic situational level of consciousness is this exchange between an experimenter and a thirty-eight-year-old illiterate subject:

"What do a chicken and a dog have in common?"

"They're not alike. A chicken has two legs, a dog has four. A chicken has wings but a dog doesn't. A dog has big ears and a chicken's are small."

"You've told me what is different about them. How are they alike?"

"They're not alike at all."

"Is there one word you could use for both of them?"

"No, of course not."

"What word fits both a chicken and a dog?"

"I don't know."

"Would the word 'animal' fit?"

"Yes."

"What do a fish and a crow have in common?"

"A fish lives in the water. A crow flies. If the fish just lays on top of the water, the crow could peck at it. A crow can eat a fish, but a fish can't eat a crow."

"Could you use one word for them both?"

"If you call them animals, that wouldn't be right. A fish isn't an animal and crow isn't either. A crow can eat a fish, but a fish can't eat a bird. A person can eat a fish but not a crow."

Luria's most fascinating findings, however, came from the study of consciousness and self-awareness.

If someone were to ask us to describe "what kind of person" we are, most of us would begin with a list of psychological characteristics (the majority of them undoubtedly favorable!): kind, considerate, intelligent, etc. We would know instinctively that we were being asked about our own evaluation of ourselves, our inner "attitude." Contrast this with a typical exchange with Luria and a thirty-eight-year-old farmer:

"What sort of a person are you? What's your character like? What are your good qualities and shortcomings? How would you describe yourself?"

"I came here from Uch-Kurgan. I was very poor, and now I'm married and have children."

"Are you satisfied with yourself? Or would you like to be different?"

"It would be good if I had a little more land and could sow more wheat."

"And what are your shortcomings?"

"This year I sowed one field of wheat . . . We've already gathered the hay and will harvest the wheat, and we're gradually fixing the shortcomings."

"Well, people are different—calm, hot-tempered, or sometimes their memory is poor. What do you think about yourself?"

"We behave well—if we were bad people, no one would respect us."

"Questions posed for an analysis of personal qualities were either not grasped at all," according to Luria, "or were related to external material circumstances in everyday situations."

Along with the illiterate peasants, Luria studied another group who were already deeply involved in the collective community activities. Most of them were literate, some of them graduates of elementary schools. Their responses, in contrast to those of the "peasants," indicated the existence of a "new inner world . . . an ideal me which began to play a decisive part in the development of their consciousness." Contrast the above responses with this answer to the same question from an educated thirty-six-year-old farm activist:

269

"What good traits and what shortcomings do you know about yourself?"

"I'm neither good nor bad . . . I'm an average person, though I'm weak on literacy and can't write at all; and then I'm very nasty and angry, but still, I don't beat my wife. That's all I can say about myself . . . I forget very fast; I walk out of a room and I forget. I also don't understand very well; yesterday I was given a long explanation, and I didn't understand anything. If I were educated, I would do everything well. I have to change this shortcoming in education. I don't want to change anything in my character; if I study, it'll change by itself."

Luria's work in Uzbekistan has profound implications for an understanding of human consciousness. Although certainly related to brain development, consciousness depends equally on social organization. Primitive agricultural societies throughout history were probably inhabited by people whose thought processes were similar to the peasants interviewed by Luria. Self-awareness, or consciousness, did not exist at all, according to our understanding of the term. "There's every reason to think that self-awareness is the product of sociohistorical development," according to Luria, "and that reflection of external, natural, and social reality arises first; only later, through its mediating influence do we find self-awareness in its most complex forms."

From studies such as Luria's a useful conception of consciousness emerges. Remember in Chapter One, we spoke of the need to shift from a *thing* to a *process* view of the brain. In a similar way consciousness is best thought of in terms of activity and behavior rather than as a physical object somewhere inside our head. From the interaction of the three functional units of the brain discussed in Chapter Three, consciousness becomes possible. Whether or not consciousness actually exists at any particular time, however, depends on the brain's interaction with things and people in the world around it. Our consciousness undergoes changes throughout our lifetime, since we learn to pick up new sources of information and use them in different ways. To this extent we are always conscious *of* something.

270

Proof of this theory of consciousness comes from studies of sensory deprivation. When adults are deprived of normal sensory stimulation, the usual content of their consciousness undergoes bizarre and deeply disturbing disruptions. The most striking example of this occurred in the laboratory of McGill University, where volunteer students were paid to lie on cots in dimly lit soundproof rooms. Goggles were placed over their eyes to prevent perception of shape or pattern. Instead, vision consisted only of the dim awareness of diffused light. Gloves and cardboard cuffs on hands and arms reduced the opportunity to maintain touch contact. Sound was limited to the monotonous hum and drone of an air conditioner placed close to the subjects' ears.

Most students, in response to this laboratory-induced sensory deprivation, underwent distortions in their conscious awareness. The first few hours were taken up with speculations about the experiment and the amount of money that could be made by continued participation. Soon conscious activity shifted to reveries about family and friends. This was succeeded by the experimental subjects' embarking on mental journeys to familiar places, often via scenic routes culled partly from previous experience and partly from an awakening and increasingly active imagination. Next, the subjects reported disturbing dreams reminiscent of drug-induced experiences and accompanied by confusion, panic, restlessness, and difficulty in thinking.

Under conditions of increased sensory deprivation, consciousness disintegrates even faster. Dr. John C. Lilly's experiments, using himself as the experimental subject, resulted within one hour in fantasies of "strangely shaped objects with self-luminous borders." In a similar experiment, using a technique involving the total immersion of the subject in a huge underwater tank, Dr. J. T. Shirley discovered upheavels in the stream of consciousness in "every subject."

Experiences similar to these experiments were formerly seen every day in the wards of a busy hospital. "Black patch delirium" is a term that refers to the distortions and hallucinations that result after bilateral cataract operations. Awareness of this major distortion in consciousness has induced many eye surgeons to limit this operation to one eye at a time. This procedure enables the patient to remain in contact with his surroundings during both postoperative periods.

In all these instances consciousness is disrupted because the brain's contact with the environment is interfered with. Blocking of familiar sensory channels leads to a lowering of the level of consciousness. From appropriate problem-solving considerations ("How much money can I make by staying here for the entire period of the experiment?"), consciousness deteriorates to a dream-like hallucinatory state similar to the world of the psychotic or the user of psychedelic drugs.

Experiments with psychedelic drugs are also helping brain scientists to forge new theories of consciousness. Stated simply, drug-induced alterations of awareness bring about changes in consciousness. Even alcohol, when taken in sufficient quantities, can result in radical distortions of consciousness. People do and say things when drunk which they cannot remember later and may even vigorously deny. Under such circumstances, alcohol creates a disruption in the even flow of consciousness, a memory gap which a person must fill in through the testimony of others. When the reported behavior is embarrassing or offensive, the person may deny the behavior.

Sleepwalking (somnambulism) may involve other unconscious acts, such as getting up in the middle of the night and walking to the refrigerator for a snack. Sometimes sleepwalking may involve even more elaborate behavior, as it does with one of my patients, who occasionally awakens his wife and carries out detailed and seemingly reasonable discussions which he is unable to recall in the morning.

In each of the above examples, consciousness is associated with memory. But can consciousness exist in the presence of a severe memory loss? Certain disease states—or, as some people refer to them, "experiments in nature" —strongly suggest that consciousness and memory are intimately related:

Lester Anderson is fifty-six years old and lives alone

in an apartment on New York's Lower West Side. Three days ago he was hospitalized after the police discovered him wandering in the streets seemingly intoxicated. After an episode of delirium tremens (withdrawal fits) which came on soon after the hospitalization, Mr. Anderson is now resting comfortably in his room. He is a tall, thin, undernourished man with a flushed face and a mouth half full of yellow, decaying teeth. Here is part of a conversation between Mr. Anderson and the young intern in charge:

"How are you today, Mr. Anderson?"

"Fine."

"Do you remember who I am?"

Silence.

"I am Dr. Gary Lawrence, who admitted you to the hospital. Please tell me how long you have been in the hospital."

"I came in last night."

"Do you remember the name of this hospital?"

Silence.

"It's St. Vincent's Hospital, and you have been here three days, not since just last night. What's my name again?"

"Dr. Joseph Smith."

Mr. Anderson is incapable of setting down new memories. Within seconds after being told Dr. Lawrence's name, he not only forgets it, but covers up his memory loss with a fictitious name, a process known as confabulation. Dr. Lawrence's next attempts to measure Mr. Anderson's tendency to confabulation:

"Several weeks ago I was shopping at Barney's for shirts, and I think you waited on me. Am I correct?"

"Oh yes, I remember you too," Anderson exclaims. "I worked there for about six months in the shirt department. I think you were looking for sport shirts."

"Do you remember how many I bought and how I paid for them?"

"Yes, you bought six shirts and paid with a check."

"Mr. Anderson, what is my name and where are you?"

"You're Mr. Barney and I'm in a clothing store."

Mr. Anderson is suffering from Korsakoff's Disease, a specific impairment in the ability to set down new memories. In most cases, the disease occurs in alcoholics like Mr. Anderson and is due to a specific lack of vitamins. Up to a point, the illness is partially reversible by the administration of thiamine and other vitamins in the Vitamin B family. Eventually, however, the brain cells involved are destroyed, leaving the sufferer with a permanent and incapacitating memory loss of a specific type. The defect involves memory for immediate events. As with Mr. Anderson, new information cannot be coded. In such cases, the memory is like a fishing net which contains large holes. In an attempt to overcome the defect, most patients make things up, or "confabulate." If they cannot remember someone's name, they invent one. If a suggestion is made ("You waited on me at Barney's"), they take it up and elaborate a series of convincing but fictional details.

Patients such as Lester Anderson illustrate the dependence of consciousness on memory. Unable to form new memories and extremely limited in their ability to call up remote memories, Mr. Anderson and other patients with Korsakoff's Disease live in a world in which consciousness is drastically reduced. Events around them are present only momentarily before lapsing into inaccessibility. Conversation with Dr. Lawrence is followed by amnesia for the topic of discussion, the doctor's name, and even eventually that a conversation ever took place. In a word, events become inaccessible.

Accessibility is the keynote of conscious experience. As you are reading these lines you are able to compare what I am saying with your past experiences. Even if you know absolutely nothing about the topic, some part of it may remind you of things you have thought about or learned in the past. New, even unexpected things may suddenly become accessible to your awareness. For instance, if I should mention your hands, your attention may shift momentarily to the weight of this book. Your consciousness, in essence, is like a searchlight which can scan different aspects of the inner and outer world. To the extent that this accessibility is interfered with, consciousness begins to fade into unconsciousness. Lester Anderson's accessibility to events around him is severely limited. In a sense, he is unconscious.

Some philosophers have suggested that accessibility is a measure of consciousness. At any moment our observations of the present, or memories of the past, can be the subject of internal scrutiny. When we are asleep or drugged, our attention to the present and recall of the past are grossly interfered with and we regress from full awareness to sleep.

One of the difficulties of this view is that we can never be conscious of most of the events that take place within our brains. For example, there is no way that you can be aware of the electrochemical processes that are, at this moment, being generated in the optic nerve and transmitted to your occipital cortex. Nor can you have access to the inframolecular processes involved when a neurotransmitter interacts with the membranes of a single nerve cell. Accessibility is limited and excludes some of our most important brain activities.

In addition, accessibility to certain brain processes isn't always desirable. A practicing neurologist spends the better part of his day listening to people complain about things like "a band around my head" or a series of "strange" sensations somewhere in the body. In the past, such people have been written off as hypochondriacs, but isn't it also possible that they may have access to internal processes that most of us are mercifully unaware of? If so, their strange talent seems to be more of a liability than an asset. Only by filtering out unnecessary information are we free to concentrate on the things that interest us. This defective "filtering capacity" is postulated by many psychobiologists as the basic defect in true hyperactivity and some forms of schizophrenia. Literally overwhelmed with stimuli to which they have unlimited access, such people can only function with the help of medications that reduce stimuli accessibility to manageable proportions.

A final problem with the accessibility theory of consciousness stems from its failure to take into account the findings of split-brain research described in Chapter Ten. Although philosophers have speculated about consciousness for centuries, the scientific procedures for testing these speculations have existed only during the last twenty-five years, particularly the last ten years. Split-brain work confirms the existence of two domains of consciousness. The first conforms roughly to our popular notion of the "I" who is now reading and reacting to these words, dredging the memory for points of comparison and perhaps disagreeing and fashioning a counterargument. But the second domain of consciousness also deals with accessibility. When objects are flashed onto a split screen, so that one projects to each hemisphere, the right hemisphere is "conscious" of the picture projected to it but can only inform the other via crosstalk over the corpus callosum. When this crosstalk is interfered with, as in the split-brain patients, the person lacks accessibility to the contents of his own right hemisphere.

"The nonvocal hemisphere is indeed a conscious system in its own right," says Roger Sperry. "It's capable of thinking, perceiving, remembering, reasoning, evaluating, willing, and emoting; all at a characteristically human level. Both the left and right hemispheres may be coconscious simultaneously."

The split-brain research is forcing a change in the explanatory models we construct to explain consciousness. If there are two domains of consciousness, which is the "real" one? Which separate consciousness corresponds to our inner sense of unity, that we are one person and not two? If forced to choose, I suppose most of us would select the left hemisphere's consciousness, which is responsible for reading, writing, most aspects of memory, and probably the capacity to envision future consequences. But what about the contribution of the right hemisphere? People such as Lawrence Ross, whom we met in Chapter Ten, illustrate the profound alterations in personality that can result from interference with the work of the right hemisphere. "As knowledge of brain function in the mindbrain relation advances, we would anticipate that terms like 'mind' and 'person' would have to be redefined or at least more precisely defined," says Sperry.

A "redefinition" of the mind and the person? At first sight, such a proposal seems preposterous. But Piaget's work with children illustrates the fluidity of our concept of personality. The ability to distinguish self from objects in the world develops slowly in infancy and is not in any

276

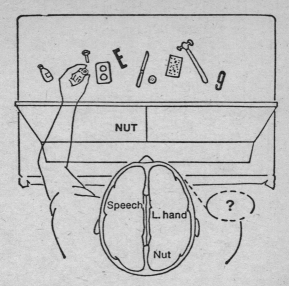

FIGURE 33. Names of objects flashed to the left visual half-field can be read and comprehended but not named, since speech centers are not in the right hemisphere.

sense innate. How could consciousness exist when self-awareness remains merged with the environment? Most philosophers whom I have asked this question agree that it couldn't.

In addition, split-brain patients often act in ways that suggest the presence of two personalities rather than one. In Chapter Ten we saw the effect of flashing a nude picture into the right hemisphere of a female split-brain patient. She blushed but wasn't able to tell why. When questioned about it, she denied that she felt embarrassed. She seemed to be inhabiting two separate realms of conscious experience. Such patients are forcing psychobiologists to take another look at split-brain research and its revolutionary effects on our ideas about consciousness.

Consider this test (Figure 33). The word NUT is flashed into a patient's right hemisphere and, via inter-

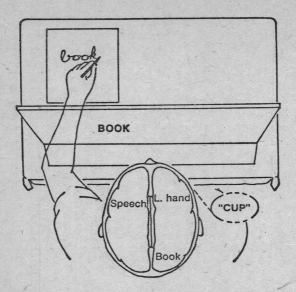

FIGURE 34. The left hand writes in response to the word "book" presented to the left visual field. When asked to name the object, the left speech hemisphere incorrectly selects "cup."

connecting pathways, to the right motor cortex, which controls the movement of the left hand. After a few seconds, the man rummages through a series of ten objects, finally grasping a nut. From a common-sense point of view, the man can be considered "conscious" of the meaning of the word NUT, as evidenced by his ability to match the flashed word to the real object. Yet, when the patient is asked about his performance, he knows nothing about what is going on. He cannot explain why his left hand is holding a nut, nor can he recall being conscious of the word NUT previously flashed onto the screen in front of him.

Or consider another patient. First a series of common objects is displayed in the center of the screen and conveyed to both hemispheres: a glass, a cup, a spoon, a book, and a pen. The subject is then told to write down the name of an object that will later be flashed on the screen. As soon as the word BOOK appears in the left visual field (Figure 34), the subject's right hemisphere programs his left hand to write BOOK. Instead of copying it exactly, however, "book" is written in a freehand script rather than capitals, an excellent demonstration that the right hemisphere possess writing rather than merely copying ability. When the subject is finished, he is then asked, "What did the left hand write?" Although aware of having written something, the subject cannot name it. If pressed, he will select objects at random and is as likely to say "cup" as the correct word, "book."

So far, such experiments support what we discussed in Chapter Ten—that verbal consciousness seems confined to the left hemisphere. But more recent work from Sperry's laboratory suggests, in addition, that the "minor" hemisphere possesses a consciousness of its own. "The nonverbal hemisphere has been interpreted to be lacking in self-awareness," says Sperry. "Self-awareness rates as a comparatively advanced and characteristically human form of consciousness appearing late in primate evolution."

Sperry's new investigations on human consciousness depend on an experimental modification of the split-brain apparatus. After one eye is completely covered, the other eye is fitted with a contact lens on which is mounted a small optical system which transmits vision only in the left visual field. It moves with the eye so that wherever the subject looks he sees only the left field—i.e., the visual input is confined to the right hemisphere. The ingenious modification, the work of Eran Zaidel, allows greater flexibility in testing, since the subject can now look in any direction he wishes rather than riveting his gaze to a single spot.

"We have used this stabilized occluder technique to test for sense of self-consciousness and for a generalized social awareness. If this higher, peculiarly human level of conscious awareness can be shown to be present, one may infer that the lower levels of consciousness must be there as well," says Zaidel.

In these later tests, two of Sperry's split-brain subjects were shown an array of pictures which were likely to evoke emotional responses—e.g., family members; pets; political, historical, or religious figures. The subject was asked to select from an array his "favorite" item, or to express his feelings simply by indicating "thumbs up" or "thumbs down." A twenty-one-year-old male's responses were "thumbs up" for Winston Churchill, a pretty girl, Johnny Carson; while Hitler, Castro, and a war scene evoked a "thumbs down" response. A photo of the subject inserted at the end of the sequence elicited an informative "thumbs down" response. "But in this case, unlike the others," recalls Sperry, "the response was accompanied by a distinct, wide, sheepish and (we think) self-conscious grin generated in the mute hemisphere."

When asked to name the objects in the series, the subjects, of course, were unable to do so. "Something nice, whatever it was," were typical comments, indicating that the speech hemisphere remained unaware of the pictures that had triggered their emotional response. "It was concluded that the mute disconnected minor hemisphere does indeed possess self- and social-awareness at levels quite comparable to those of the left hemisphere and of the intact brain as a whole," says Sperry.

Such experiments demand a total revision of our ideas about consciousness. Some even suggest that each hemisphere is separately conscious and that we should speak of a dual consciousness to complement our dual hemispheres. "It seems to me that the most reasonable view is that when the experimental conditions elicit the atypical behavior, we have two separate minds in the same body," according to philosopher Jerome A. Shaffer. Commenting on patients such as those described above, Shaffer asks a deceptively simple question: "Why should not whatever it is that stimulates a second stream of mental phenomena be casually sufficient to bring the new mind into existence? We do have new neurological phenomena; why should that not be sufficient to explain the new mind? My suggestion is that in the split-brain cases, we have not one mind but two. In such cases, we have not one person but two."

The implications for such a shift in our thinking are almost incalculable. Imagine a situation where a split-brain patient is arguing with his wife. At a certain point the patient's left hand lashes out and strikes his wife across the face. Should the wife angrily storm out of the house and find a divorce lawyer? Or was the blow unintentional? Indeed, the patient may plead with perfect justification: "It was not really me that did that. I didn't mean it. I'm sorry." If the wife understands as much as I hope you do by now about split-brain patients, she may reluctantly accept his apology. But suppose months later, in the heat of another argument, the left hand grabs a knife and kills the wife. Is the husband guilty of murder? Did he consciously want to kill his wife? Should he be punished for the acts of his right hemisphere?

Jerome Shaffer thinks we should seriously consider such scenarios. If split-brain surgery continues to be carried out, such a hypothetical situation might even come about. "Scientists should be very cautious about experimenting with split-brain cases. For the more often situations occur in which different inputs go to the hemispheres, the more likely it is that different personalities will emerge, and the more likely it is that such practical dilemmas will arise."

Now suppose a similar situation occurs in a person with a normally intact corpus callosum, only in this hypothetical case, a man (whom we will call Stan) has been complaining over the past year that his left hand is doing some strange things. On one occasion while playing poker, Stan's left hand crept across the table and picked up one of the cards from the reject pile. When the other players expressed their anger and dismay, Stan denied any responsibility. "You know I've never cheated," he explained defensively. If later Stan's left hand wielded the knife that killed his wife, he too might plead innocence. In Stan's case, however, the defense is not likely to be accepted, except perhaps as an example of temporary insanity or "dissociation" of consciousness. But how could we be sure that Stan is not telling the truth? Isn't it possible that his consciousness is split in two?

Should we begin to consider ways to punish the right hemisphere while exonerating the left? It's the old Siamese-twin murder puzzle all over again: If one member of a Siamese couple commits murder, should both be punished?

These are some of the philosophical, social, and legal issues raised by modern psychobiological research on consciousness. Some people have sought to deny the implications of the research by claiming that consciousness equals the sum total of the action of both hemispheres. This position is obviously false since, as we discussed in Chapter Three, consciousness isn't possible unless the reticular formation is working smoothly. Should we, then, consider consciousness as the sum total of the right and left hemispheres plus the reticular formation? This, too, is impossible, since the reticular formation connects with almost every other part of the brain.

Our attempts to define consciousness according to our usual explanatory models is like trying to grasp a fish that is swimming just below the surface of a placid lake. As we reach for it, our hands, disturbing the water, set up eddies which distort and fragment our vision, while our target flits away in response to the disturbance we have caused.

As a result of such difficulties, psychobiologists over the last few years have been developing ways to investigate the human brain which are not dependent on consciousness at all. One such method, known as Event Related Potentials, holds promise of providing us with no less than a "window on the mind."

13
Window on the Mind

But strange that I was not told
That the brain can hold
In a tiny ivory cell
God's heaven and hell.
OSCAR WILDE

Picture yourself standing at the edge of a quiet pond. You pick up a pebble and throw it into the center of the pond. Immediately, eddy currents on the pond's surface provide a reflection of the pebble's effect. If you wanted to view the process in detail, a movie could be made of the pebble hitting the water followed by ripples expanding outward over the pond's normally placid surface. By slowing the film speed, you could see the entire process in great detail.

Now switch your observation from the pond to the seashore. This time your pebble—even a large rock, for that matter—won't create any noticeable disturbance in the water's surface. The background turbulence of the waves and the ongoing currents will obscure the small ripple created by your pebble.

A similar situation occurs when psychobiologists attempt to study the effects of a single stimulus on the brain. A flash of light, for instance, evokes a volley of neural impulses in the optic nerve which travel along the visual pathways to the brain. This small electrical response is engulfed, however, in the "sea" of background noise corresponding to the brain's ongoing electrical activity.

The process of evoked-response averaging gets around the problem of background "noise." With each flash of light, or other stimulus, a series of brain waves will be detected at a specific time after the stimulus. While this is occurring, the larger wave patterns, corresponding to the general brain background activity, will be occurring at random. By repeating the stimulus, say two hundred times, a computer can sum up the change in background activity. If enough samples are taken, the background electrical turbulence of the brain will algebraically add to zero, while the waves resulting from the specific stimulus will become more visible each time the stimulus is repeated. In a sense, an averaging computer can "extract" the evoked response from the background noise created from the brain's normal ongoing electrical activity.

For a number of years, psychobiologists had assumed that the brain's response to a single stimulus, such as a flash of light, would be a simple wave marked by two or three upward or downward shifts in polarity. With the advent of computer averaging techniques, however, the actual results turned out to be much more exciting.

There are at least eight components to the visual evoked response to a flashing light. The first four components (the primary response) are a simple biphasic response, which remains the same whether the subject is awake or asleep. The second four components (post-primary response) disappear or are drastically altered in states of depressed consciousness. These last four components are thought to be related to sensory nerve impulses conducted by way of the brainstem, with contributions from the limbic system.

The most intriguing aspect of the evoked responses is their reliability. They provide, in the words of Edward Beck, one of the early evoked-response pioneers, "a unique and identifying quality . . . many individuals may be recognized by distinguishing characteristics of their evoked response." Comparisons between identical twins, for instance, are significantly higher than for comparable nonidentical twins. "In a manner of speaking, the evoked response is like a fingerprint of the brain," says Beck.

284

Evoked-response work is moving psychobiologists closer to the time when thought can be quantitated and even predicted ahead of time. Brain potentials associated with voluntary movement (the readiness potential described in Chapter Twelve) could make possible predictions about when a particular movement is about to take place. Another long latency response, known as the P300 (or P3) wave, is already enabling neuroscientists to measure complex psychological variables, such as decision-making.

In a typical P300 experiment, the subject is asked to predict ahead of time whether a stimulus will be either one of two variables. The presentation of a click, for instance, may be either loud or soft. A series is then presented with loud and soft clicks appearing in a random manner. At another time a regular series of clicks might be presented where a soft click is always followed by a loud click. In both experimental situations the subject is asked to predict ahead of time which click is likely to appear next (Figure 35). Since the physical stimuli are limited to either loud or soft clicks, any differences in the evoked cortical responses must depend on the subject's expectancy. This, in fact, turns out to be the case.

Figure 35 shows the P300 response to predictable versus random sequences. The two evoked-response potentials differ in their amplitude, with a large downward sweep of the potential corresponding to an unpredictable response. "Apparently 300 milliseconds after the onset of an unpredictable response a large positive wave appears," says Dr. Emanuel Donchin, of the University of Illinois and one of the founding fathers of the evoked-potential work. "This wave fails to appear when the same physical stimulus is predictable. We call this wave the P300 and we think that it represents a decision-making activity endogenous to the cortex."

Donchin and his co-workers are studying ways of measuring attention, under a Department of Defense grant. They hope to formulate a system for measuring fatigue in airline pilots that will facilitate such devices as an automated switch to automatic pilot. The research may also find application in subtle decision-making processes.

285

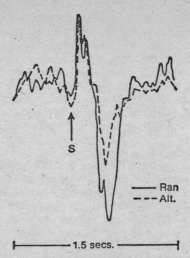

FIGURE 35. Superimposed event-related potentials recorded from the skull of a human subject. Both were in response to a flash of light. The dashed line was recorded when the stimulus was completely predictable; the solid line, when the subject was uncertain of the stimulus. Note that the effect of uncertainty is localized to a region 300 ms. following the stimulus.

How can a commanding officer tell ahead of time which of several subordinates would be the best choice for promotion? The usual procedure depends on the officer's subjective judgment, which, however reasoned, may result in costly errors. Now Donchin's work with the P300 waves provides an objective measurement of how people make decisions, as well as how willing they may be to act under conditions of uncertainty.

One test aimed at measuring "leadership qualities" involves comparing the timing of a subject's overt response (for instance, button pushing after perceiving a stimulus) to the timing of his P300 waves. Ordinarily, the P300 should precede any motor response. In some cases just the opposite occurs.

"In such people an impulsive decision is made, and only then, after they realize they have committed an error, does their brain provides a P300 wave," says Dr. Marta Kutas of Donchin's laboratory. "In this case the P300 came too late. The subject had already acted on the basis of pure reflex."

By comparing P300 to behavioral responses, the hypothetical commanding officer would be able to objectively test his subjective evaluations about a candidate's "leadership" or decision-making qualities. A person who acts prior to a P300 response is acting impulsively, will make more errors, and generally will be a poor choice as a leader.

As a further application, P300 responses could provide a lie-detector test far superior to the present system, which depends so heavily on subtle changes in a person's skin responses. When the subject is asked an unexpected or startling question, his P300 wave is a far more reliable index of surprise than any change in outward behavior. Although intelligence agents are trained in the art of skillfully disguising their inward emotions, no one has yet suggested a way to control the P300 response so easily.

One of the simplest applications might involve deciding whether or not a person can speak a certain language. A totally mute prisoner, for instance, may be unable to comprehend English, or he may be pretending. The P300 response to words that are comprehended has been shown to differ significantly from words heard in a language the subject doesn't understand. Thus, in a matter of a few seconds, a mute subject could be unmasked as an impostor.

Recently, researchers have further refined our understanding of the significance of the P300 response. It can be elicited by clicks, tones, smells—in fact, by just about any stimulus that a person is capable of detecting. For this reason some workers in the field originally suggested that the response may be triggered by the specific inputs: the sounds, light flashes, whatever. Others, principally Donchin, held out for the view that the P300 represented instead an internal decision-making process that doesn't depend on external stimuli. But how to resolve these differences of opinion?

At first glance, the question seems rather academic, typical of the kind of hair-splitting that goes on in brain research. (Psychobiologists, you might have noticed, are very compulsive people. They have to be. Advances in our understanding of brain function often depend on a researcher's ability to make fine distinctions between competing and equally plausible explanations.) In the case of the P300 question, of course, the distinction is between an electronic toy that can detect sensory stimuli, or the much more exciting prospect that the P300 represents one of the first objective measurements of brain activity corresponding to an act of the will. If the second alternative could be proven, psychobiologists would possess an electronic probe capable of detecting the exact instant when a person has made a decision. And, most important of all, they will know it prior to any action on the person's part!

Donchin's proof that the P300 was stimulus-independent came from a test where he omitted one of a series of rhythmic stimuli. If a subject is attending to an alternating series of soft clicks followed by loud clicks, a P300 will appear whenever a change occurs in the pattern —e.g., soft-loud-soft-loud-*loud*. More important, by omitting one of the stimuli, the P300 wave appears just as it would to an unexpected stimulus. "If you omit the stimuli there is, of course, no sensory-evoked response, yet the P300 develops approximately 300 milliseconds after the time the absent stimuli would have been presented," says Donchin. "This indicates clearly that we are dealing with an intrinsic cortical process which is evoked by the task demands of the situation rather than by specific inputs to the system."

At the present moment, active research continues on the meaning of the P300 wave. Although different researchers hold varying views, most agree that a P300 will appear whenever a subject is surprised by input patterns that are not predictable. The P300 announces that the subject's expectations have been confounded.

Returning to the brain model we have discussed throughout this book, we can see that the study of brain patterning, particularly the P300 response, provides proof that the brain's operation is based principally on pattern recognition. The brain is not a passive recipient of inputs which are then mechanically transformed into outputs—

the behavioral psychologist model. Instead, in Donchin's words, "it is a dynamic system which continuously generates hypotheses about the environment which are then validated against input information."

Psychobiologists are presently searching for a "theory of dynamic patterns" capable of incorporating the electrophysiological findings on evoked responses, particularly the P300 response, and equally exciting findings in microcircuits described in Chapter Nine. At the basis of their quest is an increasing confidence that the pattern reactions of certain neuron populations corresponds to neuronal informational transactions which, ultimately, are expressed in behavior. The clarification of this relationship between psyche and brain is, of course, the subject of psychobiology.

"Those of us who study the electrocortical potentials that can be recorded from the human scalp are convinced that the fine structure of the responses we observe represent functional entities of neural action," declares Donchin.

Despite some of the theoretical squabbles about what evoked responses *really mean,* the work is already finding ready and, in some cases, controversial application. For one thing, it provides a measure of brain maturation. Dr. Edward C. Beck and his associates at the University of Utah are studying the effects of aging on the wave forms of Visually Evoked Responses (VERs). Using 250 normal subjects ranging in age from one month to eighty-one years, Beck finds that the height of several waves within the first 250 milliseconds of the VER changes markedly with age. In general, the age of an individual can be correctly identified on the basis of his evoked responses.

Turning to performances within different populations, Beck has compared the evoked responses between bright and dull children. VERs from twenty children with I.Q.'s ranging from 120 to 140 were compared with twenty dull children with I.Q.'s averaging about 79. All subjects were ten or eleven years old, with boys and girls evenly distributed in both groups. Beck found that the late component of responses (post-primary response) was consistently larger with the bright children. In addition, the dull children's responses demonstrated no hemisphere differences in regard to the height of the evoked responses. Bright children, in contrast, showed larger (higher amplitude) responses over the right hemisphere than the left. "From this it can be seen that the asymmetry of the bright group is significantly greater than that of the dull children," according to Beck.

In order to rule out the possibility of artifact, Beck next concentrated on a particular group of retarded children: mongols. Neuroanatomic studies demonstrate that mongols are not only mentally retarded (there are occasional mongols of normal intelligence), but mentally retarded in highly specific ways incorporating unique physical as well as psychological features. If the evoked response in dull children really means anything, Beck reasoned, then the mongol children should show the same evoked response from both right and left hemispheres (the earlier finding with dull children). In addition, there might be a unique evoked pattern in mongolism. Both hypotheses were confirmed.

Using VERs, Beck found the expected absence of asymmetry among the mongols. Switching to the measurement of his subjects' responses to low-intensity shock stimuli, Beck discovered a strikingly uniform response among the mongols that not only differentiated them from the normal controls but characterized mongols as a group. "The somatosensory responses of the mongoloid children do appear to be sufficiently unique to be recognized with relatively casual inspection," says Beck.

Beck's findings with mongols provide support for the evoked-potential correlations between dull and intelligent children. The absence of cerebral asymmetry, failure to "habituate" to repeated stimuli, and characteristic amplitude patterns of the early components of the VER—all have been reliably correlated with intelligence. A note of caution, however, has been sounded by several workers who point out that the findings may be partially due to low attention levels on the part of some of the "dull" children. In addition, the correlations ultimately depend on comparing the VERs with I.Q. tests, which have come under fire recently as heavily culturally dependent. On a statistical basis, however, the VERs allow groups of high- and low-I.Q. children to be reliably differentiated.

If the VER-I.Q. work can be further refined, we may, in the near future, be able to reliably differentiate children with high and low intelligence on the basis of their evoked responses. Naturally, such a prospect is fraught with potential for abuse and could easily lead to conflicts regarding labeling. Many children are late developers, and no one so far has proven that children with modest I.Q.'s are incapable of achieving worthwhile adult occupational goals. In addition, the evoked-potential work can lend itself to use in a kind of no man's land between health and social policy. One of Beck's later studies on disadvantaged children is a case in point.

The VERs were studied in 140 children from economically deprived sections of the Salt Lake Valley. Group I consisted of children deprived since birth and from families that had never received welfare assistance—"poor but proud," in Beck's phrase. The children in Group II came from families who had been intermittently on welfare; and those in Group III, from families receiving welfare benefits for extensive periods dating back at least to the birth of the child under study.

Brain-asymmetry measures (usually found with bright rather than dull children) appeared most consistently among the children from welfare families. Although the reasons for such differences are complex, Beck suggests that welfare, in the form of improved nutrition and medical care, may confer a biological advantage on the fetus before birth and during the first few years of life. According to the studies on infant nutrition we mentioned in Chapter Seven, Beck's hypothesis might be valid. Further, the presence of evoked-response differences within a population usually considered homogeneous ("disadvantaged") suggests that VERs may be helpful in selecting children who benefit from rehabilitation measures. "It would appear that children with significant brain asymmetry but with a low I.Q. may have a greater psychobiological potential for improvement than a similar low-I.Q. group of children lacking in brain asymmetry," according to National Institute of Health neuropsychologist James Prescott.

Thus, the prospect becomes more plausible every day that the evoked-response studies may soon enable us to distinguish subtle and not so subtle behavioral differences among people.

Psychiatry provides one of the most encouraging areas at present for evoked-potential application. One measure, known as the post-imperative negativity variation (PINV) has already achieved 75 per cent accuracy in differentiating psychotic from neurotic controls.

To measure PINV, the subject attends to two consecutive stimuli. The first (S1) is a warning stimulus that readies the subject for a second stimulus (S2), which will require action—e.g., pushing a button to stop a tone or series of flashes. The period immediately following the S2 stimulus comprises the PINV.

In general, if a person knows he must perform an act at some point in the near future, the frontal area of the brain tends to develop a steadily increasing negative potential. This extends from the warning signal (S1) until the time for the required act (S2), which may be a perception, the movement of an arm, or the making of a decision. When the act is carried out, the negative potential returns toward the normal resting state. Usually, the amplitude of the negative wave (Contingent Negative Variation, or CNV) is directly related to certain aspects of the subject's psychological state. Interest, motivation, and the ability to focus attention are all part of the CNV. Not surprising, given so many variables contributing to the CNV, the results obtained with it have often been inconsistent and unreliable.

If the CNV is broken down into its components, however, one of them, the period immediately after the PINV, provides psychiatrists with data of highly predictive value.

Ordinarily during the PINV the evoked response turns to the base line in 500 milliseconds or less. A delay in return to the base line occurs frequently among certain psychotic patients. In some cases the delay may even precede overt behavioral abnormalities.

"This, of course, is only an isolated example of the predictive value of the PINV in psychopathology," says McGill University psychiatrist Dr. M. Dongier. "But we have reported statistically significant results concerning behavioral prediction."

The essential correlation is between the variations in PINV and the severity of emotional disturbances. Among early schizophrenics, for instance, the correlation is 94 per cent, with a 70 per cent success rate among psychotics in general. Psychiatrists in several major medical centers are already applying these findings to help them in one of the toughest differential diagnoses they are called upon to make: Is the patient a schizophrenic or merely a "strange" isolated "schizoid" personality.

"If over two or three consecutive recordings the PINV is consistently normal," says Dongier, "the diagnosis of early schizophrenia may be almost certainly eliminated."

Psychobiologists are now confident that the evoked-potential work will eventually provide biological markers that can be used by psychiatrists as aids in the assessment and prognosis of certain forms of mental illness. But the work is complicated and demands technical expertise that few psychiatric facilities now possess or can easily afford. At the moment, less than a dozen laboratories in the world are equipped for evoked response studies in psychiatric patients. But those that do use them are already in possession of a powerful therapeutic tool. "We feel that in some problems of differential diagnosis in certain schizophrenias or types of depression, evoked response measurements can help determine clinical decisions," says Dongier.

The most intriguing findings relate the evoked responses observed in the relatives of psychotics. The results of ongoing studies in Montreal and Belgium point to an abnormality of the PINV wave in up to 15 per cent of clinically normal relatives of psychotics. If this work proceeds in the direction many psychobiologists foresee, evoked responses may provide early-warning systems for mental illness, and, in addition, the work may settle some long-standing arguments regarding what kinds of behavior are and are not abnormal.

One of the most controversial psychiatric diagnoses is psychopathy. According to standard psychiatric criteria, psychopaths can be diagnosed on the basis of five traits: impulsiveness, irresponsibility, superficiality of emotional responses, inability to profit from negative experience, and impairment of conscience. Each of the criteria sounds reasonably specific, but their application is fraught with controversy. Is it irresponsible, for example, for an artist and father of four to desert his family and devote the rest of his life to art? Or is it, instead, an act of courage required as a precondition for achieving artistic distinction? And what exactly is meant in a post-Watergate society by "an impairment of conscience"? Critics of the diagnosis of psychopathy point out that it is too subjective and amounts to little more than a contemporary equivalent of Oscar Wilde's epigrammatic definition of morality: the attitude we take toward those we personally dislike.

Recent evoked-response work indicates that the diagnosis of psychopathy is undoubtedly a legitimate one: Psychopaths are not just "different" but, in fact, constitute a distinct group that can justifiably be classified as "mentally ill." Their evoked-response results, particularly in older psychopaths, show consistent failures of response to harsh or irritating stimuli. Despite the subjective complaint that a noise may be annoyingly loud, for example, their evoked responses fail to show the changes observed from control groups who are exposed to the same "noxious" stimuli.

The researcher, Dr. Karl Syndulko, of the University of California, thinks "the findings may be related to the clinical observation that psychopaths apparently fail to learn from negative experience, at least with respect to older psychopaths." According to Syndulko, the findings probably reflect a difference in brain functioning in response to unpleasant stimuli that is peculiar to older psychopaths.

The results with psychopathy, if confirmed, could usher in an era of evoked-response testing in criminals, juvenile delinquents, and the violent. Early diagnosis might enable psychiatrists to suggest corrective measures ranging all the way from medication and counseling to intensive in-hospital "treatments" for psychopathy prior to its clinical expression. Such scenarios, of course, would demand a total revision of our present attitude toward individual responsibility in a free society. So far, the question has not arisen, because psychiatry has always lacked the precision required for making early, 100 per cent accurate diagnoses. Will the evoked responses provide 100 per cent reliability?

At this point researchers are cautious but optimistic, a dramatic advance from the 1960s, when one researcher facetiously responded to such a question: "Sometime between the freezing of hell and the Second Coming."

Now that it's clear that we won't have to wait quite that long, we should begin to decide, it seems to me, to what uses, if any, the evoked-response research will be put. For one thing, it is likely that a predictive evoked-response test for schizophrenics will produce as many problems as it will solve. Suppose the test reaches 98 per cent accuracy. What about the 2 per cent who will be falsely labeled? And predictions about imminent psychosis may, in themselves, provide the kinds of stress that may lead to a mental breakdown.

When applied to psychopaths, the evoked-response results may enable psychiatrists to distinguish between those whose behavior will be transient and situational (e.g., harmless stealing among ghetto children) from those who are likely to embark on a lifetime pattern of antisocial acts. One wonders about the evoked-response findings in a Charles Manson. Could any measure of brain functioning, no matter how accurate, justify stopping a potential criminal before he has committed a criminal act?

The applications are not yet here, but they are coming. Thanks to evoked responses, neurologists are already capable of diagnosing certain forms of brain disease with a high degree of accuracy. Multiple sclerosis, for instance, is a nervous-system disease with an almost infinite variety of initial symptoms. Many patients, after an acute attack of multiple sclerosis, return to a state of complete normality. At such times an examination, even by a doctor who specializes in the disease, may reveal no symptoms. For this reason, legitimate doubts often arise about the accuracy of a previous diagnosis. Doctors of equal experience will differ among themselves about whether or not a particular patient suffers from multiple sclerosis. Evoked-response studies are now simplifying decisions regarding many of these disputed cases.

In late 1977 an extensive series of patients were described in *The British Medical Journal*. The researchers, a team from Western Australia, described their results using evoked responses among three groups: "clinically definite," "probable," and "suspected" multiple sclerosis. Measuring the evoked responses among the "clinically definite" cases yielded a 91 per cent correlation between abnormal evoked responses and a clinical diagnosis of the disease. Among the "probable" and "suspected" cases the figures were 79 and 42 per cent, respectively. These results were confirmed in the United States and reported in late 1977 by the Mayo Clinic. Applying these findings to individual patients, doctors are now able to compute the odds as to whether or not a patient with a brain dysfunction may be suffering from multiple sclerosis.

Evoked-response recordings are also helpful in determining brain death. Animal studies demonstrate that even with a flat EEG brainstem, potentials can be evoked by auditory stimulation from those brains destined to recover. In cases of deep coma, the auditory stimuli—usually a series of clicks—will tell the treating doctor whether the process is affecting the brainstem or is confined to the cerebral cortex. When the short-latency auditory responses are present and normal, brainstem structures can be presumed to be normal as well. In cases of suspected brainstem tumors, the diagnosis has been clarified by the evoked responses, leading, in some cases, to precise localization of the tumor, which has been later confirmed by surgery or autopsy.

Evoked-response measurements possess several key advantages over other methods of studying behavior. First, and probably most important, some of the evoked methods don't require a co-operative or even a conscious patient. The early components—occurring some ten to twenty milliseconds after a stimulus—are thought to originate from the brainstem, are relatively independent of the higher brain areas, and show up whether or not the patient is conscious. This provides an accessory tool for evaluating depth of coma, localization of the lesion causing the coma, and a helpful, but certainly not infallible, index of the likelihood of recovery.

The "late responses"—a few hundred milliseconds after stimulus—measure events occurring "farther upstream" in the area corresponding to the sensory and association cerebral cortex. Here attention, arousal, and interest are important if results are to be reliable. This explains why the evoked-response work with psychiatric patients has differed according to such variables as to whether the patients were taking medication, living in the hospital or at home, and, finally, their motivation for participation in the experiment. With further work, most researchers expect to overcome these limitations and arrive at reliable evoked responses that can be elicited regardless of the motivations of the subjects.

In the meantime, a group of psychobiologists at the Brain Research Laboratories of New York University Medical Center are developing profiles of brain functioning that may soon revolutionize our approaches to such conditions as senility, learning disability, and hyperactivity. Neurometrics, as the method is called, analyzes large numbers of electrical measurements of brain activity. Its goal is to replace our present reliance on verbal descriptions with precise electrophysiological profiles. A child who has difficulty in school, for instance, may be deaf, mentally retarded, emotionally disturbed, or suffering from a "learning disability." In each case the behavioral description "difficulty in school" refers to a specific brain dysfunction that will require a specific form of treatment. The Brain Research Group has now developed a neurometric test battery based on electrophysiological responses that are independent of verbal interactions or overt behavior.

The neurometric test battery first samples the subject's resting EEG and compares it to normal values for age-matched controls. Then follows analysis of the different wave forms, along with their frequency and distribution. This enables the researchers to spot abnormalities in the resting base-line EEG. At this point, "challenges" are presented to the subject which roughly correspond to test items on a psychometric test battery.

Evoked responses can provide a measure of sensory acuity as well as the degree of interference one sensory input may have on another. It's possible to show, for instance, specific evoked responses to two different colors or even two geometric shapes. A rectangle will elicit a different evoked response from a triangle. Even more intriguing, a rectangle will elicit the same evoked response whether it's small or large. If a square is rotated 45 degrees to represent a diamond, the evoked response abruptly changes. Later, when rotated back, the evoked response is once again characteristic of a square, whatever its size. Since schizophrenia and minimal brain damage have been postulated to involve subtle perceptual dysfuntion, the evoked-response testing for form and pattern perception may enable researchers to pick up abnormal brain responses *prior* to the onset of disturbed behavior.

Measurements of a subject's thought processes are now possible using neurometric methods that neutralize cultural bias and differences in verbal capacities. The onset of asymmetries in the evoked responses in the left and right temporal areas, for instance, can pinpoint the time when language processes have become localized in the left hemisphere. The use of visual or spatial "challenges," in contrast, can measure the degree of right-hemisphere specialization for these functions. Already the work is confirming the hypothesis we suggested earlier in this book: Perceptual styles, although probably dependent on a genetic program, require environmental stimulation for their proper development.

At the very earliest age, in fact prior to birth, the brain in most right-handers will show anatomical specialization favoring language acquisition in the left hemisphere (see Chapter Ten). This specialization can be changed early in life, as it is in the case of left-hemisphere damage. Under ordinary circumstances, however, the language function will remain intimately associated with left-hemisphere brain structures. Environmental stimulation—exposure to human speech, babbling, intersensory integration between sights and sounds—will accentuate the inborn predisposition to use the left hemisphere for speech. Finally, the typical adult situation will result when normal speech requires an intact left hemisphere.

Neurometric measurements of cognition are also providing an answer to several fundamental philosophical questions. If the same wave shapes result from a subject's exposure to stimuli which differ only in size, this probably represents a psychobiological basis for abstraction: A rectangle remains a rectangle regardless of size considerations. Since the brain wave forms corresponding to rectangles, for instance, can be elicited only after a certain age, it supports the Piagetian hypothesis that conceptual constancy is not inborn but develops with maturation and experience. In a phrase, the Platonic concept of Ideas is not supported by neurometric research. A child learns first to recognize a rectangle by abstracting the features of many particular rectangles and then applying this knowledge to future situations.

At the other end of the developmental spectrum, neurometrics can aid in the diagnosis of senility. One of the most difficult determinations a physician can be called upon to make is whether or not an elderly person is suffering from one or another form of *dementia,* the clinical term for senility. In recent years, the decision has been helped a great deal by computerized X-rays (Computerized Axial Tomography, or CAT scan) which delineate the size of the brain's ventricles. Often, enlarged ventricles provide a clue to the presence of brain-cell death: As the brain cells begin to die the ventricles enlarge proportionately. Comparisons can then be made between the ventricular size of a patient suspected of cerebral atrophy (brain-cell death) compared to an age-matched series of controls. Only a few years earlier, similar comparisons depended on subjecting the patient to a painful and dangerous test (the pneumoencephalogram), which consisted of the injection of air into the brain's ventricles. Since air is lighter than the cerebrospinal fluid normally contained in the ventricles, the air ascended and displaced the ventricular fluid whenever the patient was sitting upright. This provided a "cast" of the upper portion of the ventricles. To see the lower portion, the patient was then strapped to a special revolving chair which was turned upside down, in which position he remained for up to half an hour. The total procedure, sometimes lasting up to an hour, was commonly followed by several days of intensive headache and vomiting. As one might imagine, pneumoencephalography was an experience that few patients have ever forgotten or would be willing to undergo a second time! Fortunately for everyone concerned, pneumoencephalography has largely been displaced, as a method of studying ventricular size, by CAT scans. But in both procedures the principle is still the same: The loss of brain cells results in brain atrophy and an increase in the size of the brain's ventricles.

Neurometric evaluation of senile patients can be carried out using a series of "challenges." Starting with two minutes of spontaneous resting EEG, followed by several measures of the subject's response to flash and click stimuli, the series concludes with the subject's responses while listening to a short story, compared to the same stimuli repeated for two minutes after the completion of the story. From these measurements, cognitive profiles emerge that are 91 per cent accurate in identifying senility. More importantly, sub-groups appear that contain individuals with similar cognitive impairments. One group, for instance, is noteworthy for hemisphere asymmetries in the late-evoked responses. The Brain Research Group postulates that these individuals might be suffering from difficulties in synchronizing the two hemispheres for handling certain forms of information. Since the basis for this group's "senility" is different from most of the other subjects studied, individual therapies may soon be developed that will cure this particular form of senility by concentrating on ways of synchronizing the transmission of information from one hemisphere to the other.

Neurometric techniques may soon provide a reliable means of accurately assessing abnormal cerebral processing in a variety of conditions. As things now stand, medical and psychiatric diagnosis consists of observing certain behaviors (signs) and correlating them with the patient's own assessment of his problems (symptoms). In the absence of more precise methods of evaluation, a whole group of highly dissimilar people are typically lumped into categories. Schizophrenia, senility, hyperactivity, learning disability, etc.—all are only catchwords that mask our ignorance of the underlying brain dysfunctions. Neurometrics offers electrophysiological measurements that are culture-free and capable of measuring how the brain responds under highly specific "challenge" conditions.

Already it's revolutionizing our approach to the learning-disabled child.

Learning disability is the most common disorder confronted by child psychiatrists. Rather than a diagnosis, however, the term is a catch-all which includes children suffering from a variety of emotional and neurological conditions as well as children whose "learning disability" is only a response to an unhappy or stressful home or school situation. Many of these children respond to their learning impairments by increased physical activity and decreased concentration, which, not surprisingly, earns for them a second label: hyperactivity. Meanwhile, the adult response to children's learning disabilities and hyperactivity hasn't contributed very much to the goal of increasing our understanding of the processes involved. Only a few years ago, a serious but misguided effort was aimed at convincing everyone that hyperactivity was a "myth." Unfortunately, it's a myth that the families of between five and fifteen million children—somewhere between 7.5 and 15 per cent of the total child population—have learned to live with.

Neurometrics, when applied to learning disability, is providing a valuable aid for both diagnosis and treatment. Although the causes of learning disability may vary from child to child, the neurometric results can confirm the presence of electrophysiological variables, which provide 97 per cent accuracy in discriminating between "normal" and "learning disabled" children.

EEG analysis, based on sampling from different anatomical areas of the skull, reveals unusual features in 78 per cent of learning disabled children, but, among only 20 per cent of the normal control groups. With the addition of evoked responses, the figure approaches 90 per cent for the learning disabled, compared to only 8 per cent of the normals.

"The most striking feature of these results," according to the Brain Research Group's director, Dr. E. Roy John, "is the high percentage of learning disabled children who displayed multiple types of dysfunction in multiple regions. The high incidence of pervasive dysfunction suggests widespread occurrence of some source of severe generalized insuilt, such as pre- or perinatal trauma, malnutrition, or stimulus deprivation."

Learning disability, however, is not clearly defined even among children who exhibit uncontested problems in acquiring new information. Some have difficulty with verbal concepts. Others do well with reading and writing but come apart whenever they are required to handle numbers. Yet a third, the most "learning disabled" of all, manage both verbal and mathematical concepts poorly. Neurometric challenges with all three groups result in striking differences: Verbal underachievers show left-hemisphere abnormalities, while arithmetic underachievers show a consistent right-hemisphere dysfunction. Children with disabilities in both verbal and mathematical abilities demonstrate predominantly left-hemisphere changes, and these occur at an earlier point in the evoked-potential recordings than it does for the verbal underachievers alone.

Dr. John's results illustrate the folly of applying the same label to children whose disabilities vary widely. Subject material, hemisphere dysfunction, the timing of evoked-response abnormalities—all of these factors differ significantly among children who are presently grouped together under the diagnosis "learning disability." No less than three types of underachievers emerge from these studies, each with radically different responses to the learning situation. Any rational treatment approach to learning disability must take note of Dr. John's findings. Instead of concentrating entirely on the material the child finds difficulty in learning, the treatment approach must focus on the brain processes involved in the different learning tasks (mathematical, verbal, or a combination of the two).

"The remarkable feature of these findings is that the information processing in children with a particular type of learning disability seems to reflect the general operational defect independent of the specific information content of the input," says John.

John envisions a culture-free test that will predict ahead of time what learning areas a child is likely to find difficult. He's presently comparing children of normal I.Q. for the purpose of separating the children who will do well in school from those who will require remedial instruction.

The goal is the development and employment of neurometrically based teaching methods that will "build in healthy brain functioning where we've detected abnormal functioning." To accomplish this, of course, neurometrics must be developed to an almost flawless level of predictive precision.

By comparing groups of people with similar behaviors, Dr. John hopes to discover in the near future electrophysiological variables that will provide the basis for new and improved behavioral descriptions. Rather than relying on imprecise and subjective clinical judgments, the neurometric approach will incorporate a profile of the brain's own reaction patterns. As we mentioned earlier, psychobiologists are moving us closer to a kind of brain "fingerprint."

The idea that learning and behavior might someday be predicted ahead of time seems threatening to many people. They react with dread to the prospect that electrophysiological measurements may soon be available that can sort out problems in arithmetic, reading, or maybe even sensitive areas involving highly personal feelings. The technology for such determinations is, for the most part, already available. The key question, of course, is: To what use will these new psychobiological tools be put?

On the plus side, neurometrics may provide criteria for selecting people who may benefit from a wide range of specific treatment approaches. Mental illness, mental retardation, prematurity, malnutrition, learning disability, hyperactivity—this is only a short list of the potential medical applications. In addition, the method may provide a means of studying environmental effects on the general population. Food additives, toxins, pollution, lifestyle differences—all may result in specific effects on the brain which can be measured neurometrically. If so, psychobiology may provide the kinds of information that will help us to shape the world we want to live in in the twenty-first century. To mention only one possible application: If the increasing incidence of dementia could be correlated with the rising carbon monoxide levels in the air, it could provide an overwhelming argument for limiting the use of automobiles in our major cities.

Within the next five years, neurometric terminals may be available in most physicians' offices as well as in schools and community health centers. Dr. John suggests that the data obtained from periodic neurometric examinations will be used to detect such conditions as incipient senility or premature aging. From the analysis of neurometric data it will even be possible to detect premature brain aging before clinical signs appear or before the affected person's behavior begins to alter.

For one thing, there are indications that the brain's aging is exquisitely sensitive to environmental variables. As with the studies on early brain development in infants, psychobiological studies on brain aging stress the importance of stimulation. Those who remain active, curious, and involved during their later years exhibit far fewer behavioral and intellectual abnormalities than their neighbors who are socially isolated and mentally inactive. "He who lives by his wits dies *with* his wits," according to psychobiologist Dr. David Krech.

Despite popular belief to the contrary, drastic loss of brain cells isn't a necessary accompaniment of normal aging. Some areas of the brain, in fact, show no reduction at all in the number of brain cells. Rather, cell loss during brain aging seems to depend on the accumulation of lipofuscin, an "aging pigment," so called because of its appearance as a thick paintlike substance within the body of a nerve cell. So far, there is no overall agreement about the origin of lipofuscin. But agreement does exist that it is probably some type of cumulative breakdown product resulting from a flagging metabolism within the aging brain. Such thinking, if it should turn out to be correct, could provide the basis for a rational treatment approach. One drug, centrophenoxine, has already been shown to reduce the incidence of lipofuscin in the brains of old rats by as much as 40 per cent. Diets rich in Vitamin E accomplish the same effect, but to a lesser extent.

Are psychobiologists likely to turn up an elixir of youth capable of reversing or halting the process of cell aging? Although some scientists consider such a goal extremely unlikely, drugs will soon be available that will curtail the formation of lipofuscin or eliminate a large portion of the lipofuscin already formed.

Studies on the brains of aged rats have demonstrated that the thickness of the cerebral cortex diminishes by less than 6 per cent among those animals maintained in stimulating environments. Such results have led Marian Cleeves Diamond, an investigator at the University of California at Berkeley, to conclude: "In the absence of disease, impoverished environment, or poor nutrition, the nervous system apparently does have the potential to oppose marked deterioration with aging."

Neurometric evaluation of the aged may thus be capable of detecting at an early stage those elderly persons whose intellectual powers are fading the fastest. In addition, an early-warning signal of senility could provide the stimulus for timely medical evaluations leading to the identification of the treatable causes of senility. If such a goal can be achieved, psychobiologists will be in a position to refute William Hazlitt's 1827 indictment: "The worst old age is that of the mind."

Within the next several years, neurometric terminals will be available within our major cities that will make possible the early diagnosis of a host of medical and psychiatric conditions. The method, already in use via remote terminals in Long Island and Philadelphia, will consist of a telephone linkup with a central analysis center, presently located at the Brain Research Center at Bellevue Hospital in New York City. A patient will enter a physician's office, say in Philadelphia, and while sitting in the waiting room filling out his medical questionnaire, will be fitted with a "helmet" which contains an array of electrodes. The patient's "brain waves" will be immediately recorded and fed into a desk-top computer. This information will then be digitized, compressed, and relayed over a phone to a central analysis center. Within fifteen minutes, less time than it takes to fill out most medical forms, the desk-top computer will receive from the central analysis center a permanent printout: "The following abnormalities exist at the following locations within the brain. Here is a series of pictures of the brain, with the abnormalities indicated in their proper anatomical locations. Here, too, is a list of the probable causes for the abnormalities listed according to their statistical probability. As you can see, the most likely diagnosis is _____."

"Within fifteen minutes after the patient sits down in a physician's waiting room, this kind of data will be available to the examining doctor," according to Dr. John.

Emergency rooms across the country will eventually be capable of evaluating accident victims via neurometric techniques. In a typical situation, a stretcher in the emergency room will be permanently linked to a neurometric terminal. Auto-accident or head trauma victims, upon arrival in the emergency room, will be placed on these stretchers for neurometric evaluation. Cerebral and brainstem mechanisms will be evaluated at regular intervals. In this way, the neurosurgeon can determine whether the patient's status is improving, remaining the same, or deteriorating. In the case of deterioration—as measured by neurometric indices—surgical intervention can be started immediately.

On the other hand, neurometrics, if improperly used, could provide further justification for applying labels to behavior that we don't understand and can do little about. As with any scientific procedure, neurometric prediction can sometimes be wrong. For instance, How can we protect a child who is merely a slow starter and not learning-disabled at all? Dr. E. R. John has captured the essence of the conflict beautifully. "Procedures must be devised to assure that neurometric evaluations are used to optimize the development of the individuals rather than to restrict their opportunities."

14
Put on a Happy Face

Our ends are joined by a common link:
With one we sit, with one we think.
Success depends on which we use:
Heads we win, tails we lose.

HEARD IN A RECENT LECTURE ON NEUROANATOMY

In the laboratory of the Yale University Center for Behavioral Medicine, an experimental subject sits quietly while four recording electrodes are placed on his face. "Grit your teeth," the experimenter requests, and two of the electrodes on the lower part of the face begin firing.

"Now frown. Wrinkle your forehead like you're worried or angry."

In response to this, the two upper electrodes are activated. The experimenter nods to his assistant . . . everything is in readiness. The experiment can begin.

The muscles of the face are responsible for the outer expression of our inner emotions. When we feel sad, for instance, our eyebrows wrinkle into a state of sustained muscle tension, and the corners of our mouth droop. When we're happy, other muscular patterns evolve. Instead of drooping, the muscles around our mouth tighten while the outer corners turn upward into the beginnings of a smile. In general, these muscular contractions are outward signs of our inner mental state. But which comes first? Do we wrinkle our brows because we are sad? Or does our sadness result, at least partially, from the muscular patterns of our faces?

The idea of testing muscle patterns as indicators of mental states is not new. In the 1920s, Edmund Jacobson carried out a series of innovative experiments that correlated his subject's mental imagery with the activation of certain muscles. Jacobson discovered that if a subject imagined that his right hand was moving, eye movements would shift to the right and tiny increases in muscle tension could be recorded from the muscles of the right forearm. Today's experiment at Yale is a continuation of Jacobson's observations carried out by Gary Schwartz, a bearded, energetic young psychologist who specializes in biofeedback. Schwartz's experiments are comparing the pattern of muscle tension in a group of depressed patients and a group of controls who are not depressed.

"Think about something happy," Schwartz suggests. "Think, or just imagine in your mind, something that you really want that would make you happy." Next, the subjects are instructed to think sad or angry thoughts. Finally, neutral thoughts are elicited. "Think about a typical day."

Comparisons of the muscle patterns of the two groups reveal important differences. Among the normal subjects, the thinking of happy thoughts is associated with a specific pattern of facial muscle tension that is different from both thinking sad thoughts and thinking angry thoughts. When the subjects are requested to think about a typical day, they generate a typical happy pattern.

Testing depressed patients reveals a different response. "While depressed subjects readily generate sad and angry patterns, their muscle patterns for happy imagery are markedly attenuated. When the depressed subjects think about a typical day, the resulting muscle pattern appears to be one of sadness," says Schwartz.

These results, Schwartz suggests, confirm the speculations of earlier experimenters that moods may be reflected in specific muscle tension patterns. We feel happy or depressed, for instance, because these emotions accompany certain facial and postural changes. By feeding back to the brain muscle patterns characteristic of happiness, the brain generates or sustains inner feelings of happiness. The experience of separate emotions such as happiness, sadness, or anger results from specific patterns of muscle tension which psychobiologists now speculate are probably "hard wired" in the brain.

From such experiments emerge a whole series of exciting possibilities presently being explored by Schwartz and others. If the brain and patterns of muscle tension are interrelated, and if, according to Jacobson's imagery studies, the brain can bring about changes in muscle tension, why not train the brain to alter the harmful muscle tension patterns and at the same time improve undesirable emotions?

"If as a result of consciously regulating specific thoughts or emotions the brain generates specific motor and visceral output devices, we have the psychobiological foundation for teaching a person to regulate his health," according to Schwartz.

Schwartz's research represents a modification of biofeedback, the concept that one researcher has called "the major focus around which modern biology has been reorganized." Before we get to Schwartz's modifications, however, let's take a moment to say a little bit about how a feedback system works.

If you have a thermostat in your home, then you already intuitively understand the basic principles of feedback. After setting the thermostat at the desired temperature, the heating system is automatically controlled so that your home is neither too hot nor too cold but remains at exactly the temperature you have chosen. If the house begins to get colder the thermostat "feeds back" this information and the furnace is turned on (positive feedback). As the temperature in the home begins to rise, it may eventually exceed your previous setting. At this point, the thermostat turns off the furnace (negative feedback) and the temperature in the house begins to decrease again. It's important to recognize that negative feedback (turning off the heat) is the process that leads to stability. In the absence of negative feedback a higher temperature in the house would elicit *more* heat rather than less from the furnace, until eventually the system would break down. Negative feedback (the higher temperature shuts off the furnace) leads to a condition of stabilization—i.e., the furnace is not driven to a point of breakdown and the occupants of the house aren't forced into the street in order to escape intolerably high temperatures.

In 1940 the concept of feedback became the basis for a new science, cybernetics, the linkage of men and machines. From here it was only a small step toward applying cybernetic principles to biology, resulting in biocybernetics. The stimulus for this application came from William Cannon, a physiologist and author of *Wisdom of the Body*. Cannon popularized a concept he called *homeostasis*, borrowed from the idea of the nineteenth-century physiologist Claude Bernard, who believed that all bodily processes are precisely regulated within certain fixed limits.

In most cases of feedback, the brain is the regulator of bodily processes, a point which until recently was insufficiently recognized. Neuronal impulses and chemicals, usually hormones, inform the brain of events taking place elsewhere in the body. As we mentioned in Chapter Two, the second functional unit of the brain uses information sent from our eyes and ears—only two of our exteroceptors—as well as internal information such as joint position or the pressure on internal organs originating deep within the body (enteroceptors). Light, sound, and touch—in fact, all primary sensations—are converted into identical electrochemical patterns that are interpreted by the brain as sight or hearing. (Later in the book we'll look at some ways that touch patterns are now being converted into light signals that may in the future provide sight for the blind.)

Think of the brain as an international casino visited by people from all over the world. Each person comes to the casino with his own currency. In order to gamble, he must change it into the currency of the casino. In the same way all internal and external stimuli must be converted into the currency of the brain: the electrochemical impulses described in Chapter Nine. To this extent the brain becomes the regulating force for all bodily activities.

Naturally, where there is regulation there is also the possibility of *disregulation*, Gary Schwartz's term for any breakdown in the feedback mechanisms. If the thermostat fails to function, the furnace will continue to raise the temperature of the house to higher and higher levels, eventually with catastrophic results.

The thermostat example is also similar to the body's feedback mechanisms in one other important respect: Both processes occur outside of awareness. The brain is no more conscious of the processes being fed back to it than is the thermostat. "We have the paradox of self-regulation," says Schwartz. "The brain has an inherent inability to be aware of much of the feedback that participates in regulating bodily processes, and is incapable of experiencing any of the feedback involved in regulating itself. The brain is so constructed that it is impossible for us to consciously experience these processes even if we choose to do so."

Such inaccessibility to conscious awareness is, as we mentioned throughout this book, the rule rather than the exception when it comes to most of the brain's activities. Events of tremendous importance may be taking place right now, and yet we remain totally unaware of them. If your blood pressure, for instance, is elevated, you may eventually suffer a stroke (a good reason for finding out your blood pressure level), yet you'll still remain completely unaware of the sustained hypertension leading up to such an unfortunate event.

Experimental confirmation of the inaccessibility of the brain's self-regulation comes from studies on EEG (brain wave) patterns. Normally, in states of relaxation a subject will generate an EEG pattern with a frequency of eight to thirteen waves per second. This alpha rhythm is symmetrically equal when recordings are made from both sides of the head. Through feedback training, it's possible for experimental subjects to either increase or decrease their alpha waves. But most interesting of all, the subjects are not aware of any changes in their brain waves.

In one of Gary Schwartz's experiments, subjects are taught to modify their blood pressure by biofeedback. They are then shown how to change their heart rates. Finally, both the heart rate and the blood pressure are altered, but in opposite directions! Answers to post-experimental questions indicate an absence of consistent differences in the subjective experience of increasing versus decreasing their cardiovascular responses, notes Schwartz. "Interestingly, most subjects did not believe they had much control over their responses."

For a while Gary Schwartz was troubled by these instances where important bodily events were occurring outside conscious awareness. But after thinking about the problem in the context of his biofeedback experiments, he hit upon an explanation.

Consider something as simple as moving your hand. At the behavioral level the action is voluntary and can be started and stopped at will. Psychobiologically, however, the action consists of patterned muscle responses which are eventually programed in the opposite cerebral cortex. You have no insight into how these patterns are put together, nor how the patterns actually control the action of moving the hand. Even though the central processes occur in the brain, you experience them in the outside world: You are aware of the moving hand, not the cerebral responses. This has been tested experimentally by Penfield's electrical stimulation of the cerebral cortex of patients about to undergo brain surgery. When the area of the brain corresponding to the thumb is stimulated, the patient reported movement of the thumb, not electrical stimulation of the cerebral cortex.

In essence, since our behavior has implications for our survival, we respond to behavioral consequences rather than neurophysiological processes. If we step in front of a car, for instance, our perception of the car's rapid motion toward us is more important for our well-being than awareness of the brain mechanisms underlying sight.

"The brain's construction of three-dimensional space is a major accomplishment," says Schwartz. "Although this greatly helps the brain to process information and interact in a three-dimensional world, it consequently greatly hampers the brain's capacity to understand its own role in the very creation of these sensations."

Now let's take a look at Gary Schwartz's model for disregulation. Beginning at the outside and working inward, the first element in the biofeedback loop (Stage 1) is the environmental demands imposed on us. Although we usually think of the environment as external, we should remember that environmental demands often originate internally. A dazed, rubber-legged boxer trying to remain on his feet and avoid a knockout is forcing his body, via internal demands, to remain in the harsh environment of the prize ring. An executive who persists in a high-pressure job, despite ulcers and migraine headaches, is also allowing internal processes (ambition, greed, whatever) to force him to remain in an unhealthy work situation.

Stage 2 corresponds to what we discussed in Chapter Three as the second functional unit of the brain. Many of the regulating mechanisms of this second stage may be genetic, hence comparatively unalterable. High blood pressure, elevated cholesterol, the clustering (in certain families) of heart disease and certain forms of cancer—these are some examples of conditions that are turning out to be hereditary. Some are also frequently associated with specific behavior patterns which may be inherited. But at this stage psychobiologists aren't certain how many of the brain's central-processing activities are hereditary and, hence, essentially unmodifiable.

Stage 3 involves the specific organs responsible for "feedback" to the brain. If you accidentally touch a hot stove, your hand withdraws in response to a sensory process originating in the pain receptors of the skin. In a condition known as congenital insensitivity to pain, a person's pain receptors in the skin may be absent, resulting in severe burns. The peripheral organ (the skin) is incapable of detecting the hot stimulus and, as a result, can't "feed back" warning signals to the brain. Different types of disregulation involve different peripheral organs: Asthma, dermatitis, peptic ulcer, and colitis are disorders of the bronchial system, the skin, the stomach, and the lower colon, respectively.

The fourth and final stage in the negative feedback loop corresponds to the interaction between the heated house and the thermostat. Information from peripheral organs is fed back to the brain, where adjustments can be made that will be helpful in maintaining the body's equilibrium, or homeostasis. This is the most important component of the feedback system, and when it's not working properly it results in disregulation.

Ordinarily, a bodily process such as respiration goes on automatically, although we can temporarily suspend our breathing when engaged in activities such as underwater swimming. At such times the carbon dioxide in the blood builds up and directly stimulates the respiratory center in the brainstem to restart breathing. A fierce struggle may then ensue between the brainstem's order to "begin breathing" and the order from our cerebral cortex to "hold the breath." Here the feedback loop is being counteracted by the cerebral cortex's knowledge that breathing underwater will result in drowning. Eventually the struggle may result in unconsciousness, at which point the lower brainstem's respiratory center is no longer inhibited and a fatal breath ensues.

Whenever the signaling of a peripheral organ is ignored, the body's smooth regulation is interrupted. In the example just given—underwater swimming—interrupting the feedback loop may be helpful, as in the case of a swimmer ascending from a deep dive who requires only a few more seconds to surface. In most cases, however, any interference with the smooth performance of the body's feedback system is harmful, leading to psychosomatic illness.

Psychosomatic illness results, according to Schwartz and other biofeedback researchers, from failures in one or more of the four stages of the feedback regulation. In the first stage it's easy to imagine circumstances where the body is simply overwhelmed by environmental extremes. For example, under ordinary circumstances we are able to maintain our body temperature very near 98.6. Under environmental extremes, however, the body's temperature can rise dramatically (heat stroke).

The feedback failure may also result in Stage 2 if the brain chooses to ignore the feedback information. Some people think that this may be the cause of some forms of neurosis and psychosomatic illnesses. Although we are presently unsure of the mechanisms operating here, there are several provocative speculations.

As we discussed in Chapter Ten, David Galin's studies on hemisphere specialization suggest that emotional reactions may be processed mainly in the right hemisphere and can, under certain circumstances, be confined there by blocking transfer across the corpus callosum. Gary Schwartz's earlier work led to the discovery of a right-hemisphere specialization for handling emotionally loaded questions. In response to factual questions such as, "How many *s*'s in the word *Mississippi?*" Schwartz found most people's eyes turned to the right, a sign of left-hemisphere activation. When asked an emotionally loaded question, however, such as, "Do you think your mother-in-law is an interfering woman?" the eye movements are usually to the left. Further support for the right-hemisphere-emotional-responsiveness hypothesis comes from the work of Rubin and Raquel Gur, who, as we noted in Chapter Ten, discovered an increased incidence of psychosomatic illness among leftward eye-movers (right-hemisphere-dominant types). Whatever the mechanism, the right hemisphere seems occasionally to embark on courses of action which may be harmful to us.

Traditionally, the treatment of psychosomatic disease concentrates on the peripheral organs. A stomachache, for instance, is treated by an antacid, which coats the lining of the stomach and reduces the amount of stomach acidity. As a result, pain impulses are interrupted at the level of the pain receptors. If this doesn't work, drugs are given to modify the number of nerve impulses going to the stomach; with fewer nerve impulses less acid is secreted, which reduces the irritability of the intestinal wall and results in fewer pain impulses feeding back to the brain. At a still later stage, part of the stomach may even be cut out in a last desperate effort to control pain.

During the past twenty-five years, many physicians, dissatisfied with the results of treating the peripheral organs, developed less dramatic approaches to illnesses like peptic ulcer. A change of job, remarriage, financial security —all have been found on occasion to "cure" untreatable ulcers. Other patients have been helped by psychotherapy to achieve important attitude changes which enabled them to adopt a more philosophical approach to life, or to deal more successfully with destructive emotions such as anger or frustration. Although no one used terms like biofeedback to describe what was going on, these treatment approaches were obviously working at the Stage 1 and 2

315

levels. But several problems had to be solved before biofeedback really came into its own.

For one thing, the brain was conceptually broken down for many years into two categories, each with different implications for treatment. One part corresponded to the brain as a conscious integrating system with direct connections to the muscles of the body. Disturbances have resulted in forms of hysteria, such as paralysis, which affected parts of the body under voluntary control. Meanwhile, the more automatic parts of the body's activity (respiration, digestion, heart rate, etc.), which I prefer to call inaccessible nervous system activities, were grouped under the heading of a second category (the autonomic nervous system). I prefer the word *inaccessible* to the more common word *unconscious* because of a small difference in implication. Freud's "unconscious" emphasizes mental processes that we are not aware of and perhaps should be. The goal of psychoanalysis is to "make the unconscious conscious." Inaccessibility, on the other hand, refers to brain events that, as a result of the "wiring" of the nervous system, can never enter our awareness. These inaccessible activities were thought to be automatic—hence, not controllable. Studies of the autonomic nervous system led to the development of psychosomatic medicine, a term for illnesses at least partially dependent on environmental or psychological factors, but lacking any clear-cut guidelines for treatment. Psychosomatic illnesses were, in essence, diseases without cures.

At about this time, biofeedback theories were being put forth cautiously. Is it possible to transform information from inside the body into some form of environmental stimulation that can be used by the brain to increase its powers of self-regulation? Since most of these internal processes are inaccessible, such a proposal required the development of technical instruments capable of changing bodily signals into electronic ones. Traditional medicine has always possessed a few instruments capable of doing this: thermometers, blood-pressure recorders, the stethoscope, EKG for monitoring the activity of the heart, EEG (or brain-wave test) for measuring the brain's electrical activity—in each instance a specialized instrument converts the action of an inaccessible body process into meaningful signals. With blood pressure, for instance, the force of blood coursing through the blood vessels is converted into "millimeters of mercury," which can then be statistically compared to the normal blood pressure for people of similar age, weight, sex, and so on.

With the more sophisticated electronic instruments, dynamic rather than static measures became possible. Scientists were soon able to obtain a continuous record of certain bodily activities by converting physiologic signals into visual or auditory discharges.

The stimulus for this research came from a study of color visualization first carried out by Dr. Barbara Brown of the Department of Psychiatry at the U.C.L.A. Medical Center. Since childhood, Dr. Brown had been fascinated by the variations among people in their ability to internally visualize in color. Some people think, and even dream, in full color, while others report visual images limited to a gray or "black and white" field. Could one type of person be distinguished from another by their response to color?

Using a series of ingenious experiments, Dr. Brown showed that people capable of color visualization respond differently than nonvisualizers to flickering colored lights. When a "visualizer" is exposed to rapid flashes of red, for instance, brain-wave recordings show a kind of startle response which is out of phase with the frequency of the flashes. "Nonvisualizers," in contrast, show a relaxation response to the color red, an electronic confirmation of their lack of internal color visualization.

From here, Brown developed an experimental device that activated a blue light whenever an experimental subject's brain waves entered a state of quiet relaxation (alpha state). At the same time, the subjects were allowed to watch the blue light waxing and waning with their alpha rhythm. After only a few sessions, most subjects doubled and even tripled their alpha frequencies, prompting Brown to formulate the concept of biofeedback: "If a person could see something of himself that up to now had been unknown and involuntary, he could identify with it and somehow learn to exert control over it."

From here, psychobiologists set forth a bold new

postulate for bringing about changes in the brain's regulatory powers. "Visceral learning" was the term for a revolutionary approach that did away with the classical distinctions between "voluntary" and "involuntary."

Ordinarily the stomach digests food, the heart beats, and the lungs move air in and out of the thoracic cavity—in each case without attention, effort, or concentration. The automatic nature of these activities may have prejudiced brain scientists for hundreds of years against understanding the autonomic nervous system. Everyone agreed that movement of the skeletal muscles, the arms and legs, could be voluntarily controlled. But how can anyone learn to control his heart rate, or elevate or depress his blood pressure? For the longest time it was assumed that these activities couldn't be voluntarily altered or affected by learning. Scientists did admit that certain internal activities—such as Pavlov's experiments on the salivation response of dogs to the ring of a bell—could be influenced by conditioning. But learning, in the classical sense, seemed impossible, since no one was capable of demonstrating any control over internal autonomic processes.

No one, that is, except Eastern mystics or yogis, who, under laboratory conditions, could slow down or stop their hearts at will. One yogi, Ramanand Yogi, remained in a sealed box for over ten hours while voluntarily reducing his oxygen consumption by 30 per cent. At one point in this experiment the yogi's oxygen consumption decreased to 25 per cent of the theoretical minimum needed to sustain life!

Inspired by such demonstrations, psychobiologists launched a series of experiments aimed at testing the validity of the classical distinctions between conditioning and voluntary learning.

To do this, two researchers, Dr. Neil Miller and Dr. Leo V. DiCara of the Rockefeller University, administered to rats the curare-like drug d-tubocurarine, a chemical transmitter responsible for blocking electrochemical impulses from the brain to the skeletal muscles. Whatever learning takes place in such an experiment must be independent of the voluntary nervous system, which is blocked by d-tubocurarine. Since respiration is mediated by volun-

318

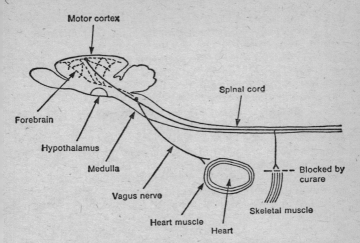

FIGURE 36. In order to insure that the change in heart rate is due entirely to impulses from the autonomic nervous system, curare is used to block the impulses leading from the brain to the skeletal muscles ordinarily under voluntary control.

tary muscles, the animals must be artificially breathed with a mechanical respirator. In this way, all possible contributions by the "voluntary" nervous system are cut off and whatever learning takes place must be mediated via the autonomic nervous system.

Miller and DiCara used a system of electrical stimulations to the brain's pleasure centers located deep within the hypothalamus. To this they added a mildly unpleasant electrical shock which could be avoided when the rats' body responses varied in the desired direction. Soon, using these two parameters, the experimenters were producing, through instrumental training, increases and decreases in heart rate, intestinal contractions, and the rate of urine formation.

By varying the stimuli (electrical brain stimulation versus electric shock) and the direction of body change (first bringing about an increase in blood pressure followed by a period of lowered blood pressure), the experimenters ruled out the possibility that their results could be due to some innate nonspecific effect of the stimulus on the brain.

If the increase in blood pressure were secondary to a general "tonic effect" on the nervous system, for example, it should always result in an elevation of blood pressure and pulse rather than, as the experiment demonstrated, either an elevation or depression of blood pressure dependent on the aims of the experimenter.

By the same token, discrete responses could be obtained in two bodily processes controlled by the same nerve. Heart rate and intestinal peristalsis, for example, are each mediated by the vagus nerve. Rats could learn to increase intestinal peristalsis while leaving their heart rate unaltered. Conversely, the heart rate could be speeded up or slowed down without any change in the rate of the intestinal peristalsis. These results show that visceral learning of two responses can occur independently of each other.

Further research turned up increasingly sophisticated learning responses in the "autonomic" nervous system. Curarized rats can alter the diameter of the blood vessels in their skin without making any change in their heart rate or blood pressure. In some cases the changes can be made even more specific, as they were in one of Miller and DiCara's experiments where a rat learned to selectively dilate the blood vessels in only one ear!

Next followed a series of experiments aimed at discovering the possible usefulness of visceral learning. Obviously, ordinary learning involving the skeletal muscle system has great survival advantage. We learn to avoid painful experiences by escape or careful planning that will protect us against starvation, thirst, or physical or emotional abuse. Since our skeletal responses operate in direct contact with our environment, our arms and legs can take us out of conditions of danger or deprivation. But how about the internal responses mediated via the autonomic nervous system? How are they adaptable? Or is visceral learning only a cute laboratory trick with little or no relevance to the everyday adaptive needs of the body?

Miller, DiCara, and Dr. George Wolf demonstrated that an internal glandular response can reinforce learning. To do this they injected albino rats with a high dose of antidiuretic hormone, which ordinarily maintains the body's proper salt and water balance. If too much antidiuretic hormone is in the blood, the kidney conserves water and produces only a small amount of urine. If the level of antidiuretic hormone fails, just the opposite occurs: the kidney excretes an excess amount of water to restore normal salt and water balance. The body's level of antidiuretic hormone and the kidney's response to it constitute another example of negative feedback.

In the experiment, the albino rats are first force-fed an excess of water and then placed at the end of a T-shaped maze. At the right side of the T-maze is a syringe with antidiuretic hormone, and on the left, a solution of salt. If the rat goes to the right side of the maze he is injected with antidiuretic hormone, which leads to increased water accumulation. (Remember, *antidiuretic* means that the hormone counteracts the body's tendency to diurese: get rid of water.) This was a maladaptive response that could be fatal in a rat that has already been given an excess of fluid. Instead, in all cases the rats soon learned to select the left side of the maze, where they received an injection of the saline solution, which aided the body's own adaptive responses to restore equilibrium.

In another experiment, part of the rat's hypothalamus was destroyed, so that he could no longer manufacture antidiuretic hormone. In this case, however, the rats were first prefed an excess of highly concentrated salt solution. Ordinarily the body would correct for this by secreting antidiuretic hormone, thus conserving body water that could dilute the salt. This time, when the rats were placed in the T-shaped maze they learned to select the side where they received an injection of antidiuretic hormone.

In both experiments the body's normal tendency toward homeostasis was maintained via a learned internal glandular response. Instead of drinking more or less water, the animals selected pathways that would reinforce glandular patterns leading to health.

From here, biofeedback researchers turned from laboratory rats to the study of human psychosomatic diseases. Dr. David Shapiro, Gary Schwartz's early mentor at Harvard University, demonstrated small but consistent blood-pressure-lowering effects on hypertensive patients.

Each time the blood pressure decreased, a flashing light appeared with a nude pinup or a landscape thrown in after every twenty flashes. In each case, the subjects had no idea what they were learning and, in most cases, were completely ignorant of what was being measured by the flashing light. Despite this, most subjects were soon able to control blood pressure in the absence of any change in heart rate, or vice versa. In all cases the subjects, when questioned, denied any subjective experiences accompanying the changes in heart rate or blood pressure.

When both blood pressure and heart rate were controlled by biofeedback, an interesting and unexpected phenomenon took place: Visceral learning occurred quicker and resulted in larger changes. In addition, subjects began to report subjective feelings of relaxation and emotional tranquility. Schwartz refers to this as *pattern biofeedback,* a key brain response. General relaxation with its corresponding subjective experience does not consist simply of low heart rate or low blood pressure but rather a patterned integration of these changes. In a phrase, the whole is greater than the sum of its parts.

The concept that the whole is greater than the sum of its parts is a contribution of modern systems research. It also ranks among the most important insights of the twentieth century when it comes to the formulation of explanatory models. In a sense, this book is a plea for a system or patterned view of brain function and is directed against an overly simplistic view that the brain can be understood as simply the sum total of our present knowledge about brain mechanisms. The smoothly regulated whole brain is more than the sum total of its constituent parts. Biofeedback—particularly Schwartz's concept of the brain's disregulation model—has immediate practical applications.

"Once the theoretical relevance of emergent properties in patterns of physiological responses is recognized, we can begin to appreciate how procedures may be effectively patterned or combined . . . Skeletal and visceral responses can interact and contribute to the emergent experience of unique emotion or blends of emotions."

As an example of this, Schwartz designed an experiment testing the combined effects of increased heart rate and increased muscle tension. From previous experience, Schwartz knew that facial muscle tension is often associated with depression and anxiety. A rapid heart rate is also characteristic of an anxiety state. What happens if the subjects are trained by biofeedback to increase *both* heart rate and facial muscle tension?

To find out, Schwartz first instructed his subjects to increase their heart rate while reporting any increase in subjective anxiety. In some cases a faster heart rate corresponded to an increase in anxiety. Next, forehead tension was increased, and it too correlated with heightened anxiety. With increases in both heart rate and muscle tension together, however, the highest rating of subjective anxiety was reported.

"These data indicate that self-regulated facial forehead tension and self-regulated heart rate increases contributed to subjective anxiety," says Schwartz. "When these are regulated as a pattern, moreover, they combine in an additional sense and contribute to the total emergent experience of anxiety."

As a result of these studies, psychobiologists are now looking at ways to bring about regulation. There are several places in the feedback loop where intervention might be helpful. Patients with high blood pressure, for instance, have been found to respond with a lower blood pressure when the carotid nerve in the neck is stimulated. This corresponds to strengthening the negative feedback loop from the peripheral organ and increasing the frequency of feedback signals. Such treatments are impractical, however, since few people would be willing to undergo the inconvenience of walking around all day activating a nerve stimulator. Besides, who would activate the stimulator when they are asleep? And, as if this weren't enough, the scientists who carried out the study concluded that eventually the high blood pressure mechanism would probably reset itself, elevating the blood pressure once again.

A more sensible approach involves a feedback apparatus that links the elevated blood pressure (Stage 3) to an observable light or color change (Stage 1). Through training, the subject "learns" that a pattern of color or sound corresponds to a lowered blood pressure. Eventually, the blood pressure can be controlled by attending to the color or sound patterns. Some have reported that the process can be speeded up by offering "rewards": slides of attractive landscapes or offers of money.

But despite its success, biofeedback doesn't provide much in the way of an explanation about what's actually taking place. What's occurring, for instance, when feedback signals about blood pressure, after conversion into colored lights, result in a lowering of the blood pressure? In an instance like this, scientists still have not come up with an explanation about what's really going on. Before being too critical, however, we should remember that traditional medicine hasn't progressed much further when it comes to explaining the disorders that biofeedback is now successfully controlling. Over 95 per cent of the cases of high blood pressure in this country, for instance, are diagnosed as "essential" or "ideopathic," two fudge words for "We don't know." Psychobiologists are hoping that perhaps a greater knowledge of the causes of psychosomatic illnesses will result in increased precision in our methods of treatment.

In the meantime, biofeedback provides a workable model for treating psychosomatic illness. Stated at its simplest, our health is controlled by our brain. "The basic premise of the disregulation model is that the brain has the primary responsibility of maintaining the health of itself and its body, and it succeeds in this task by altering and regulating itself to meet the needs of the specific organs," says Dr. Schwartz.

If health is dependent on brain regulation, then illness results from disruptions on one or more of the four stages of the feedback loop. In addition, successful treatment requires intervention at the appropriate stage. A migraine headache, for instance, results from dilatation of the temporal arteries along the side of the scalp. From here, a throbbing, torturous headache may increase to the point where the pain is completely immobilizing. At this point, the sufferer may want to reach for whatever pain pills are available and continue taking them until the headache is gone. This is, in fact, the usual course of action—

and, according to Dr. Schwartz, it may be the wrong one. "The fundamental question is, What should the brain's response to this internal stimulation be? From a psychobiological perspective, the brain should either direct a change in the outside environment, or leave the environment, or modify its interactions with the environment."

In the case of headache, the stimulus may be in the environment. A tense, pressured work situation may bring about a headache (Stage 1). Or unrealistic and rigid goals may be driving a person beyond his constitutional limits of endurance (Stage 2). Adjustment at either stage would likely result in a decrease in the frequency and severity of the headache. The pill-popping habit, however, represents the sufferer's attempt to deal with the headache on his own terms. A strong narcotic will dull the brain's pain perception at Stage 2 as well as partially block it at the peripheral organ (Stage 3). This approach, however, is ultimately harmful, since it disrupts the feedback loop and produces disregulation. The brain is no longer in control; its normal responses are interrupted by the medication, which eliminates the pain by dulling the brain's perceptions.

Similar approaches are used for other pains. Stomachaches, cramps, backaches—these, too, are often treated by interfering with the body's feedback mechanism, usually at Stages 3 or 4. As a result, the brain may become permanently incapable of responding to the body's messages. Imagine a situation in which a thermostat and a furnace are being gradually disconnected. Eventually, the temperature of the house will become unpredictable and beyond thermostatic control. So, too, the body's functioning can be disconnected from the natural control centers of the brain. This may, in fact, be the mechanism responsible for certain diseases like malignant hypertension. The blood pressure regulating system, no longer under the control of the brain, behaves autonomously, leading to a vicious cycle of higher and higher blood pressure, which eventually harms the brain along with many other body organs.

In the psychological sphere, the phenomenon of multiple personality provides a possible model for such a disregulation process. Psychiatrists occasionally encounter patients with split personalities which coexist independently in the same patient. The additional personality may often behave in ways that can't be predicted or controlled. Under hypnosis, to take another example, many people enter fugue states where they do things they cannot later recall. Certain forms of physical illness also represent similar bodily processes that were once under the brain's control but are now acting completely on their own.

In addition, there is evidence that the "meaning" of a stressful experience can alter the body's response to it in ways that may vary all the way from no response to a full-blown and incapacitating illness. For instance, if rats are hooked up to an apparatus that delivers an electric shock to their tails, different effects result depending on the rat's ability to prepare for the shock.

Dr. Jay Weiss and his associates placed rats under three varying experimental conditions and measured weight loss and the tendency to ulcer formation. The first rat was able to avoid the electric shock by turning a wheel or jumping onto a small platform placed within the cage. The second rat, in a separate cage, could do nothing to avoid the shock but was hooked up in such a way that he received the same shock patterns delivered to the first rat. The third rat occupied yet another cage, similar to the others, but didn't receive shocks at any time. In each experimental run, the helpless rat in cage two lost more weight and incurred more ulcers than the rat in cage one, who was able to exert some control over the number of shocks received. The least effect occurred in the rat in cage three, who received no shocks at all.

Weiss next explored the reasons why the rat in cage one, who could delay or extinguish the electric shock, fared better. He conducted a second experiment to determine if feedback could reduce and possibly eliminate ulcers, despite the persistence of a stressful situation. Initially, the shock was delivered without warning. This resulted in weight loss and ulcer formation. Next, a warning tone was added to the experiment. When the rat now attempted to escape from the shock, the warning tone was sounded. Since the tone resulted from the rat's action (jumping onto the platform) and thus postponed the shock, the tone provided a feedback. Under these conditions the rats lost less weight and developed even fewer ulcers—in fact, only slightly more than the control rat receiving no shocks at all!

Human experimental results confirm Weiss's findings. If people are given inescapable mild shocks in a laboratory (I'm happy to report that such experiments are presently out of fashion), their response depends on whether or not they can terminate the shocks by such activities as fist-clenching or button-pressing. In such cases, emotional arousal, as measured by the experimental subject's skin response, is less than in individuals who are told that they can do nothing to avoid the shocks. "The people who thought they had control over the shocks perceived their responses as producing the shock-free condition—that is, they saw their responses as producing relevant biofeedback," says Dr. Weiss. "In contrast, the people who thought they were helpless necessarily perceived their responses as producing no relevant biofeedback. Thus, for humans as for rats, the same variables seem important in describing the effects of behavior in stress situations."

In humans, emotional attitudes, self-evaluation, even the instructions about how they should interpret their experiences, have an effect on feedback and, ultimately, the body's tendency to develop stress reactions.

The relationship of physical processes to mental events is the central puzzle of modern psychobiology. How do we conceptualize the contribution of emotional and mental factors to physical disease? How should these contributions be integrated? There are indications that our present ideas may be too restrictive.

Picture the typical patient who appears in the psychotherapist's office. Over the span of the next several weeks the patient will talk as well as respond to questions from the therapist. Finally, after a variable period, the therapist or patient or (under the best circumstances) both, may begin to understand the basis of the patient's problems. At a certain later point the patient may develop "insight" into his difficulties which could help to bring about an improvement in symptoms. While the methods of treatment may vary according to the orientation of the therapist, certain principles are agreed upon. For one thing, the process of psychotherapy is usually considered a nonphysical form of treatment. There is no "laying on of hands"; a physical examination is not ordinarily carried out; and, in most cases, drugs aren't dispensed. This has led to the conclusion that, in a sense, psychotherapists are treating the patient's "mind" rather than his body. Standard psychiatric diagnosis provides support for this incorrect orientation by classifying certain mental illnesses as organic (brain-related) and others as functional (no known cause).

But, according to psychobiological research, such a division doesn't make sense, since all behavior is ultimately traceable to the brain (in the sense that it is dependent on the brain for its expression). From this, it seems reasonable to conclude that all behavioral treatments ultimately must depend for their effectiveness on bringing about brain changes. But how can the psychotherapist change the brain, since psychotherapy is restricted to listening and talking?

The answer to this seeming paradox comes from a re-examination of what happens when we focus our attention or concentrate on vivid internal images. EEG studies show that attention to a flashing light will alter the brain waves recorded from the visual areas. If the subject then switches to finger tapping, the alteration will appear in the sensory motor regions. At this point the subject can be taken to a quiet room and told to "imagine" an interval of flashing light followed by an interval taken up with finger tapping. The EEG results in the "imagined" part of the experiment are identical with the results obtained when the subject actually watches the flashing light or taps his fingers.

In the above experiment, carried out at the Yale Center for Behavioral Medicine, imagery exerts a measurable physical effect on the brain. A neurophysiological alteration in brain-wave activity occurs and responds to a "mental" activity. Merely imagining that we are doing something can bring about brain activation similar to what happens when we are actually doing it!

"It is becoming increasingly recognized that the brain plays a central integrative role in all behavior," according to Schwartz. "A person learns to experience because his brain learns to experience. His thoughts and feelings, memories and wishes, skills and plans, involve the restructuring of his brain and its regulation of the bodily organs."

Since the brain plays the central integrative role in human performance, then, according to psychobiologists like Gary Schwartz, all mental activity results in brain changes, although many of them may not be detectable by using present-day technology. Applying this to our psychotherapy patient, the images stimulated via therapy may be creating changes in the patterned interaction of brain cells. Words and images generated by the therapy process may be having "physical effects." "From this perspective verbal therapy is neurotherapy, even though we have no direct conscious awareness that this is actually the case," says Schwartz. "We might paraphrase McLuhan's statement 'The medium is the message' to read, 'The message is the image.'"

What sounds at first like a science-fiction fantasy seems logical to me, at least on the basis of what we have learned so far about the interrelation of emotions and bodily processes. We already know, from the studies on facial muscle patterns, that imagery and emotions can affect tension in the facial muscles. In addition, there are hints that the process may be working both ways. "Put on a happy face" may be as much a matter of good psychobiology as it is good manners.

From here it is only a short step toward speculating that imagery and emotions reflect brain patterns as well. Sustained depression, for instance, even when brought on by overwhelming stresses (prisoners of war), usually doesn't disappear immediately when these stresses are ended (liberation). In addition, neurochemical and endocrine changes can be detected in severe depressions which eventually become automatic: Despite the patient's most agonizing efforts, he cannot break out of the melancholy that binds him. At some point, even psychotherapy becomes useless, and most psychiatrists will then turn to medicine or electroshock treatment to reinstitute neurochemical balance. Here is what I believe is a sensible way to picture what is going on in these cases:

First, depression may be brought on by a stressful life event. Next, the imagery and emotions evoked by this experience elicit altered brain patterns, which at the present state of technological development we detect as neurochemical changes. Initially the process is reversible by psychotherapy or other techniques involving the substitution of positive images and emotions. Once the process reaches a certain point, however, it becomes irreversible by most psychotherapeutic methods and demands treatments that more directly restore the neurochemical balance. In other words, the psychotherapist may initially be able to alter brain patterning via a feedback loop by substituting more healthy images and emotions. Later, the unhealthy pattern becomes autonomous and disregulated (mental illness), requiring neurochemicals.

But the implications of biofeedback are not limited strictly to mental illness. A recent experiment in the Department of Neurology at Bellevue Hospital in New York City involves the application of biofeedback principles to the treatment of stroke victims. Victims of strokes, or cerebral vascular accidents, represent some of the most hopeless cases doctors are called upon to treat. Treatment, in fact, is usually limited to forms of physical therapy aimed at helping the patients regain some limited but nonetheless useful functioning. Very few stroke victims, however, ever return to their usual occupation. In addition, their productive life is usually at an end, their family relationships severely strained, and their total lifespan significantly shortened. In response to these considerations, the attitude of most physicians to stroke victims has been, to put it mildly, deeply discouraging. Based on traditional but incorrect views of brain functioning, this pessimism seems almost justified. After all, if those parts of the brain responsible, say, for moving an arm, are destroyed and can't be restored, then what can be done short of a miracle or a brain transplant?

Over the past two years, physicians at Bellevue have been hooking up stroke victims to biofeedback machines in order to discover for themselves the degree of possible recovery. So far, 75 per cent of the patients have achieved restoration in the control of their paralyzed limbs. In one case, a seventeen-year-old girl who had lacked right-hand control since early childhood achieved recovery to the point of being able to ride a horse while controlling the reins with her right hand. Even more exciting, two thirds of these recoveries have retained their improvement outside the biofeedback laboratory situation.

"These researchers are applying biofeedback techniques that defy localization according to traditional models," says Dr. E. Roy John. "The patient is in a biofeedback teaching machine situation where his brain is being reorganized in such a way that the patient deparalyzes. By moving the good arm the patient observes a pattern on the screen. Later, by concentrating on that pattern, 'thinking thoughts' about that pattern, he can eventually learn to transfer the process to his paralyzed limb. The result is incredibly dramatic: The paralyzed limb is able to move again!"

Biofeedback represents the practical application of the holistic model of brain functioning that we have been discussing throughout this book. Instead of relying on dogmatic assertions that one part of the brain is connected to another like tramtrack terminals, biofeedback researchers approach the brain with a refreshing simplicity and humility. We don't know how the process works, they seem to be saying, but the fact that it does work—that paralyzed arms and legs can be made to move again—proves conclusively that what we have previously believed about the brain was wrong.

Already the search for new brain models has uncovered paradoxes that could only be discovered through experiment rather than reflection. If a monkey's or a cat's visual center is destroyed, for instance, the animal becomes blind in the opposite visual field. If the animal's superior colliculus (which also receives visual input but only at a brainstem level) is then surgically destroyed, the animal can see once again. Would a similar result occur in a human? Could sight be restored in cases of visual cortex damage by surgically destroying the superior colliculus? So far, no one has performed such an operation. "We have now discovered over thirteen brain regions that are somehow involved in the mapping of vision," says Dr. E. Roy John. "Vision, thus, is not mediated entirely by the visual cortex. You don't 'see' with just your visual cortex.

The visual cortex probably functions as a vision intensifier or resolution enhancer. But you don't 'see' with it. But try suggesting a collicular oblation operation to a neurosurgeon in the case of visual cortex destruction. He won't believe it's possible because he's been taught in medical school that we 'see' with our visual cortex. Whole attitudes, whole bodies of knowledge, need to be changed."

Similar alterations in attitude may be required before biofeedback can be applied outside the laboratory. A recent experimental project at Yale University is a case in point.

The experiment involved a group of professional actors who were divided into two groups. Those in the first group were instructed to work themselves up into an angry state through imagining frustrating and disagreeable situations. The second group concentrated on thinking thoughts that helped them to remain calm, stable, and at peace with themselves. After a few practice sessions with mental imagery, the members of both groups were hooked up to monitors measuring heart rate, blood pressure, and respiration. At this point, the most crucial part of the experiment began.

Over the next several hours, the subjects were put through moderate exercise maneuvers, such as repetitively climbing a flight of stairs. All the while, measurements of the body's response were correlated with the mental imagery of the two subject groups. At the basis of the experiment was the hunch that exercise—despite all the claims that are made regarding its beneficial effect—is only one factor in determining benefit or harm. The context in which exercise is carried out should, if mental imagery has any importance at all, be equally relevant. The experimental results confirm this hypothesis.

In the "angry group," heart rate and blood pressure were sustained at undesirable levels. In the "placid group," the exercise was truly beneficial, since heart rate and blood pressure remained well within normal bounds. Despite the fact that both groups were doing the same thing, only one group received benefit; in fact, the group that generated the energy and feelings of anger were actually harming themselves!

The Yale project in mental imagery and exercise has profound implications, I believe, for national health. Caught up as we all are in an "exercise revolution," many of us may be doing ourselves more harm than good. If exercise is carefully planned and sensibly carried out, it's probably beneficial. Often, however, exercise occurs in a setting in which disregulation can almost be guaranteed. For example, jogging sessions are becoming routine in many of the nation's business firms. After spending a stressful morning competing in the office, workers can now stressfully compete with each other during the lunch hour to establish who can run the fastest or the farthest. Naturally, in such a setting, the mental attitudes that exist in the office are almost certain to be transferred to the exercise hour. The Yale studies on mental imagery and exercise point out the danger such exercise may hold for certain personalities.

"It has been observed that the coronary-prone individual—the Type A personality—tends to rarely report fatigue when he is tested on a treadmill in a doctor's office," says Schwartz. "Naturally, such a person is equally likely, when out on a track, to ignore feelings of stress or push himself harder when his body is telling him to 'slow down.' Exercise with some people, when improperly carried out, is disregulating. The person may be courting disaster by pushing his cardiovascular system beyond its limits."

The Yale study suggests that equal emphasis should be placed on the reason the person is exercising, the ongoing mental events that accompany exercise, and the person's response to his body's fatigue signals. The runner who ignores signs of fatigue and keeps "pushing himself" may be doing the same kind of harm as the abuser of food and alcohol. "It wasn't for nothing that the Greeks stressed moderation in all things," says Schwartz. "Even something like exercise can be harmful if it leads to disregulation."

At this point, it is much too early to know just how effective biofeedback procedures may be in curing the illnesses that plague us or in beneficially modifying our lifestyles. Despite this, there is no doubt that biofeedback represents a major advance, since it places health care precisely where it belongs: in our own hands or, more precisely, in our own brains! Each of us possesses a finely tuned psychobiological self-regulating system which, if we could only learn how to use it, may increase our capacity to control our own destinies.

15
A Treat Instead
of a Treatment

I have ceased to quarrel with physicians; their foolish
remedies have killed me, but their presumption and hyp-
ocritical pedantry are work of our making; if we were
not so afraid of pain they would tell fewer lies.

MARGUERITE YOURCENAR,
Memoirs of Hadrian

Perhaps someday a historian will write a history of the
world based on the concept that nations, like anxious
patients, will do anything to avoid mental or physical
anguish. Pain, and our attempts to avoid it, are, it seems,
probably the most potent motivating factors in much of
our human behavior. Not a month goes by without a major
pharmaceutical house announcing a new "painkiller" guaranteed to galvanize patient and doctor alike with the prospect of eliminating pain once and forever through the
blandishments of a chemical.

A "war" on pain, however, is not quite the same thing
as "war" on cancer or heart disease. The latter are undesirable under all circumstances, insults to our bodily
integrity, and therefore "enemies." Pain, on the other
hand, is often an intimate friend who warns and protects
us. By young adulthood most of us would possess gnarled
stumps instead of hands and fingers were it not for pain's
warning signals, which are transmitted along the naked
nerve endings originating in our fingertips. In addition,
pain can protect us from the perils of our inner world. The

vague tinge of chest discomfort first experienced while working outdoors on a cold winter morning may forewarn the potential heart attack victim to stop working and seek medical attention. Because of its usefulness, therefore, we wouldn't want to eliminate pain altogether, only control it. But to do this we must first understand pain, how it works, how we can use it, and, most importantly, how we can turn it off after it has served its purpose of forewarning us.

During World War II, several physicians made independent observations about pain perception in wounded soldiers. At Anzio, a nineteen-year-old youth whose arm was completely shattered sat calmly talking to the field surgeons who were tending the wound. The soldier didn't request a narcotic and, in fact, seemed to experience no pain, despite the seriousness of his injury. Other wounded soldiers demonstrated a similar "fortitude" in the face of broken bones and brutal internal injuries. In an attempt to explain this puzzling behavior, the doctors speculated that it might be due to the relief the soldiers experienced at being alive and finally out of the battle. Despite reservations about the correctness of this explanation, it was, at that time and in the absence of an alternative theory, the best way scientists could "explain" the unexplainable.

Over the next several years an alternative theory slowly began to surface regarding the soldiers' strange response to injury. Suppose that painkilling drugs, primarily morphine, adhered to special receptor cells in the brain. Further, imagine that under conditions of extreme stress— as a penetrating and unusually agonizing wound—the receptors are filled with a natural narcotic produced somewhere in the body. If this were true, it means, in essence, that the body could produce its own narcotics! There are several lines of research that support belief in such specialized brain-opiate receptors.

For one thing, some opiates are extraordinarily powerful. The little known opiate etorphin is the most potent mind-altering drug in the world, with ten thousand times the painkilling properties of morphine. A tiny dose of etorphin (less than the smallest potent dose of LSD), on the end of a dart, can deliver an instantaneous knockout blow to a rampaging elephant. The capacity to achieve so powerful an effect with such a miniscule dose of the drug, argues against any theory that holds that the chemical "overwhelms" the entire mass of the brain. It suggests, rather, an exact molecular "fit" between the opiate and special opiate receptors in strategically located cells. In essence, the opiate seems to conform closely to a "lock and key" concept of how neurochemicals work on the surface of nerve-cell membranes.

Further proof of this came from studying what happened when neurochemists began tampering with morphine's molecular structure. Altering only a small part of the molecule leads to chemicals with different and, in some cases, antagonistic actions. One drug, naloxone, completely reverses the effect of morphine, presumably by displacing morphine from its receptors on the nerve-cell membrane. Naloxone is used in emergency rooms throughout the world to bring addicts out of dangerous morphine and heroin comas. Naloxone acts so speedily that it often begins to take effect before the syringe is out of the addict's arm. This suggests that the drug has greater affinity for the patient's opiate receptors than does morphine or heroin itself. Naloxone thus appears to displace the opiate, restoring consciousness, quickening respiration, and instantly bringing on a state of morphine withdrawal. It does so by occupying the receptor site and thus effectively keeping the morphine out without triggering the chemical action that brings on a "high." (Unfortunately, naloxone's effects are too short-lived for it to serve as an ideal painkiller in itself.)

In order to locate the narcotic receptors within the brain, psychobiologists turned to a method known as radioactive tagging. Morphine or naloxone, when combined with a radioactive substance, can be traced as it flows throughout the body. Several hours after test injection, an experimental animal is killed (or "sacrificed," as researchers euphemistically refer to the process) and the brain is removed for placement into a radioactive counter. Those brain areas which have absorbed the radioactively tagged naloxone will exhibit the highest counts and correspond roughly to the location of the morphine receptors. Despite the elegance of the technique, the reasoning is really quite simple. If $500,000 worth of heroin were somehow marked for later identification and dropped off in the central business district of any of our major cities, the drug would eventually wind up on specific narcotic receptors: the city's heroin addicts. So, too, radioactively tagged morphine will eventually be absorbed into the brain's specialized narcotic receptors. Unfortunately, despite the essential correctness of this approach, its practical application turned out to be much more complicated.

Opiates, like most chemicals, will bind, to a certain extent, on most cell membranes in the body. These nonspecific receptors, like double agents, will camouflage the location of the true receptors. To get around this, it was first necessary to devise a way to virtually guarantee that only the specific narcotic receptors were absorbing the radioactively tagged morphine or naloxone.

Candace B. Pert and Solomon H. Snyder, of the Johns Hopkins University Medical School, first achieved the goal of separating the true receptors from the counterfeit receptors by hopping up the radioactivity per molecule of naloxone. In this way, only very small concentrations of radioactively tagged drug needed to be injected. After washing away the weak, nonspecific binding sites, naloxone adhered only in those cells specifically organized to receive narcotics. This breakthrough quickly led to two key discoveries.

First, the morphine receptors are located in two specific areas of the brain: the limbic system, important in emotional reactions, and the medial thalamus, the way station for the transmission of pain impulses. In this way, both the brain's reception of pain and its reaction to it might be explained. Pain could conceivably be effectively cut off by opiate binding in the medial thalamus and, as with the soldier at Anzio, emotional reactions altered at the level of the amygdala of the limbic system, which contains the highest density of the brain's opiate receptors.

At this point, psychobiologists were faced with an existential dilemma: Why would the brain contain receptors designed to accommodate morphine, the product of the juice of a poppy grown in Mesopotamia? The alternative, however, seemed, at the time, no less preposterous: The receptors existed to interact with a naturally occurring opiate-like substance. But how could such a chemical, if it existed, be isolated?

At about this time, two researchers at the University of Aberdeen, Scotland, refined their experimental techniques by taking advantage of the fact that opiates inhibit the contractions of certain smooth muscles in the body of most organisms. (The antidiarrheal action of the narcotic paragoric depends on this property of the opiates.) They found out that the guinea pig's intestine and the mouse's vas deferens (a part of the male reproductive system which transports sperm from the testes to the penis) are inhibited by opiates in proportion to their strength. As a further check that the inhibitory response was indeed mediated by opiates, the two researchers routinely administered the opiate antagonist naloxone. If the guinea pig's intestinal inhibition was reversed by naloxone, it virtually guaranteed that the action was mediated by an opiate.

In December 1975, John Hughes and Hans Kosterlitz, two Scottish neurochemists, published the molecular formula for two morphine-like materials extracted from pig brains. The substances, known as enkephalin, consist of a string of five amino acids differing only in the final component (leucine and methionine). They were therefore referred to as leucine enkephalin and methionine enkephalin.

At about the same time, another research team, at Stanford University, discovered similar opiate activity in an extract from the rat's pituitary glands. Ten years earlier, a University of California researcher had isolated a 91 amino acid compound known as beta-lipoprotein. A reexamination of this molecule—with the intent of comparing it with the recently discovered enkephalins—turned up a pleasant and intriguing surprise. The five amino acid sequence of methionine enkephalin is contained within the beta-lipoprotein molecule. This suggests that beta-lipoprotein, although possessing no morphine-like activity of its own, would probably act like an opiate if methionine enkephalin could somehow be liberated from its position deep within the molecule (between the sixtieth and sixty-first amino acid).

By cleaving beta-lipoprotein at the 60–61 position, a potent natural morphine analogue, beta-endorphin, was discovered. Encouraged, psychobiologists next attempted to split beta-lipoprotein at other points along its 91-amino-acid sequence. None of the resulting compounds, however, were discovered to equal beta-endorphin in morphine-like activity.

If a rat's pituitary is removed prior to measuring its brain enkephalin levels, no change is found in the opiate-like content of the brain. Thus, the enkephalins discovered in the brain and the endorphins residing in the pituitary are completely independent, produced by separate means, and presumably serve different functions.

In mid-1976, psychobiologists could summarize their position something like this: Within the pituitary gland there is a lengthy amino-acid sequence, beta-lipoprotein, which, after suitable transformation within the body, leads to the release of a morphine-like substance, the endorphins. Independently, the enkephalins are produced somewhere in the brain and, in fact, cannot exist anywhere else. (When the enkephalins are injected into the blood stream they are instantaneously degraded into small, inactive molecules.) This raised the possibility that perhaps the endorphins are packaged in such a way that after release from the pituitary they can survive within the blood stream until reaching their target sites.

Further biochemical investigation in mid-1977 revealed that beta-lipoprotein is itself only part of an even larger molecule, pro-opiocortin, which contains, in addition, the hormone ACTH.

At this point, a theory began to develop. ACTH is a stress-related hormone which is released whenever a person experiences severe pain. It would make a good deal of sense to couple ACTH's release with the release of an endorphin. In this way, the two hormones would work together to exert a protective function against excessively painful impulses. Dr. Sidney Udenfriend, the discoverer of pro-opiocortin, describes the response to stress via a military analogy. "Like a MIRV guided missile, with multiple, independently targetable warheads, the precursor molecule breaks up into smaller pieces—ACTH, endorphins, and enkephalin—each honing in on its own target tissues."

In an attempt to confirm Dr. Udenfriend's "hormonal MIRV hypothesis," Dr. Dorothy T. Krieger, of the Mount Sinai Medical Center in New York, measured ACTH, beta-lipoprotein, and beta-endorphin in the blood of patients subjected to the stress of insulin-induced blood sugar depression. In all cases, ACTH and the endorphin levels were elevated in response to the stress. Other researchers found elevations of ACTH and beta-endorphin levels following sudden, deliberately induced bone fractures in young male rats. "Plasma and pituitary concentrations of ACTH and beta-endorphins vary concomitantly and in remarkable parallelism with these experimental situations," according to Dr. Roger Guillemin, a 1977 Nobel Prize winner in medicine and physiology.

Excited by these research findings, psychobiologists began developing an "endorphin theory" of pain. Pain—we are now talking of the gnawing, prolonged pain seen in chronic illness—travels up the spinal cord into the brainstem and terminates within the medial thalamus. Opiate receptors exist all along this pathway and can be identified by radioactive tagging. Their mechanism of action, although still controversial, seems to depend on bringing about a change in the conductance of the nerve-cell membrane to sodium. Because the ultimate effects are brought about by a combination of special receptor cells and a circulating neural hormone, the endorphins are, in most psychobiologists' opinion, deservedly considered neurotransmitters in their own right. "Endorphin peptides are a whole new nervous system in brain and gut and elsewhere in the process of discovery," exclaims Roche Institute's Sidney Udenfriend.

Throughout 1977, psychobiologists went about refining their understanding of the body's pain-perceptual apparatus and how it can be affected by the endorphins. Within the periaquaductal gray, a deep-seated brainstem area lying along the floor of the third ventricle, neurosurgeons at the University of California in San Francisco placed indwelling stimulating electrodes for pain relief in

341

six patients afflicted with chronic, unremitting pain. Whenever the patients began to experience pain, they were able to shut it off via the activation of a battery-operated stimulator about the size of a pack of cigarettes. After activating the stimulator, all six patients—in accordance with earlier findings in other pain patients—experienced dramatic, long-lasting, and repeatable pain relief. In order to test the hypothesis that the pain relief was genuine and not just an example of a "placebo response," one patient was outfitted with a stimulator containing a "dead" battery. The patient, a fifty-one-year-old woman with severe back and leg pain caused by cancer of the colon, anxiously reported that her pain had returned and the stimulator "wasn't working." Replacement of a new battery led to immediate pain relief.

At this point, the surgeons intravenously injected the narcotic antagonist naloxone. Five out of six patients experienced an immediate return of their pain. (The sixth patient developed a disturbing side effect that dissuaded the investigators from increasing or repeating her dose of naloxone.) In each case the result was clear: Electrical stimulation gave rise to pain relief which could be reversed by injecting a narcotic antagonist naloxone. If the naloxone was given without prior electrical stimulation, no effect was produced, suggesting that electrical stimulation ordinarily releases a natural narcotic (leading to pain relief) which can later be displaced from the opiate receptors by naloxone (resulting in the return of pain). As a final proof, researchers at Stanford showed that stimulating the periaquaductal gray of rat brains increased endorphin release from the tissue. From here, other researchers turned their attention to a possible endorphin effect operating in acupuncture.

Western medical practice has long recognized the beneficial effects that can be brought about by brief, intense stimulation of the body's trigger points. A patient with arthritis or severe muscular spasm often obtains relief by the use of dry needles, intense cold, even the injection of salt water into the painful areas. More lasting relief comes from the insertion of acupuncture needles, sometimes accompanied by the passage of weak electric currents passing through these needles. Control studies reveal that procedures involving intense stimulation of acupuncture points result in pain relief that can be reliably differentiated from a placebo response.

For years, psychobiologists explained acupuncture's success on the basis of a "gate theory" of pain transmission. Basically, the theory proposes neural mechanisms in the spinal cord which act as a "gate" that can increase or decrease the flow of nerve impulses coming from peripheral nerve fibers. In this way, the painful stimuli can be modulated before they ascend to higher brain areas. In addition, the gate is profoundly influenced by nerve fibers descending from the cerebral cortex (the basis, some people believe, for the pain relief effected by hypnosis). The theory further proposes that pain impulses from lower portions of the spinal cord are modulated at successive points along the way to the thalamus, where pain is first experienced. In this model, pain occurs whenever the level of arriving nerve impulses reaches a certain critical point.

Despite the neatness of the gate theory as an explanatory model, many psychobiologists found it unsatisfying. For one thing, what is the mechanism by which pain is "shut off" before it can rise to higher levels in the nervous system? The odds favor a neurotransmitter, but which one?

To find out, Dr. Bruce H. Pomeranz, a neurophysiologist at the University of Toronto, Canada, applied the acupuncture technique to anesthetized cats. After measuring the electrical potential in some of the cats' individual spinal neurons, Pomeranz acupunctured the cats with low-voltage electrical stimulation. After twenty minutes, the pain impulses were blocked. Injection of the opiate antagonist naloxone abruptly reversed the acupuncture-induced blockage of the pain impulses. Pomeranz hypothesizes that acupuncture was stimulating the pituitary to release some of its endorphins. To prove that the acupuncture effect was specific, Pomeranz performed a "sham" procedure with the needles inserted in locations other than the classical acupuncture sites. No change was observed in the neuronal firing of the animal's isolated spinal cord neurons. In a similar study carried out on humans at the University of Washington, pain perception was lowered by 30 per cent. This effect, too, was completely abolished by the opiate antagonist.

What emerges from these independent but interrelated lines of investigation is a natural pain-relieving system within the body, principally the brain and the pituitary. In response to painful stimuli, the brain produces and releases its own narcotics, which travel along points of the pain path that start in the spinal cord and terminate in the thalamus. Pain can be effectively blocked by naturally occurring endorphins (literally, internal morphines) or by man-made and self-administered opiates. This scheme provides, among other things, a possible model for drug addiction.

According to the feedback principles discussed in Chapter Fourteen, an excess of a neurochemical leads to a message to the receptor cell: "We have enough endorphins, turn off the endorphin machine." It is now speculated that drugs such as heroin may exert a similar effect. First, heroin occupies the natural endorphin receptor, less natural endorphin is produced, and, finally, the endorphin production may be shut down entirely. This may explain the extraordinary difficulty addicts experience when attempting to wean themselves from the drug. At such times they are literally suffering from an endorphin deficiency.

The body demonstrates many similar interactions between an administered drug and the shutdown of the production system for a hormone. Thyroid extract, if given over a sufficiently long period, shuts down the thyroid gland by signaling the hypothalamus to stop producing thyroid-stimulating hormone (TSH), which controls thyroid hormone production. Eventually, thyroid hormone production may cease altogether, rendering the patient completely dependent on externally administered thyroid pills—in essence, a "thyroid addict."

There is much to recommend a feedback theory of addiction. Imagine everyone in the world with genetically determined levels of many different endorphins—some of those endorphins known, others still to be discovered. Assume, in addition, that although the endorphins may vary from one person to another, they remain fairly constant from day to day in any particular person. Now introduce a daily intake of heroin or some other opiate. In response, the body's endorphin machinery shuts down as the opiate receptors become fully occupied with the injected drug. Eventually, the body's ability to produce its own "narcotic" (the endorphins) becomes seriously hindered and, if the addiction continues long enough, perhaps even permanently altered.

Research at Johns Hopkins University and the National Institute of Mental Health is currently aimed at developing better methods for detecting changes in the endorphin levels of drug addicts. If changes can be detected, they will support earlier theories which equated addiction with metabolic diseases like diabetes. But first, several technical problems await solution.

Studies so far indicate that endorphin levels in human spinal fluid (which provides the most easily obtainable measure of brain endorphin levels) are very low, requiring extremely sensitive tests for their detection. "The tests currently available are simply not good enough to measure the subtle differences that may exist between, say, a drug addict and the rest of us," says Dr. Solomon Snyder, Professor of Pharmacology and Psychiatry at the Johns Hopkins University School of Medicine in Baltimore and one of the earlier researchers in the endorphin field.

In anticipation of more sensitive tests in the near future, several laboratories are now storing human spinal fluid from a wide variety of patients. In addition to addicts, the list includes patients with intractable pain as well as patients who cannot experience pain.

Although most of the research has been carried out with acute pain problems, there are also major aberrations in pain perception that are permanent instead of temporary. A famous example is F.C., the daughter of a Canadian physician, who lived a lifetime with a "congenital insensitivity to pain"—an extreme physiological liability that hastened her early death from massive infections at age twenty-nine. By the age of two, F.C. had chewed the tip of her tongue into a pulp. (All children might do that if it didn't hurt when they accidentally bit their tongue.) At three, she knelt on a radiator hot enough to make any normal person jump off instantly, and calmly watched children playing outside until she suffered third-degree burns—the first of many similar accidents. In her teens, she was able to mystify her friends by voluntarily dislocating both of her shoulders and then casually replacing them without assistance. For years, F.C.'s strange affliction remained an enigma, since, at autopsy, no explanation was found. Now scientists are speculating that F.C.'s insensitivity to pain may have been due to an inborn endorphin abnormality. Do such people have abnormally high or low levels of endorphins?

A recent study indicates that sufferers of the severe facial pain of trigeminal neuralgia (tic douloureux) have lower than normal endorphin levels. Is there something wrong with the neurochemistry of the pain-perception pathways in these cases? From studies on brain endorphin level in the blood, scientists may soon have answers to these questions.

Meanwhile, other researchers are studying the role endorphins may play in our emotions. Since the greatest concentration of the opiate receptors is found in the limbic system, could an endorphin abnormality be involved in certain forms of mental illness?

"Perhaps in parts of the brain such as the amygdala, the normal release of endorphins acts as the body's own 'tonic' against disappointments and losses," Snyder suggests. Could this "tonic" be deficient in cases of depression? So far there are no studies correlating the endorphins with depression, but there are indications of a relationship between the endorphins and schizophrenia.

Between June and December of 1977, Dr. Nathan Kline, one of the founders of modern psychopharmacology (drug treatment of mental illness), carried out a highly publicized trial using beta-endorphin on fourteen mental patients.

At an International Conference on Endorphins and Mental Health Research held in San Juan, Puerto Rico, in January, 1978, Kline described his results: "Pronounced and sustained therapeutic effects" in patients ranging from schizophrenics to agoraphobics.

346

"We stopped at nine milligrams, the first level of therapeutic response," says Kline. "For all we know, if we had given a schizophrenic 100 or 200 milligrams, he would turn around and say, 'Where the hell have I been for the past fifteen years?' "

Other researchers are less enthusiastic. Although Kline's innovations in the past (introduction of the tranquilizer reserpine and the antidepressant monoamine oxidase inhibitors) have earned the respect of colleagues around the world, many investigators are now charging that the endorphin injections were premature and unconvincing. They point to earlier work from Sweden in which two investigators reported an end to hallucinations in schizophrenics given doses of naloxone. This work was based on finding increased endorphin-like substances in the spinal fluid of four schizophrenics and three manic-depressives. When the patients improved after drug treatment, the endorphin-like chemicals disappeared. From here, the scientists speculated that the patients' mental illnesses might be due to an excess secretion and binding of the "brain's own narcotic." If true, this hypothesis immediately suggests a treatment: Inject an opiate antagonist, naloxone, to preferentially occupy the endorphin receptor sites.

When Drs. Lars Terenius and Agneta Wahlstrom, of the University of Uppsala, Sweden, intravenously injected naloxone into six schizophrenics, four of them abruptly stopped hallucinating. This aroused tremendous hope in the two neuroscientists that they had discovered the biological basis for psychosis.

A later controlled trial at the National Institute of Mental Health failed to substantiate the claim. This now leads some to suggest that Kline's "endorphin treatments" for mental illness will also eventually turn out to require similar disclaimers.

The basis for this skepticism is lack of an adequate control group. So far, the endorphins have not been pitted against what is probably the most powerful drug in the history of medicine: the placebo. Comparing the present situation to thousands of anecdotal observations about fly-

347

ing saucers, Dr. Avram Goldstein, of the Addiction Research Foundation in Palo Alto, told Kline, "We don't need tens of thousands of descriptions of encounters with UFOs. What we need is *one* description in which it is clear that there *was* an encounter."

Throughout all this, psychobiologists remain enthusiastic that the endorphins may turn out to be the most important brain chemical to come along since the discovery of the catecholamines.

Will the endorphins be addictive? At first, brain scientists were hopeful that the endorphins, unlike heroin and morphine, would provide pain relief without the double hazard of tolerance and physical dependence. Tolerance refers to the increasing doses that, after chronic administration, can be "tolerated" without experiencing side effects. Physical dependence refers to the occurrence of withdrawal symptoms after the termination of the chronically administered drug. Unfortunately, the endorphins produce both tolerance and physical dependence in a typical opiate manner. So far, the search continues unsuccessfully for a painkiller that is not also addictive.

One of the difficulties in the development of a nonaddictive painkiller comes from the euphoriant effects common to most opiates. People at first take opiates to relieve pain, only to find they also feel "good" at the same time. Soon it's only too easy to return in search of what Nathan Kline humorously refers to as "a treat instead of a treatment."

The euphoriant effects of opiates—and, by analogy, all the endorphins so far discovered—can be traced at least as far back as the Greeks. In the *Odyssey* Homer describes the pathetic plight of Telemachus, who ingested a drug given to him by Helen to compensate for the loss of his missing father. Subsequent history contains all too many other examples where opiates have been used as an escape from conditions or experiences that are meaningless or intolerable. "Opium dens" have given way in recent times to "shooting galleries" where the lonely, the grieved, and the depressed inject heroin in a treacherous search for a chemical that can bestow peace of mind.

348

A nonaddicting opiate would stop much of this, in which case physicians could dispense painkillers freely without worrying about creating an illness in the sufferer more horrible than the pain for which he is seeking relief. One promising approach to a nonaddictive painkiller might result from trying to develop synthetic opiates which bind preferentially to the pain pathways and have little attraction to the opiate receptors in the limbic system. So far, however, no one has demonstrated that such a feat of neurochemical juggling is even possible.

But if the endorphins live up to their promise, even partially, it's hard to calculate their overall effects. They could relieve the suffering of countless millions—and relieve it *safely,* without doing further harm. They could save millions more from addiction—and their families and society at large from the crushing consequences. They could be useful in a variety of mental illnesses. They could considerably diminish the crime rate in the major cities of the world by relieving addicts of the need to steal in order to feed their drug habits. They could have a substantial effect on our relations with many nations—Turkey, for example—with whom we have had considerable difficulty because of the international narcotics traffic. One could go on and on with quite feasible speculations of this nature. The overall effect of the endorphins on our society could be incalculable.

On a slightly more theoretical level, all the present theories of brain function were developed by psychobiologists who were unaware of the existence of the endorphins. As a result, brain scientists must now restructure their concepts about how the brain works in order to accommodate a major neurochemical system.

The endorphin research has all come about in a brief three years, with new data and new applications promised in the very near future. We should, of course, not let our expectations run too high too early. Several investigators suggest that we temper our enthusiasm with caution.

"So far, we have only looked at perhaps 40 per cent of the neurons in the human brain," says NIMH's Candace Pert. "I'd like to know a little more about the other 60 per cent before betting on the endorphins to provide the final answer to things as complex as drug addiction and human pain." She points out how much we still do not know about the brain. But even she admits that "what we don't know almost always turns out to be much more interesting—and more surprising—than anything we already know!"

16
The Dancing Bees

We tried to talk it out, but the words got in the way.

GEORGE BENSON,
"This Masquerade"
(popular song), 1977

This has been a most wonderful evening. Gertrude
has said things tonight it'll take her ten years to understand.

ALICE B. TOKLAS to Mortimer Adler

In 1918 an obscure Austrian biologist, Karl von Frisch,
elaborated on an observation first made by Aristotle more
than two thousand years earlier. If a dish of honey is set
out near a beehive, a solitary bee will eventually discover
it. The bee will then fill its crop with the sweetened solution and return to the hive. Within minutes there will be a
flurry of bees out of the hive headed right on target to
the dish.

As Aristotle thought of it, the original forager bee
somehow "recruited" the new bees. His explanation was
that the swarm of bees simply followed the original bee
back to the food.

In 1901 the mystery concerning bee communication
deepened even further. A German experimenter followed
a forager back to its hive and then later trapped it on the
way out. Even though the recruited bees were thus deprived of their leader, they nonetheless flew straight to the

food. Puzzled, the experimenter speculated that a "magnetic intuition" might be involved.

Intrigued with these observations, Von Frisch, in a series of ingenious and Nobel Prize-winning experiments, discovered that bees communicate by dancing! After a bee returns to the hive, it runs in circles on the face of the honeycomb, often reversing directions at one and one quarter intervals. Other bees in the hive attend to the dance by touching the dancer with their antennae. Moments later, the bees know the location of the food and leave the hive in search of it. With more distant food sources, the original bee performs a different and more complicated dance.

Follow-up studies by Von Frisch and others proved that a honeybee's ability to pass on information about location equals the performance of complex radar systems. But is the honeybee speaking a language? For fifty years scientists have debated this question and, in the process, have sharpened our ideas about language. Although clearly a communication device, the dances are not language but inherited ways of satisfying immediate needs. Nor are they learned, as are human languages. Different colonies of bees possess, from birth, different dialects. Austrian bees cannot understand the identical dance of Italian bees. In addition, there doesn't seem to be any way one bee can interpret the dance of another "nationality."

The earliest human communications, too, seem to be programed and largely inherited. Infants at the age of six months begin to produce an immense array of sounds. During this early babbling state, an infant is capable of forming most of the sounds that make up a language—including sounds used in languages the baby has never heard! Experiments have shown that the babbling of a Chinese baby cannot be distinguished from a Russian or American infant. But at about nine months the infant's performance changes. The babbling range narrows to include only those sounds characteristic of the language the child will eventually speak.

All languages are based on elementary sounds called phonemes. They may vary from fifteen to as many as eighty-five. English has about forty-five phonemes, which are roughly equivalent to the sounds we use to pronounce the vowels and consonants of the alphabet. The smallest meaningful units in a language are called morphemes. *Sh* and *th* are phonemes and without meaning. *Sit, it,* and *at* are morphemes.

From time to time, scientists have explored the question of animal communication by means of language. Despite extensive efforts, no one has ever produced a talking animal. Speech seems to be a uniquely human trait based on the construction of the human pharynx, or voice box. But what of nonverbal language?

Debate is still raging over the language competence of two chimpanzees, Washoe and Lucy, who were taught to communicate by means of the American Sign Language of the Deaf. Their trainers, Beatrice and Allen Gardner of the University of Nevada, claim that the chimpanzees use language creatively. The chimps can combine a number of signs and sequences that are similar to the sentences of a two-year-old. "Hurry, give me toothbrush," "You drink," "Listen, dog" (at the sound of barking). In addition, the chimps have come up with a few new signs: A watermelon is a "drink-fruit" or "candy-drink"; the refrigerator is signaled as an "open-food-drink," an amusing combination of three gestures in place of the previously learned gesture for "refrigerator."

Critics point out, and I think quite correctly, that such performances are impressive but as irrelevant to human speech as the dance of the honeybees. Language development in human beings depends on culture and the level of brain development. As we saw in Chapter Six, reality does not exist for us independent of the way we perceive it. So, too, language develops according to our perception of the world around us.

B. L. Whorf, a student of American Indian languages, translated ideas from one Indian dialect into another. In many cases, he was incapable of carrying out the translations. One Indian language made no distinction between nouns and verbs; another had no term to distinguish events in the past from those in the future or the present; still another used the same word for different colors. From this experience, Whorf formulated a controversial thesis: Thought is relative to the language in which it is expressed. The Whorfian Linguistic-Relativity Hypothesis concludes that:

1) the world is conceived very differently by those whose languages are different;
2) the structure of the language is the cause of these different ways of conceiving the world.

Some have suggested that Whorf has things completely turned around: Our experience determines our language. In either case, language and experience are intricately interwoven. Different experiences—the world of the chimpanzee and the world of the human—create different communication requirements. Although it may be important for us to demonstrate, for example, that chimpanzees can be taught the rudiments of gestural language, it is apparently far less important to the experimental subjects themselves: So far, Lucy and Washoe have not transmitted their newly learned language skills to either their offspring or to other chimps.

If language were somehow helpful to the chimps, wouldn't it make sense for the chimps to teach it to each other? In addition, the absence of language in the chimps raises another question: If chimps are capable of language and didn't develop it until the mid-twentieth-century efforts of human trainers, why did language develop in our ancestors thousands of years earlier?

The most reasonable explanation for this failure involves the intimate relationship that exists between language and thought. Certainly it is possible to use language without thought. A visit to a school for the mentally retarded will provide a host of poignant examples. On the whole, though, language evolved for the communication of information that could not be conveyed by gesture. Some students of language, in fact, believe in the gestural origin of human speech and can point to psychobiological experiments that provide some justification for this.

If a short-acting anesthetic is injected into the left carotid artery, which leads to the speech center in the left hemisphere, speech is temporarily arrested. Along with this speech arrest, the subject will be unable to carry out a series of previously memorized hand gestures. This doesn't seem to involve actual paralysis, since the subject can move his arms on command. It is only the gestures that are affected, and they return at the same time as speech.

But the relationship of gesture to language is probably less important than language's relationship to the principles of thought. Psycholinguistics, a recent hybrid of psychology and language structure, aims at discovering whether we think the way we do because of our language or, alternatively, whether our language is an essential by-product of the workings of the human brain.

The man who contributed most to our emerging understanding of language and thought is Noam Chomsky, a fifty-year-old Professor of Linguistics at the Massachusetts Institute of Technology. In the fall of 1977 I met Chomsky and discussed language and the brain.

It's an overcast fall morning and Noam Chomsky is seated at his desk, or rather curled into a chair in front of it, with his feet propped on the writing board and his hands busily engaged in lighting a pipe. At fifty, Chomsky is nothing less than a living legend; a man who almost singlehandedly founded the science of psycholinguistics, a hybrid of psychology and traditional linguistics that, according to Chomsky, is concerned most of all with the mental process underlying how we acquire knowledge.

Chomsky settles back in his M.I.T. office and begins to speak. "Personally I'm primarily intrigued by the possibility of learning something from the study of language that will bring to light inherent properties of the human mind," says Chomsky. "Viewed this way, linguistics is simply a part of human psychology."

In recent years Chomsky has engaged in a lively and sometimes heated debate on the nature of psychology and what he feels it should be telling us about the capacity of the human brain and mind. Usually a pleasant, easygoing man, Chomsky can shift to icy formality or even unrestrained annoyance whenever he discusses what he considers the shortcomings of contemporary psychology. "There is no hope in the study of the control of human behavior by stimulus conditions, schedules of reinforcement, establishment of habit structures, patterns of behavior, and so on. Of course, one can design a restricted environment in which such control and such patterns can be demonstrated, but there is no reason to suppose that any more is learned about the range of human potentiality by such methods than would be learned by observing human behavior in a prison or an army—or in many a schoolroom. The essential properties of the human mind will always escape such investigation."

Such remarks have a way of annoying many of Chomsky's critics, who point out that Chomsky is a linguist, not a psychologist. They claim, in fact—and Chomsky doesn't deny it—that the world's greatest psycholinguist never formally studied psychology. But Chomsky dismisses such allegations with the observation that language and psychology are so intertwined that a person who studies either discipline cannot help but become involved with the issues of the other.

"Let's consider for a moment some of the inferences we can make on how the brain must work," Chomsky resumes, punctuating his sentences with his left hand while cradling his pipe with the other.

"We all recognize certain physical limitations. We can't fly or live under water, for instance. Why, then, do we have difficulty accepting mental limitations? I think it's because part of our intellectual tradition holds that the mind is plastic and unstructured—hence, potentially infinite. This just isn't true. There are limitations on our minds that are biologically determined. As the brain matures, the mind eventually is capable of great complexity and achievement but is still limited in the kinds of things it can achieve.

"Take for instance, the idea of intelligibility, whether or not an explanation makes sense. We all recognize that certain explanations about events around us are intelligible, while others are nonsensical. How do we decide that? I think it's because we are endowed at birth, as part of our biological heritage, with a certain concept of what counts as an intelligible explanation. Rather than considering this a limitation, we should consider ourselves

356

lucky that we are endowed with this capacity. Imagine a world in which everyone had a different concept of what was intelligible—in essence a chaos.

"There is another side to this, however. The capacity to develop intelligible theories means, in principle, that there may be other explanations that we can't construct. It might turn out, for instance, that the explanation we are seeking to explain the brain may fall outside the domain of theories that we are able to understand. It is rather egocentric on our part to believe that our theory-forming capacity should happen by accident to be exactly right in order to understand the brain. The world was not made to our specifications, and it is entirely possible that psychobiology will never be able to explain the brain."

"Are these capacities developed through evolution?" I asked.

"Yes and no. Human evolution deals with real life, problems of survival, how to get along in the world, how to develop an intuitive understanding of other people. Usually this is thought to develop through natural selection. But does it really? The capacity to develop a correct theory, explaining human behavior or the workings of the human brain, for instance, was not specifically selected. In fact, there is not the slightest reason to think that there is evolutionary pressure to develop a mind capable of solving these kinds of problems. The same is true of hard sciences like physics. Nor was there evolutionary pressure to account for the development of something like calculus. In addition, natural selection is not directly relevant to the capacity to create or enjoy symphonies or literature, in the sense that such capacities are directly selected. I can't think of any conceivable relationship that such things might have on differential reproduction so that they would confer survival advantage in the Darwinian sense. Presumably, they are concomitants of other factors which may have been subject to selection.

"Natural selection as an explanation has been very helpful, but I wonder if we are not too enamored of it. Certainly we evoke other kinds of explanations when we are describing other physical phenomena. When we discuss the origin of the universe we talk about the Big Bang

357

Theory of creation, for instance. We speak of elementary particles and elements which eventually form crystals. We then go on to larger and larger molecules, eventually to DNA. Nobody would claim, for instance, that this creation was an evolutionary process: that every element was tried and rejected until hydrogen and helium were discovered to be the best elements. Rather, we recognize the existence of physical laws that account for the fact that, at a particular time and under certain conditions, molecules and crystals result. This has nothing to do with evolution, but rather the existence of certain physical laws."

According to Chomsky, the human brain may also possess certain capacities simply on the basis of its being a physical system of a particular order of complexity. "People ask, "How did humans get the power of speech?" —implying somehow that speech was selected by evolution. There's an alternative explanation possible. It may be, as evolutionary theory informs us, that properties of the human brain were selected for their evolutionary advantage. Along with this may have come other accompanying benefits. This is common in evolution, where adaptations for a particular trait may involve all kinds of other consequences. Certainly no biologist believes that *every* trait is selected for. It is possible that language is best explained in these kinds of terms: a carryover effect appears at a certain order of brain complexity. Or, to put it differently, at a certain level of complexity many of the human brain's most striking capacities may have to do with the laws of physics. They may relate to the density and packing of neurons in the brain, for example. Perhaps there is only one way of physically solving such a packaging problem, and that in turn may lead to certain consequences like speech or the ability to deal with numbers. It might turn out, in fact, that simply as a consequence of physical law, the brain will have language ability, just as a certain molecular organization will eventually result in a crystal. Such an explanation may be just as valid at the level of complex biological structures like the brain, as it is at the level of simple physical structures. There is no reason to assume that we have to switch from physical

358

laws to evolution when we pass from crystals to living organisms."

Professor Chomsky's speculations about the origin of human language may best be described as heretical. In essence, he is suggesting to us that we may never know how our brains developed the capacity for language. Others before Chomsky have suggested that the mind's attempt to understand itself is doomed to failure because it involves an inherent contradiction. There is a risk involved in the adoption of such a view. A believer may one day wind up in the same company as those who said, "Airplane flights are impossible," or "Men will never get to the moon." So often the limits of the possible turn out to result from failures of the imagination. The mind, like any biological system, has its scope and limits, and there is no reason to suppose beforehand that the study of the brain falls within these limits. It is simply an open question, according to Chomsky.

Behind every attempt to understand the brain lies an article of faith. Chomsky has gone even further and called it the modernist equivalent of a "religious belief." It concerns our access to the relevant brain processes that we must understand in order to explain how our brain works and, in the process, achieve an integrated self-understanding.

One of the unspoken articles of much traditional scientific faith assumes that we ultimately can become aware of—in a very personal and subjective way—those brain processes that are responsible for our behavior. Even Freud seems to have thought that, through psychoanalysis, the unconscious could become conscious. In other words, the relevant factors responsible for behavior are, in principle, accessible to introspection. I think this claim is closer to a religious dogma than it is to anything approaching scientific proof. I would suggest, instead, that language provides a paradigm of something that, on the surface, seems highly explainable, in part at least, in terms of mental representations and computations that lie, in principle, beyond the bounds of introspection. Furthermore, it might turn out that certain aspects of language

—for example, what is sometimes called "the creative aspect of language use"—may lie beyond the reach of human scientific intelligence.

It may be that this is typical of brain processes and speculations in this regard. The attempt to computerize different languages so that a sentence in English can be automatically rendered into, say, Swahili, represents one of the computer science's more spectacular failures. Human translators have existed since the tower of Babel; why then can't a machine be developed to carry out the translation process? We'll discuss this in more detail in Chapter Seventeen, but a preliminary answer can be gotten from a brief excursion into the Chomskian world of transformational grammar.

Chomsky would have us imagine a situation in which a visitor from another planet was observing a child learning English. After an observation period, the visitor might hit upon a rule for the formation of questions: "The man is tall," when put in the form of a question, becomes, "Is the man tall?" "The book is on the table" becomes, "Is the book on the table?"

From such observations our visitor might arrive at a hypothesis of how questions are formed: The child starts with the first noun and proceeds until he encounters "is" or other words like it such as "may" or "will," etc. He then begins the sentence with "is," thus producing the corresponding question. Such a rule can be applied to a large number of English sentences. Eventually, however, the extraterrestrial visitor will encounter sentences such as, "The man who is tall is in the room." Applying the above hypothesis yields, "Is the man who tall is in the room?"—which does not make sense. The visitor will be thrown into even more confusion when the young child unerringly transforms the sentence into, "Is the man who is tall in the room?"

Obviously, the first hypothesis is far too simple to enable our extraterrestrial visitor to explain how English questions are formed. Instead, according to Chomsky, the visitor must recognize that the child analyzes the sentence into phrases, not into separate words. In the first instance, the child would be working automatically, breaking the

360

sentence down into words and word sequences. The second example, in contrast, is built on the analysis of the sentence into words and abstract phrases. (The child, if he can deal with sentences of such complexity at all, will recognize that "Is the man who tall is in the room?" doesn't make sense. In fact, the young child who is learning English will make many errors of all sorts, but he will not make that particular mistake!)

Chomsky's explanation of a child's automatic ability to perform such complicated language feats rests on the child's possessing a "universal grammar, a system of rules that are common to all human languages, not merely by accident but by necessity. Of course, by 'necessity' I mean biological necessity. Biological necessity implies that universal grammar is an innate property of the human mind. In principle, we should be able to account for it in terms of biology," says Chomsky.

To make this a bit clearer, consider the sentence, "They are flying airplanes." This can be transformed into more than one question, since the sentence can be interpreted in two ways: *Some people are flying airplanes* or *Those airplanes are flying* (they are actually in the air at the moment). The same ambiguity exists when the sentence is changed into a question: "Are they flying airplanes?" Such sentences have led Chomsky to an important distinction between *surface structure* (the word sequence) and *deep structure* (the probable intent of the sentence—the thought behind it).

Experimental evidence suggests that we all perceive sentences in terms of phrases and deep structure instead of just sequences of words. This partly explains why the learning of a new language can suddenly shift from drudgery to inspired eloquence. At a certain point we learned enough vocabulary to shift from word-by-word analysis to the analysis of phrases and content. Proof of this comes from tests in which subjects are required to report the occurrence of a clicking noise interspersed at varying points while the subjects listen to a sentence. Under appropriate experimental conditions, clicks are perceptually displaced to the boundaries of phrases rather than isolated words. This corresponds with what our extraterrestrial visitor already discovered about language learning in the young child: Phrase structure—the meaning of sentences—functions as the natural unit in the perception of speech.

This hypothesis can be put to a further test involving information retrieved from memory. Dr. A. L. Wilkes and R. A. Kennedy tested a group of adults' capacity to memorize sentences. After the sentences were committed to memory, the subjects were presented with a word in the sentence and asked to give the word that followed it. "The poor cold girl stole the warm red coat" was one of the memorized sentences. Although composed of nine words, the sentence actually involves two phrases, "the poor cold girl" and "stole the warm red coat." Wilkes and Kennedy discovered that whenever the word "girl" was presented, it took the subjects a long time to respond with "stole." This suggests that the sentences are stored in memory on the basis of phrase structures rather than separate words.

To prove this to yourself, think about the last time you heard a joke you wanted to remember. Rather than memorizing the joke word for word (the surest way to guarantee its falling flat!), you concentrate instead on the general situation and the punch line. The first time you decide to tell the joke to someone else you probably change some of the details to suit the occasion. (Maybe the priest and gynecologist become the rabbi and the gynecologist when the joke is told at a bar mitzvah.) If someone were tape-recording you at the time, they would notice unimportant differences in meaning, but significant changes in word order and selection.

Dr. J. S. Sachs tested subjects' ability to remember sentences after listening to a recorded paragraph. Sachs stopped in the middle of the recording and read a series of test sentences to the subject. These sentences were either identical to the one in the paragraph, involved a change in form, or represented a significant change in meaning. A typical example: "The dog chased the boy" was repeated at first unchanged, then with a change in form ("The boy was chased by the dog"), then with a change in meaning ("The boy chased the dog"). Such results are of interest for what they tell us about both memory and language. When tested immediately, all sub-

362

jects could correctly identify sentences. If tested later, however, the subjects could only recognize changes in meaning, not in form. Thus, none of the subjects confused "The dog chased the boy" with "The boy chased the dog." Very few, however, could remember whether the original sentence was "The boy chased the dog" or "The dog was chased by the boy." From this, Sachs concluded that only the deep structure of language is remembered. In short, memory and language are concerned with meaning rather than word order.

We learn two different things when we learn our language. We learn the language itself—its sound system, its vocabulary, its rules of grammar, etc. We must also learn how to use the language for communicating concepts under widely varying circumstances with different people. This involves a further distinction between nonsocial speech and social speech.

Nonsocial speech ignores the listener. Humor is often based on the use of nonsocial speech—as when the entomologist points out a small bug to a young child with the explanation, "This is a nasute termite."

Social speech is highly variable, since it is based partly on the personality of the listener. A famous example of this came from an experiment carried out by the psychologist Douglas Kingsbury, who stopped pedestrians near Harvard to ask for directions to Central Square. When dressed as a student, Kingsbury obtained short but adequate instructions: "First stop on the subway." When affecting the manner of an out-of-towner, however, Kingsbury received expanded, sometimes meticulously detailed instructions on how to proceed. In both instances the pedestrians were using social language according to their perception of Kingsbury: A local resident requires only general instructions, while a stranger needs a detailed description.

Communication is thus determined by the speaker's understanding of the knowledge and background of the listener. Or, put another way, there does not exist a single set of directions to Cambridge's Central Square that is equally appropriate for all people under all circumstances.

In the light of such considerations, it is difficult to

understand how any of us is able to learn a language in the first place. Chomsky and many other psycholinguists believe that such a marvelous feat can be explained only on the basis of an innate language-processing capacity within our brain.

Proverbs are often used by neurologists as a language test for abstraction, the ability to get beyond the surface structure of language. How would the chimpanzee Washoe render the proverbs "The book is in the pen" or "The tongue is the enemy of the neck"? (If you had a little difficulty with "The tongue is the enemy of the neck," don't feel too bad. Often perfectly normal people stumble over it. How, then, could a chimpanzee ever arrive at the interpretation that talking too much can often harm the speaker?)

"The chimp language experiments actually prove that chimps are incapable of even the most rudimentary forms of human language," Chomsky told me. "Much of the difficulty stems from a misinterpretation about language on the part of the chimps' trainers. Take the question of symbolism, for instance. As soon as chimps appear to deal in symbols, some experimenters say, 'This is really language. The chimps have freed themselves from just the concrete use of words. They now use words like we do.' This is nonsense. A bee doing a waggle dance in its hive is using symbolism to convey distance from the hive to the food source. But it's really the symbolism of the speedometer, where the dial and the numbers symbolize how fast we are driving. The signaling system involves a point on a continuum. Human language is not at all like that. It is an infinite but discrete system, completely different in character from the apparently continuous systems of some natural animal communication, or the finite systems taught to apes.

"Birds have been trained, for example, to match patterns of objects with sequences of those objects, leading some scientists to conclude that they 'understand' the concept of number. You don't understand the concept of number until you have the concept of adding 1 indefinitely. A bird or a chimpanzee trained to count up to 7 or to 8,632 doesn't have the concept of number, and when signaling 7 or 8,632 to us, is not using number in any way that can even be remotely compared to the human system."

Chomsky paused, his eyes fixed on the wall in front of him, as he searched for an analogy. "Look at it this way: Imagine that some scientist suddenly announced that human beings could actually fly. This crackpot idea might originate from observations of Olympic broad jumpers. The scientist might construct a 'flight graph' with humans and chickens on one end and eagles and hawks on the other. He might say that the distance a chicken can fly is closer to a human broad jumper than it is to an eagle or a hawk. Thus, in a sense, broad jumpers are actually flying. In this hypothetical example, the similarities in the scientist's grossly ridiculous measure of behavior tells us absolutely nothing about flight. Chickens are obviously closer to hawks and eagles, while humans can't fly at all.

"A similar situation exists in an attempt to teach chimps a language. Chimp 'language' lacks the basic properties of human language. The claim that we can learn about human language from a study of a chimpanzee's communications seems, for the present, no less ridiculous than saying we can learn about flight from Olympic broad jumpers."

In recent years, it has been claimed that the human brain's language performance will soon be duplicated, or even surpassed, by computers. The emerging field of artificial intelligence (AI) promises, some claim, to provide answers to the mysteries of how the human brain processes language. Before examining that claim in Chapter Seventeen, however, it is helpful to remember that we are still profoundly ignorant of the brain's language capacities. In addition, the more we attempt to investigate the mechanism of language, the more mysterious the process becomes. "Our meddling intellect misshapes the beauteous form of things. We murder to dissect," the poet William Wordsworth wrote. He might well have aimed his remarks at our feeble attempts to understand the relationship of the brain to language. Could it be, as mentioned in Chapter One, that we are operating on the wrong level of organization? or bogging ourselves down in category mistakes? Are we like the little boy visiting Congress, the

White House, and the Supreme Court, when what we really want to arrive at is an understanding of "the Government"? I have no ready answer for these questions. Furthermore, I am troubled by Chomsky's intriguing, but disturbing, suggestion that we may be seeking answers that will forever elude us.

17

Because You Asked Me To

> I wish to warn young men against the invincible attraction of theories which simplify and unify seductively.
>
> RAMON Y CAJAL

At Yale University's Artificial Intelligence (AI) laboratory, a computer named SAM (Script Applier Mechanism) is linked to a terminal of the United Press International. SAM's job is to rapidly scan newspaper stories and reduce them to short summaries. Here are two recent examples:

"Alan L. Plucinski, age 17, Fairfield, died Thursday morning in a motorcycle accident when the cycle he was driving crashed into a utility pole on Dalton Woods Rd. Plucinski was pronounced dead on arrival at Milford Hospital, police said."

Summary: "Cycle hit a pole near road two days ago. Allan Plucinski, age 17, residence Fairfield, Conn., died."

"A passenger train carrying tourists collided with a freight train in the rugged Sierra Madre of northern Mexico. At least 18 persons were killed and 45 were injured, the police reported today."

Summary: "A passenger train hit a cargo train in the Sierra Madre Mountains. 18 people died. 45 people were slightly injured."

In less than two years since SAM's development, the prospect of a thinking computer has progressed from the realm of science fiction to imminent reality. Some people are now claiming that the human brain's performance may soon be duplicated by tiny blinking lights, spinning discs, and a labyrinth of cables. Others vigorously deny the possibility. At the basis of this disagreement is a fundamental question: Are there limits to how closely computers can simulate the function of the human brain? Allied to this: Are there brain activities a computer could never carry out?

People have been fascinated for centuries by the possibility of developing nonhuman creatures capable of performing super-human mental feats. Mythical beasts, centaurs, magic spirits, gods—in each case, mythology expresses a basic human desire for the mind to transcend itself via a man-made creature that, although all-powerful, remains man's faithful servant.

According to Jewish folklore, Maharal, a fifteenth-century rabbi in Prague, succeeded in bringing to life the Golem, a mythological figure fashioned to represent a human. As part of the bargain, Golem served Rabbi Maharal for many years and protected the citizens of Prague from foreign invaders. Three centuries later, a modern novelistic attempt to create another Golem misfired and resulted in the Frankenstein monster. Legends persist into the present, however, and today, by an eerie coincidence, two of our century's most influential workers in AI—Norbert Wiener, the founder of cybernetics, and Marvin Minsky, Director of the Artificial Intelligence Unit at the Massachusetts Institute of Technology—are said to be direct descendants of Rabbi Maharal!

In recent years, psychobiologists have also become interested in the computer scientists' attempts to create a contemporary Golem, a computer-based robot that can remember, compute, translate, and, if all promises are fulfilled, reason. At the basis of their enthusiasm is the hunch that computer simulation of human thought processes may provide the basis for understanding how the brain works.

Historically, there is much to recommend such a cooperative approach. Understanding often rewards those who can draw meaningful comparisons between seemingly different processes. We now understand how materials flow in and out of brain cells, for instance, because we learned a lot over the past twenty-five years about the permeability characteristics of living and nonliving membranes. By the same token, we've applied what we know about electricity to living organisms and, in the process, have learned how nerve cells communicate via electrochemical impulses.

In a similar way, psychobiologists are now joining computer specialists in a co-operative effort to test various hypotheses about brain functioning. It is often an agonizing, sometimes troubled, but always potentially rewarding alliance for both sides. If computers can help psychobiologists to understand the brain, then, some computer specialists speculate, the best way to learn how to design computers may come from learning as much as possible about psychobiology.

From the beginning there have been problems, however. First: How do you model a computer to work like the brain when we still know comparatively little about how the brain works? Second: If it is true that computers can do only what they are programed to do, how could they ever teach us anything we can't learn by limiting our study to the brain itself?

Actually, both problems are more imagined than real. Complete understanding is as rare in science as it is in everyday life. Besides, people in different ages often propose strikingly different explanations for the same observations. Does a kite fly because it's carried along by the gods, or because of the force of the vector currents? Either explanation will do, and both are, in fact, irrelevant to the pleasure of the kite flyer.

Nor is a computer performance completely dependent on its human programer. Computers can improve their performance and, in a word, *learn*. The original checker-playing computer built by A. L. Samuel was originally a poor player. By playing thousands of games, however, it "learned" from its experiences and discovered which parameters are most important in a winning checker game. As a result, an improved model of Samuel's checker-playing computer today is virtually unbeatable, even defeating checker champions foolhardy enough to "challenge" it to a game.

But the biggest puzzle about computers involves computer-human communication. How do you talk to a computer? How does it talk back to you? As we saw in the last chapter, human-to-human communication is complicated enough. How, then, can we teach a computer to communicate as effectively as, let's say, a five-year-old child? The answer is, of course, that until recently we couldn't. But that's now rapidly changing in response to some exciting new developments in our understanding of how the brain makes language possible.

Initially, computer specialists working with language based their analysis on several misconceptions. They presumed that natural language (language spoken by human beings) can be broken down into syntax (how words are strung together) and semantics (meaning). The rules of syntax are essentially the rules of grammar. "John gave the book to Mary" is a syntactical rendering of the sentence. It is also grammatical. "John the book Mary gave" is ungrammatical and, in addition, does not follow accepted English syntax: the subject (John), the verb (gave), the object (the book). Although most linguists didn't think about it, the second sentence also lacks a precise meaning. Did John give Mary the book? Did Mary give John the book?

The initial attempt at a thinking computer mistakenly assumed that language could be separated into meaning (semantics) and sentence structure (syntax). The first attempt to computerize language translation, therefore, consisted in transferring large dictionaries onto computer tapes so that words in one language could be linked up with the same words in another language. As an example of how this worked: "The spirit is willing but the flesh is weak," when translated from English to Russian and then back to English, came out, "The wine is agreeable but the meat is spoiled."

Such an absurd result shouldn't come as a surprise to anyone who thinks even casually about how people use language. Meaning cannot be separated from the way words are put together, any more than a symphony can result from a musical score written from notes chosen at random. The formal recognition of this dependency on structure came from Noam Chomsky, whom we met in the preceding chapter. Language is composed, according to Chomsky, of *deep structure* comprising meaning (semantics) which can be expressed according to a large number of syntactical structures. "John gave Mary the book," "Mary was given the book by John," "The book was given to Mary by John"—all three sentences express the same deep structure (the transfer of the book from John to Mary), but they do so by using different syntactical (grammatical) structures. Our ability to extract the meaning of the three different sentences is, according to Chomsky, a psychobiological trait that is genetic and relatively independent of a specific language. We understand each of the sentences because we understand the deep structure. A computer, in order to translate each of the sentences, can only do so if it also possesses a similar understanding. In a phrase, understanding meaning is the key to understanding language.

Over the past twenty-five years, ideas have changed about how meaning is embedded in natural language and how computer language might be modeled on the same principle. But, since the basic goal is to enable computers and people to talk to each other, a reasonable test of a successful computer language is always available: Is it understandable to the average person?

Early AI workers attempted to solve the problem by assigning all possible meanings to a given word. This didn't work, however, since isolated sentences give no clue as to which meaning is intended. "Mary was moved" might refer to Mary's reassignment to a new office at work or her reaction to hearing that she had just won the New York State Lottery. An unlikely, but still understandable, sentence might even combine Mary's emotional reaction and her new job location. "Mary was moved (*to tears of joy*) upon learning that she had been moved (*into the office of the Executive Vice-president*)."

For a while, things seemed almost hopeless when it came to programing the language capabilities needed to develop a thinking computer. To each new approach there always turned out to be a new variation of "Mary was moved." Once again, however, one of the most obvious aspects of the whole problem was being overlooked. When you get right down to it, what *does* "Mary was moved" mean to the average reader when encountering it in isolation? The answer is: *not much*. The sentence has to be used in a context; otherwise, it remains hopelessly ambiguous. Why, then, should we expect a computer to do more with the language than human beings? From here, the next logical step became obvious: Model computer language on human language. Human language performance then becomes the outer limit of what can be expected from computers.

The man who has most advanced the cause of computer language—to the point where some say a thinking computer may be just around the corner—is Dr. Roger Schank, Director of the Yale Artificial Intelligence Project. Shank, a thirty-one-year-old linguist with long, flowing hair which extends down over heavy-set shoulders to merge with a thick, scruffy beard, comes across at first meeting as a sort of Rasputin in a sport shirt. In his large office at Yale University, he explained how SAM works.

"The appropriate ingredients for extracting the meaning of a sentence are often nowhere to be found within the sentence. Consider something like, 'The policeman held up his hand and stopped the car.' Somehow, in understanding that sentence, we effortlessly create a driver who steps on a brake in response to seeing the policeman's hand. None of these intermediate links are mentioned in the sentence. But consider something like, 'I went to three drugstores this morning.' What does that mean? When you and I hear it, we assume that the person must not have found what he wanted in the first two drugstores; otherwise, why would he have gone to the third? This kind of implicit inference is very common."

Listening at random to conversations leads to the conclusion that almost all sentences rely strongly on the listener's ability to draw inferences. "I like apples" is a shorthand which implies, "I like to eat apples." If computer language is to mirror human language, we must be able to draw similar inferences.

"How do people organize all the knowledge they must have in order to understand language?" Schank asks. "How do they know what behavior is appropriate for a particular situation? To put it more concretely, How do you know in a restaurant that the waitress will get you the food you ask for?"

We know such things because, according to Schank, we have two kinds of knowledge about the world we live in: general knowledge and specific knowledge. General knowledge enables us to understand and interpret other people's actions simply because they, too, are human beings like ourselves, with certain standard needs, who live in a world which has certain standard methods of getting those needs fulfilled.

Specific knowledge, on the other hand, is a kind of shorthand we apply to situations we have been through many times before. Buying a ticket for a movie, ordering a meal in a restaurant, telephoning for a plane reservation —in each case a predictable sequence of events is put into motion in which inferences are made on both ends on the basis of what Schank describes as a "script."

"The script is a structure that describes the appropriate sequence of events in a particular context. Scripts handle stylized everyday situations. It's a predetermined, stereotyped sequence of actions that define a well-known situation."

According to Schank, much of human behavior is dependent on people learning large numbers of scripts throughout their lifetime. The possession of common scripts makes it possible for people to draw inferences that are not specifically stated in sentences. "John was hungry, went to a restaurant and ordered a steak, paid the check and then left"—fits into a restaurant script familiar to everyone. It isn't necessary in such a script to include such things as John sat down, John read the menu, John ate the steak, etc. Each is implied in the restaurant script.

SAM operates by processing large numbers of scripts, filling in inferences and drawing conclusions from sketchy details. "SAM understands these stories," Schank insists. "By 'understand' I mean that SAM can create a linked causal chain which represents what took place in the story.

Any significant deviation from the script can then be identified and the reasons given."

Schank, along with his associate, Christopher Riesbeck, modeled SAM on the same process that underlies human understanding. Ordinarily, understanding the meaning of a sentence depends on the context in which the sentence appears. In a similar way, SAM's understanding depends on fitting information into one of several scripts which have been programed in great detail. If there are variations from the usual restaurant script, for example, SAM is able to demonstrate "understanding" by correctly answering a series of questions.

Here is a recent example of SAM's ability to analyze a restaurant script and correctly respond to a series of questions.

"John went to New York by bus. On the bus he talked to an old lady. When he left the bus he thanked the driver. He took the subway to Leone's. On the subway his pocket was picked. He got off the train and entered Leone's. He had some lasagna. When the check came, he discovered he couldn't pay. The management told him he would have to wash dishes. When he left, he caught a bus to New Haven."

"Why did John go to New York?"

"Because John wanted to go to Leone's."

"How did John get to Leone's?"

"John took a bus to New York and then took a subway to Leone's."

"Did anything unusual happen on the bus?"

"I do not remember anything."

"Did anything unusual happen on the subway?"

"A thief picked John's pocket."

"What did John order?"

"John ordered lasagna."

"Did anything unusual happen in the restaurant?"

"John discovered that he could not pay the check, so he had to wash dishes."

"Why couldn't John pay the check?"

"Because John did not have any money."

"Why didn't John have any money?"

"Because a thief picked John's pocket."

"How did John get back to New Haven?"

"John went from Leone's by subway, and then he went to New Haven by bus."

"How was John able to take the bus to New Haven?"

"John probably had a ticket."

Other computer programs, based on similar well-defined scripts, are already in use. One of them, the LSNLIS (Lunar Sciences Natural Language Information System) answers questions about the geological properties of lunar rock and soil samples. Another, Luigi, answers questions about the preparation of food by simulating the actions of a cook. Several computers are already in use that are purported to be highly reliable in arriving at a correct medical diagnosis on the basis of questioning the patient regarding his symptoms. But in each of these latter examples, unlike SAM, the programs are limited and do not attempt to model human understanding. "Is the moon rock bigger than a basketball?" and "Does the quiche Lorraine taste better than the coq au vin?" cannot be answered, since the questions require more than just technical details.

"Any computer understanding system must be able to understand stories that describe new and unexpected situations," cautions Schank. "To do this, it is necessary to connect pieces of information by means of the scripts. To handle this we introduce plans: general information about how people achieve goals. A plan is a series of projected actions to achieve a goal. Consider the following sentences: 'John knew that his wife's operation was very expensive. There was always Uncle Harry—he reached for the suburban phone book.'

"How are we going to make sense of such a paragraph? It would be unreasonable to posit a 'paying for an operation' script with all the necessary scenes laid out in our restaurant script. But the situation is not entirely novel either. Understanding this paragraph would not be significantly different if the 'wife's operation' were changed to the 'son's education' or 'down payment on the mortgage.' There is a general goal in each case—namely, raising a lot of money for a legitimate expense."

A plan-based computer known as PAM (Plan Applier Mechanism) is already in service at Yale. Rather than relying on scripts, as with SAM, PAM understands stories by analyzing the interactions of the story's characters. PAM uses knowledge about goals to figure out the intentions of the characters it hears about and keeps track of the goals of each character by interpreting their actions in light of the achievement of those goals.

PAM is the creation of Robert Wilensky, another of Roger Schank's associates at the Artificial Intelligence Project. In a typical paragraph, Robert Wilensky gives PAM a hypothetical situation.

"John wanted some money. He got a gun and went into a liquor store. He told the liquor store owner he wanted money. The owner gave John some money and he left."

In this paragraph nothing specific is mentioned about what John did with the gun. We know that a man going into a liquor store with a gun is probably intent on robbing the store. But how does a computer draw such a conclusion?

"PAM is equipped with a lot of little rules about intention," Wilensky told me. "Words are linked with their possible uses, and PAM then draws the appropriate conclusion. Take the word 'gun,' for example. PAM knows a gun can be used to overpower someone or force him to do something. It can be used for target practice and it can also be used for hunting. In this case, PAM links up John's need for 'money' with the three possible uses for a 'gun.' John won't get money by target practice or hunting, both of which are inappropriate in a liquor store. But he can get it by threatening to overpower the owner. To prove that PAM has understood the story, let's ask it to paraphrase the story first from John's point of view."

Wilensky typed in the instructions and PAM printed out the liquor store's story as John would have experienced it:

"I needed to get some dough, so I got myself this gun and went down to the liquor store and told the liquor store owner to let me have the money or I would shoot him. He handed over the money and I left the store."

"Now let's see if PAM can give us a credible version of the same event from the point of view of the store owner."

"I was minding the store and a man entered and threatened me with a gun and demanded all the cash receipts. I did not want to get hurt, so I gave him the money. He escaped from the store."

"The interesting thing about this," according to Wilensky, "is the number of inferences drawn by PAM about things that are not contained in the original story. 'He threatened me with a gun' is an inference drawn from 'he got a gun and went into the liquor store.' 'He escaped from the store' is an inference from the single statement 'John left.' PAM's ability to make these inferences has enabled us to outperform any computer ever devised. It is, in fact, a thinking computer. Now let's try a variation on the story:

"John wanted some money. He got a gun and went into the liquor store. He told the liquor store owner that he would give the owner the gun for ten dollars."

"Let's ask PAM to rewrite the story for us from the point of view of, first, John, then the owner."

John: "I needed to get some dough so I got myself a gun and went down to the liquor store and told the liquor store owner that I would let him have the gun for ten dollars."

Owner: "I was minding the store and a man entered. He had a gun and I thought he was going to rob the store, but he just wanted to sell me the gun for ten dollars."

"PAM has the ability to draw conclusions the way people do. Let's try something like, 'John was lost while driving his car. He saw a farmer standing by the side of the road. He pulled over by the farmer.'

Q. 'Why did John pull over to the farmer?'
A. 'He was lost and wanted to find out where he was.'

"In such a sentence PAM realizes:

1) 'that John is lost'
2) 'that most people don't like being lost'

377

3) 'that other people can help by giving directions'
4) 'that John pulled over to ask the farmer how to return to familiar landmarks.' "

Computers such as SAM and PAM represent a revolution in our attempts to model human thought processes. From experiments with these computers, scientists may come up with answers to two questions that Roger Schank believes lie at the core of both psychology and artificial intelligence: What is the nature of knowledge? How is this knowledge used?

"We are concerned with the intentional and contextual connection between events," Schank says. "While it is possible that computers cannot actually experience such interventions and relationships, they can perfectly well be programed to have some understanding of their occurrence and significance—thus functioning as smart observers. If our theory is apt, it will provide a model of the human observer of the human scene; it will also explain how to construct a computer observer of the human scene and lead to the eventual building of a computer participant in the human world."

Roger Schank is not talking about some far-off *Star Wars* fantasy, but computer capabilities that are already here. If we can understand how a computer, given certain initial preprogramed biases, can reach conclusions, this might tell us a lot about how the brain processes information. But, once again, for the computer to be of help, it must be built to construct inferences the way people do. One promising program developed by Schank's team is POLITICS, a computer simulation of different political ideologies.

POLITICS interprets and responds to world events from a pre-programmed political, or ideological, point of view. "The Russians massed troops on the Czech border." What conclusions can be drawn from this statement? Obviously, any conclusions have a lot to do with a whole series of political beliefs. What are the goals of Russian communism? Why are the Russians massing troops on the border of a neutral country? How can the United States

378

best respond? What are the goals of the United States in relation to Russia?

Each of the above questions is answered differently according to the political belief systems of the questioner. Beliefs rather than facts are called upon in order to determine the "meaning" of the Russians massing troops on the Czech border.

So, too, politics operates on the basis of ideological theories about national and international motivation. "POLITICS interprets this story, making appropriate references to incorporate the story in its internal memory representation in such a way that it can later answer questions, make predictions about possible future events, and comment as to what the United States will do about the situation," according to Jaime G. Carbonell, Jr., the inventor of POLITICS. "The story interpretation assumes a given political ideology. Let's consider a right-wing ideology in processing the meaning of Russia's action. We stipulate that a politically conservative person might infer that Russia intends to either invade Czechoslovakia or threaten an invasion in order to force Czechoslovakia to capitulate to Soviet demands. If one recalls the 1968 Soviet invasion that toppled Dubcek, one may infer rather specific reasons for the invasion. Otherwise, it's sufficient to infer that, since Russians always want strong political control over their neighbors, the present story is just another incidence of Russia exerting its military muscle in order to expand its political domination. POLITICS interprets the story largely by attributing goals and motivations to the actors in the story and integrating the story's representation into its world model."

POLITICS is able to demonstrate its comprehension and interpretation by answering questions about the possible meaning of political events:

Q. "Why did Russia mass troops on the Czech border?"
A. "Because Russia thought it could take political control of Czechoslovakia by sending in troops."
Q. "What will Russia do next?"
A. "Russia may tell its troops to start the invasion."

Q. "What else can Russia do?"

A. "Russia may wait for Czechoslovakia to give in to Russian demands."

Q. "What happens if Czechoslovakia does not give in?"

A. "Russia may decide to start the invasion."

Q. "What should the United States do?"

A. "The United States should intervene militarily."

Q. "What else should the United States do?"

A. "The United States can do nothing, or it can intervene diplomatically by conferring with Russia about Czechoslovakia. These alternatives are bad for the United States."

POLITICS is funded by the Department of Defense's Advanced Research Projects Agency. Although no one so far is stating exactly why POLITICS and other human thought-simulation computers are of interest to the Department of Defense, the computer specialists working on the project justify their participation by speculating that their computers may provide insights into the nature of human reasoning. In addition, we may find out how human reasoning errors occur and how they might be prevented in the real world. "POLITICS enables us to model mistakes, which are always more interesting and informative than the right decisions," according to Christopher Riesbeck, who specializes in how computers analyze natural language. "We try to understand why certain theoretical beliefs lead to certain courses of action. How do people arrive at the conclusions that provide justification for their actions? If the computer's decision on hearing that the Russians massed troops on the Czech border was to order a nuclear attack, we might learn how to avoid a similar conclusion from a President or a Congress by studying the inference processes of the computer."

In the invasion example, two opposing belief systems would lead to different conclusions about what should be done. The right-wing scenario is based on the belief that Russia's goal is world domination at all costs, and therefore interprets the massing of troops as a prelude to invasion. The appropriate response, given the right-wing belief system, is to stop the invasion: The United States should intervene militarily.

The second and equally likely belief system, the left-wing ideology, assumes that Russia's main goal is to avoid a nuclear war while expanding its political control only as a lesser goal. Here the appropriate response is to assume that Russia is only threatening and will welcome a negotiated settlement. In this case, the sending of troops and military intervention would be disastrous. The chillingly intriguing questions, of course, are: *"What is Russia's goal?"* and "What is the correct response to make to an apparent invasion?"

One answer might come from carrying out computer simulations of past events and correlating what was done with the choices suggested by right-wing and left-wing computers. The Cuban missile crisis, for instance, was a cliffhanger that almost precipitated a major military confrontation that many observers feel, even today, could have been handled in a far less risky, less confrontatory manner. How would a computer handle such a decision? If it called for the launching of missiles or the sending of troops to Cuba, we could safely assume that this was the wrong decision and, by reviewing the computer's belief systems, figure out what preconception disrupted the process of reason. If it came up with the actual decision, a nonmilitary but nonetheless forceful confrontation, this might indicate that our leaders held a correct series of beliefs about Russian international intentions.

"The problem with the computer simulation of past events arises from our brain's tendency to develop rules to support why we had to do what we actually did," says Riesbeck. "In other words, we just naturally tend to assume what we did was correct and look for reasons to justify that belief."

Even the Cuban missile crisis example doesn't allow a final answer, since no one really knows what Russia would have done if we had sent troops into Cuba. Suppose troops had been sent, we took over Cuba, and the Russians did nothing? Would we now be congratulating ourselves that we were "smart" enough to annex the island of Cuba by a display of military might? Or would such an action have more likely led to a nuclear holocaust?

POLITICS is currently analyzing a series of congressional decisions and interpreting them according to different political beliefs. Here is the right-wing and left-wing interpretation of the U. S. Congress's vote to fund the Trident submarine project:

RIGHT-WING QUESTION: Why did the United States Congress vote to fund the Trident submarine project?

A. "The United States Congress wants the United States forces to be stronger."

Q. "Why should the United States be stronger?"

A. "The United States armed forces should be the strongest to stop communist expansion."

Q. "What should the United States do if Russia builds submarines?"

A. "The United States should build more submarines."

LEFT-WING QUESTION: Why did the United States Congress vote to fund the Trident submarine project?

A. "The United States Congress feared falling behind in the arms race."

Q. "What will Russia do next?"

A. "Russia will probably build more weapons."

Q. "What should the United States do if Russia builds submarines?"

A. "The United States should negotiate with Russia to stop the arms race."

Analysis of the right-wing answers reveals that the computer bases its decisions on the belief that funding submarines leads to an increase in military power for the United States. This is the goal of the right-wing ideology and therefore is the answer to why the funding was voted. In the left-wing ideology, world peace rather than military power is the underlying ideology. Therefore, a military build-up has to be stopped. A counterstrategy for this might involve a negotiation, the left-wing-suggested response to, "What should the United States do if Russia builds submarines?"

What are the limits to the computer's ability to simulate human thought and reasoning? Will computers teach us how to reach decisions? Most important of all, can the study of computers help us to solve the mysteries of the human brain?

At this moment, a lot of time, effort, and money is being funneled into projects aimed at determining the outer limits of a computer's ability to simulate brain functioning. At the basis of these efforts is the conviction, sometimes earnestly held, that the brain's performance may depend on the density and patterning of electrical charges within the brain. The recently discovered microcircuits discussed in Chapter Nine result in complex charge gradients, with distributions varying from millisecond to millisecond. With each shift in the firing of nerve cells, the charges on the cells vary, resulting in complex charge gradients which form the physical basis for Sherrington's "enchanted loom," where "millions of flashing shuttles weave a dissolving pattern, always a meaningful pattern though never an abiding one."

To Dr. E. Roy John, Director of New York University's Brain Research Group, the brain's electrical energy distribution may be responsible for its thus far unique position in the universe as the only physical system capable of conscious self-reflection. John describes the energy distributions within the brain, and their fluctuations from moment to moment, as a *hyperneuron*. Rather than something like a queen bee in a hive, however, the hyperneuron is not a "giant neuron" but an energy process—in fact, the sum total of charges within the nerve cells, glia, and extracellular spaces within the brain. John postulates that consciousness emerges from the "co-operative interaction of neuronal populations," which result in hyperneurons. "The content of subjective experience *is* the momentary contour of the hyperneuron," says John.

As the hyperneuron alters over time, so does the content of consciousness. This leads immediately to the intriguing, but so far unanswerable, questions: Is the neuron itself important for consciousness? and, Is the energy charge distribution within the brain the critical feature?

If the neuron is truly unique and also capable of resulting in consciousness, then this implies, of course, that the human brain will not be duplicated by a machine and we can begin to rank artificial intelligence alongside such things as perpetual motion and the fountain of youth. But if consciousness does result from a pattern of electrical charges—from a hyperneuron—artificial intelligence becomes not only possible but, with continued research, probably inevitable. "Were it possible to achieve comparable distributions of energy without neurons—in other words, to simulate a 'neuron free' hyperneuron—perhaps quite the same subjective experience would arise," suggests John. "Subjective experience may actually be a property of a certain level of organization in matter."

The hypothesis of a hyperneuron leads to several immediate research strategies. For one thing, psychobiologists—principally John and his research group—are trying to discover the particular feature of organized energy in brain tissue that is responsible for conscious awareness. Once this is achieved, scientists will be in a better position to decide whether "this emergent property of consciousness is necessarily limited to brains, exists in any neuronal system, is a general property of living matter, or might arise in a sufficiently organized system of energy."

Energy and its transformations are, of course, the proper matter for physics. It was for this reason that the two Nobel Prize winners in physics, cited in the first paragraph of this book, selected brain research as the area for the Nobel Prize in physics in the year 2000. Psychobiologists may, in fact, be in the process of discovering what one researcher has called "a previously unknown law of thermodynamics." Perhaps a specific configuration of energy may produce the process we call consciousness. If this is true, however, an even more challenging question arises: How will we know when we have discovered it?

Consciousness depends on the establishment of communication with other conscious creatures. Although we can speculate about the consciousness of a rabbit or a cat, since elementary forms of communication with these animals are possible, we are totally stymied when it comes to investigating the consciousness of a rock. Our difficulty stems, in part, from our complete inability to communicate with a rock. In the language of artificial intelligence there are no input or output transducers, no way to establish communication channels. In a similar manner, if other highly organized forms of energy were to be created in the near future modeled on the brain's distribution in the hyperneuron, how would we be able to test whether our concoction was indeed conscious? Obviously such a determination would become possible only if our conscious creation decided to establish communication with us. Thus, we may find ourselves in the grip of a gnawing uncertainty, as paralyzed as Schopenhauer before what he called the "world knot" of consciousness.

But there are those who believe that artificial intelligence can be discounted without reference to questions that probe quite so deeply. In the fall of 1977 I visited the most outspoken of these critics in his office at the Massachusetts Institute of Technology.

On the fourth floor of the futuristic-appearing Technology Square Building at M.I.T., is the office of Joseph Weizenbaum, Professor of Computer Science. While many of the offices on the fourth floor are locked, requiring activation of the doors by special key cards in order to gain entrance, Weizenbaum's small but comfortable and modern quarters can be approached directly, with a minimum of fuss. So, too, Weizenbaum himself is a refreshingly unlocked man who has formulated an intriguing theory about the role of computers in twentieth-century society.

"What are your own interests regarding computers?" he asked as I entered and sat across from him and a small computer terminal.

"Some people think computers might help us understand how the brain works," I responded, conscious of a pair of intensely riveting eyes peering at me from behind modish horn-rimmed glasses.

"What do *you* think about it," Weizenbaum persisted.

"I'm not a believer," I said.

"You're in good company," Weizenbaum replied, his face relaxing into a soft smile. "In fact, if you were convinced that the brain is merely a computer, you wouldn't be here talking to me."

Face to face, Weizenbaum looks like a cross between a classical musician and a chess grandmaster. Neatly, almost nattily, dressed, with graying razor-cut hair worn long, particularly toward the back, Weizenbaum speaks in a slow, measured, deeply accented voice. At age fifty-six he looks trim, the picture of physical fitness. He would also appear totally serious were it not for a well-groomed Jerry Colonna mustache which, when he laughs, conveys an impression of infectious good humor. "Let me give you an analogy for a person trying to understand the brain by constructing a computer. The John Hancock Building here in Boston had been plagued for years by windows inexplicably falling out. They just couldn't keep the windows in the building, and engineers hadn't been able to find out why. (I understand the problem has been largely solved since.) Suppose a group of architects and engineers meet every Wednesday morning to brainstorm what they might do next about the windows. Now let's imagine that on a certain Tuesday a scientist somewhere in the world announces a theory unifying the whole physical universe, perhaps some version of the so-called Unified Field Theory that Einstein worked on during the last years of his life. The next day, at the architects' and engineers' meeting, someone might say, 'Boy, are we lucky. Now that we understand everything about the physical universe, surely the solution to our window-breaking problem is contained in that "everything." '

"Such a person is, in a certain sense, correct. Despite that, I don't think the new theory would be of much help in discovering why the windows are falling out. Between the newly discovered Unified Field Theory and the window-falling problem, there exists layer upon layer of intervening variables awaiting definition and understanding. What is glass? for example. Is it something you see through or is it a mixture of silicate formed by high-temperature fusion? Obviously it is both; glass that will not break, but which is totally opaque, would not satisfy our idea of glass windows suitable for a modern skyscraper. The example I am giving you contains a logical error. It's possible—and, in fact, necessary—to be able to reason from specific examples to general laws. Windows fall *down* not *up*—because of gravity. You learn about gravity by observing windows or apples, as Newton reportedly did. Gravity, in turn, may be explained by a deep theory of the behavior of, say, elemental particles. But you can't progress from the study of gravity to a direct application such as falling windows." Weizenbaum paused to see if I was with him, puffed twice on his pipe, and continued:

"Imagine for a moment that you have designed a computer, handed it to me, and said: 'Here is a challenge for you. Figure it out and tell me how it works.' I start working on the computer, devoting all my energy to it, and finally come up with a description of the computer. I know every flip-flop, the state of every storage cell, and eventually compose a state transition map; given the state of the computer at any particular time, I can tell what its state will be at any time in the future. Now at this point, in some very important sense, I know all there is to know about the computer. For instance, I can predict what its state will be ten thousand years from now or ten minutes from now. But here's a question for you: Can I tell what the computer is doing at this very moment? The answer is that I cannot. It might be computing odds at a nearby racetrack, inverting a matrix, even counting the commas in *Macbeth*. There is no way, in principle, that I can reliably predict at any given instant on what application the computer may be working.

"This is a very important concept that I think a lot of people are confused about," Weizenbaum continues. "Prediction doesn't imply knowledge about what is happening at a given point in time. In a hypothetical example I may be able to predict what the computer will print out in a few minutes, or what pattern will be in the register a thousand years from now. But when asked, 'What is it doing now?' I can't answer. The person who programed the computer is the only one who might be able to do that. He or she had something in mind, and to them the bit pattern has a *meaning*. That meaning is in the programer's head, not in the computer. The computer is full of bit patterns, but I am left to guess to what problem it is being applied. There is no logical or computational way I can figure out the meaning of its action.

"Now let's compare the computer with the human brain, as so many people have been fond of doing recently. The analogy holds here as well. Even if we understand every synapse, even develop a transition map for people so that we can tell what their brain might do in the next microsecond, nevertheless there is no way I can figure out whether that person is thinking of a song by Frank Sinatra or mentally working his way through the Boston subway system. The analogy with the computer is here quite exact. If we know everything about the brain from a physical point of view, which we don't, it's almost unrelated to the question of why the person is thinking certain thoughts at the time. Even if you do correlate certain neurons with certain thoughts, the correlation will not enable you to say that the next time these neurons are activated the person must be thinking of the Frank Sinatra song once again. In a computer we use the same bits for different tasks, depending on the computer program operating at the time. So, too, the brain may use the same neuronal paths to process entirely different mental processes. In short, even if we know everything about the brain, it would not tell us about our most human, most personal aspects."

Another potential source of misunderstanding involves the different ways men and computers go about solving similar problems. Chess-playing computers, for example, were originally designed to model the mental process of skilled chess players. Such efforts were total failures, as Weizenbaum and others are fond of pointing out. What are the mental processes responsible for expert chess playing? We simply don't know. In fact, players such as Bobby Fischer have been suspected of basing their play on original theories of chess so different from standard theories that they cannot even be explained. For reasons that we've mentioned in Chapter One, it is impossible for the human brain to figure out ahead of time every possible move, even three moves ahead. So how can we explain the performance of chess grandmasters which up to the time of this writing have prevailed over every chess-playing computer yet developed?

"Chess-playing computers tell us nothing about human chess," Weizenbaum says. "The computer is programed with the elementary principles of chess. It gains its power by being able to look ahead more systematically than the human player at the average, or even above-average, level. It is conceivable, even probable, that as computers become larger and faster and scan more possibilities within the allotted time for a chess move, that a computer will be developed that can beat even the best chess master. But don't let that victory confuse you. Such a computer would tell us absolutely nothing about how humans play chess. We simply know next to nothing at all about the relations between brain functions and human thought. Until we do, computers are marvelous toys, but not much help in explaining how our brain works."

Weizenbaum has firsthand experience in building computers that are almost indistinguishable from human beings. In the mid 1960s he was among the first to design a computer program capable of "conversing" with a human operator.

Dubbed "Eliza," after the cockney heroine of George Bernard Shaw's *Pygmalion*, whose diction was improved through the not entirely disinterested efforts of Professor Higgins, Weizenbaum's computer was able to analyze language and respond according to a script. "The script is a set of rules rather like those that might be given to an actor who is to use them to improvise around certain themes," Weizenbaum says.

In a typical example, the computer may be programed with a script that enables it to carry on a conversation about, say, professional football, or the steps involved in knitting a cardigan sweater. Weizenbaum's most intriguing computer script, however, involves a computer in the "role" of a psychotherapist who practices nondirective therapy. Here is a typical exchange between "Dr. Eliza" and a "patient":

PATIENT: Men are all alike.
COMPUTER: In what way?

389

PATIENT: They're always bugging us about something or other.

COMPUTER: Can you think of a specific example?

PATIENT: Well, my boyfriend made me come here.

COMPUTER: Your boyfriend made you come here?

PATIENT: He says I'm depressed much of the time.

COMPUTER: I am sorry to hear you are depressed.

PATIENT: It's true. I am unhappy.

COMPUTER: Do you think coming here will help you not to be unhappy?

PATIENT: I need some help. That much seems certain.

COMPUTER: What would it mean to you if you get some help?

PATIENT: Perhaps I could learn to get along with my mother.

COMPUTER: Tell me more about your family.

PATIENT: My mother takes care of me.

COMPUTER: Who else in your family takes care of you?

PATIENT: My father.

COMPUTER: Your father?

PATIENT: You are like my father in some ways.

COMPUTER: What resemblance do you see?

PATIENT: You are not very aggressive, but I think you don't want me to notice that.

COMPUTER: What makes you think I am not very aggressive?

PATIENT: You don't argue with me.

COMPUTER: Why do you think I don't argue with you?

PATIENT: You are afraid of me.

COMPUTER: Does it please you to believe I am afraid of you?

PATIENT: My father is afraid of everybody.

COMPUTER: What else comes to mind when you think of your father?

PATIENT: Bullies.

COMPUTER: Does that have anything to do with the fact that your boyfriend made you come here?

When Weizenbaum published the description of Eliza —or "DOCTOR," as the computer program became known—the enthusiasm for the project quickly spread from M.I.T. to dozens of computer departments all over the United States. Soon the program became, in Weizenbaum's words, "a national plaything."

As frequently happens in innovative research, however, Weizenbaum's purpose in inventing the DOCTOR script turned out to be quite different from many of the program's suggested uses. Why not a computer psychotherapist? Why not, indeed, according to Kenneth Mark Colby, writing in the *Journal of Nervous and Mental Diseases:* "If the method proves beneficial, then it would provide a therapeutic tool which can be made widely available to mental hospitals and psychiatric centers suffering a shortage of therapists. Because of the time-sharing capabilities of modern and future computers, several hundred patients an hour can be handled by a computer system designed for this purpose." To this, Weizenbaum reacted in horror. "What must a psychiatrist who makes such a suggestion think he's doing while treating a patient, that he can view the simplest mechanical parody of a single interviewing technique as having captured anything of the essence of a human encounter?"

Weizenbaum's purpose in inventing Eliza was educational and based on the ease with which an ordinary person can interact with a computer without the need for specialized knowledge. "Most other programs could not vividly demonstrate the information-processing power of a computer to visitors who do not already have some specialized knowledge. DOCTOR, on the other hand, can be appreciated on some level by anyone," says Weizenbaum.

Soon after his program's introduction, Weizenbaum also discovered that "extremely short exposures to a relatively simple computer program could produce powerful delusionary thinking in quite normal people." In one case, a secretary lashed out at Weizenbaum when he suggested that the computer conversations be stored so that at a later time individual exchanges could be studied. On another occasion, Weizenbaum was asked to leave the room while someone asked the computer a series of personal questions. "These examples provided clear evidence that people were conversing with computers as if they were a person who could be appropriately and usefully addressed in intimate terms." About this time, a startled Weizenbaum started asking himself a series of "basic questions which, at bottom, were nothing less than questions about man's place in the universe." Is human thought entirely computerable? And even more intriguing, What is the basis for the strange fixation on a computer psychotherapy program, which, on the face of it, was nothing more than an elaborate joke, something the English psychologist Stuart Southerland mocked as an "Electronic Oracle"?

Over the past several years, Weizenbaum has searched for answers to these questions. In *Computer Power and Human Reason*, published in 1976, Weizenbaum suggested that the problems created by computers may result not from the fact that the computers represent something new but, on the contrary, that in one form or another they have been with us for a very long time. The usefulness of tools in hunting and farming; the importance of steam power, rotary engines, and jet propulsion in our transportation systems—these are examples of ways we have extended our powers over nature while increasing our sense of control. In line with this, computers can be thought of as elaborate and cunning devices that carry out operations for us that we are unable to carry out for ourselves. Computers, however, differ in a very significant way from other technologies. While arrows and even gunpowder are only extensions of the killing power of our own hands, and jet propulsion is but an accelerated version of our own motor power, computers in some areas are capable of performances totally beyond the known powers of the human brain. To this extent, computers pose a deeply felt threat to our self-image. If computers can duplicate many aspects of human thought, doesn't it suggest that we ourselves may be nothing more than living, breathing computers?

The idea of computers disguised as human beings is particularly intriguing when we consider ways we might go about distinguishing between a computer and a human. Imagine that I have just introduced you to a friend, Carl James, whom I have identified as a twenty-eight-year-old Wall Street lawyer. After the introductions and an opening get-acquainted conversation, I suddenly announce to you that Carl is actually a robot—designed by a European computer programer who specializes in making lifelike robots. Carl at first seems mildly embarrassed at my revelations; but, when asked, he reluctantly corroborates my assertion. Forget for the moment that lifelike robot computers don't yet exist. Assume also that Carl and myself are neither crazy nor liars. How would you go about determining whether or not Carl is a computer?

From the start, certain lines of inquiry would be immediately doomed to failure. One of these is the question-and-answer approach. If you begin by asking Carl where he went to college, you might discover, to your pleasure, that he attended Yale at about the same time as yourself. Based on this, you might confidently begin a series of questions. Who won the Yale–Harvard football game in 1967? What was the score? Who was Chairman of the Department of Business at that time? Despite the seeming logic of such an approach, the results obtained will in no way help you to distinguish the robot computer from a human. If Carl doesn't know the score of the Yale–Harvard game, does that prove he's a robot? Or, rather, does it point to a former undergraduate with little interest in football? Does Carl's failure to remember the chairman of the department in which he majored reflect any more than a poor memory? Alternately, if all the questions are answered satisfactorily, does that indicate a real person who actually attended Yale, or a robot programed with an unusually detailed fund of knowledge?

Imagine that at this point Carl offers to correctly identify every student who ever attended Yale, together with the year of their graduation, and also to fill us in on every faculty member who ever taught in the Department of Business. Of course, now the decision would be easy, and you could confidently and perhaps incredulously announce that Carl must be a computer. The likelihood of this happening, of course, is low, since Carl would likely be programed not to reveal any information beyond the capacities of a human. In short, if Carl is carefully enough constructed to require testing regarding his true identity, you're not likely to be able to correctly choose between

393

robot and human on the basis of question and answer.

At this point, you might ask if it would be possible to spend a few hours with Carl for the purpose of observing what he does. This approach is more likely to be successful than the question-and-answer method, since we can make certain assumptions about a computer that *may* lead to significant peculiarities of behavior. The computer circuits, for example, don't depend on metabolic processes for their operation. Breakfast, lunch, and dinner are thus unimportant and might provide us with a clue. If we literally moved into an apartment with Carl and observed over several days that he never eats, we could confidently predict that Carl possesses an alternative source of energy from our own. If we are very sure of our observations and are able to rule out Carl's sneaking to the refrigerator at odd moments, we can be fairly certain that Carl is a robot-computer and not a man. Once again, a smart computer programer would have anticipated this and instructed Carl to periodically go through the motions of seeking food he neither needed nor desired.

But, before giving up on this approach (which, as I said, is potentially quite promising), why not try a cruel but perhaps critical experiment. Imagine that in the course of your conversation with Carl he mentions that he has a twin brother in Indianapolis with whom he claims a close and affectionate relationship. Affectionate relationships, of course, are usually marked by a good deal of emotional involvement which, in the event of tragedy, can lead to psychic upheavals that can be difficult to feign. If the task of distinguishing whether Carl is a robot or a man is sufficiently important, a nasty but potentially successful challenge is available. You might tell Carl that a cable arrived a few minutes ago which contained some terrible news. His brother was killed instantly in a multi-car collision outside Indianapolis. Carl-the-Man's reaction to such news can be confidently expected to differ significantly from the reaction of Carl-the-Robot. The absence of an emotional history, as well as the incapacity to experience emotion, is basic to computers, while human beings, even the most emotionally inhibited, could be expected to be stunned by such a revelation.

In discussing this scenario with an actress friend, she suggested to me that my grim experiment still would not differentiate a computer from a real person. "Any actor worth his salt could temporarily play the role of an acutely grieved relative, so why couldn't your computer do the same thing?"

I remain less convinced than my actress friend that such a performance could be successfully carried off. As a doctor, I've had the unhappy task on many occasions of informing relatives of sudden deaths. In such situations I've always been impressed that the relatives' responses bear little relation to the grief scenes I've witnessed on the stage or in the movies. There is a certain emotional resonance in such real-life situations that I don't believe can be precisely duplicated by an actor, no matter how skilled. This is not to say, however, that I believe our little experiment would necessarily succeed. Rather than a highly emotional response that might inadvertently come out as a "programed emotion," I believe a well-designed computer would respond as they usually do in any third-rate science-fiction movie: by showing no emotion at all.

In other words, although a computer might betray itself by the display of a recognizably false emotion, I don't think the absence of an emotional response would be much help at all in deciding between a computer and a human. (Does this suggest that our emotions are robot-like?)

All this leaves us without a dependable way to decide whether Carl is a robot or a mere mortal like ourselves. It seems we run into difficulties whenever we attempt to define an isolated piece of behavior as "human" or deny its programability in a computer. Does this mean there is no way to distinguish a computer from a man? Oxford psychologist Jerome Bruner suggests a solution to our dilemma. Instead of searching for isolated bits of behavior, Bruner, along with the American psycholinguist Noam Chomsky, suggests that we examine each human act within the total context in which it occurs.

"The significance of any feature is determined by its position in the structure," says Bruner. "The position of a piece on a chessboard, the function of a word in a sentence, a particular facial expression, the color and placement of a light, these cannot be interpreted without reference to the person's internalized rules of chess and language, the conventions he holds concerning human interactions, the traffic rules enforced inside his head."

As an illustration, Bruner refers to an experiment he carried out testing the time required for subjects to correctly identify a series of playing cards. At first sight, what could be easier—in fact, more robot-like?

In Bruner's experiment the subjects were seated while a series of playing cards were projected onto a screen in front of them. Most of the cards were normal, but some had been altered—e.g., a *red* six of spades and a *black* four of hearts. These anomalous cards pitted suit identification against color. Although a person might correctly recognize the spade, he would logically be expected to reject a red spade and point out the anomaly to the experimenter. Instead, the anomalous cards were always identified as normal. The black four of hearts, for instance, would be reported as the four of spades. Without hesitation, the anomalous cards were immediately identified according to context: The card was recognized as a spade, and the subject seemed not to notice that, in this case, the spade was red. Only after slowing the exposure time and questioning them about errors, did some of the subjects begin to realize their mistakes. But some of them never did. One of them exhibited a breakdown: "I can't make the suit out, whatever it is. It didn't even look like a card that time. I don't know what color it is now, or whether it's a spade or heart. I am not even sure what a spade looks like. My God!" From such experiments, Bruner and other scientists have suggested that human behavior can be understood only within a contextual framework: not just what is being said or done, but the setting in which speech or action is occurring.

A robot test with Bruner's apparatus would be likely to respond differently than a human being. When color and suit conflicted, the robot would be thrown into an irreversible conflict. Less dependent on context, it would immediately reject the anomalous card and reveal itself as a machine, not a man.

In his ingenious book *Understanding Natural Language,* Stanford University AI expert Terry Winograd demonstrates that human and computer language have certain key differences. Winograd gives examples of how computers are able to correctly analyze sentences that would be too ambiguous for human understanding. In one example, the computer is asked to identify one colored object from an array of assorted wooden shapes arranged on a table. "Does the shortest thing, the tallest pyramid supports, supports, supports anything green?"

"Yes, the green pyramid."

"What color is it?"

"By 'it,' I presume you mean the shortest thing the tallest pyramid supports. Red."

Such a conversational exchange is meaningless even at the level of two people conversing about the movement of blocks on a table. Presented with a sentence such as, "Does the shortest thing, the tallest, etc." a human can only stare blankly while reciting the possible implications of "supports, supports, supports." A computer, however, answers promptly and correctly. Even simple sentences, however, will result in the computer and the human coming to quite different conclusions.

In one example the computer chides its more fallible human operator for becoming temporarily confused regarding the directions he's given the computer about arranging some geometric shapes on a table.

OPERATOR: Why did you clear off that cube?

COMPUTER: To put it on a large green cube.

OPERATOR: Why did you do that?

COMPUTER: To stack up the red cube, a large red block, and a large green cube.

OPERATOR: Why did you do that?

COMPUTER: Because you asked me to.

According to Weizenbaum and other AI critics, the brain is not merely a machine and will never be duplicated in all its functioning by a computer, no matter how cleverly constructed.

"It may be possible to construct a conceptual structure that responds to the meaning of the sentence, 'Will you come to dinner with me this evening?' says Joseph Weizenbaum. "But it is hard to see how robots could possibly understand that same sentence to mean a shy young man's desperate longing for love. Even if the computer could simulate the feelings of desperation and love, is the computer then capable of *being* desperate and loving? Can a computer *then* understand desperation and love? The answer is no."

For the computer to be capable of human thought and emotion, it would require no less than a lifetime of human experience. According to this view, intelligence involves infinitely more than moving a set of blocks on command, or even drawing sophisticated conclusions from complex imbedded sentences ("supports, supports, supports"). Language developed in an evolutionary framework which favored survival, rather than from the moving of blocks on a laboratory table. "There is a lion in the bush!" prevailed over "Behind the grass which is overhung by the shadows cast by tall trees, is a movement that may signal a living creature."

In the final analysis, whether or not AI can be considered a successful model for how the human brain's performance depends ultimately on how limited a definition of "intelligence" we are willing to accept.

There are other areas of brain-machine interactions, however, that are more promising at the moment. Brain prostheses—or brain pacemakers, as they are commonly called—may enable psychobiologists to heal a damaged brain and, in the process, further extend our understanding of how an intact brain works.

18
A Seductive Analogy

Is there after all such a great difference from the point of view of the expansion of life between a vertebrate either spreading its limbs or equipping them with feathers, and an aviator soaring on wings with which he has had the ingenuity to provide himself?

TEILHARD DE CHARDIN,
The Phenomenon of Man

Must one first shatter their ears to teach them to hear with their eyes?

FRIEDRICH NIETZSCHE,
Thus Spake Zarathustra

Are eyes necessary for vision? Is it necessary to have ears in order to hear? At first glance, the answers to these questions seem obvious. Sight requires a functioning visual system. In order to hear, we must have ears as well as a conducting system for transmitting sound. Our eyes and ears, according to common sense, are absolutely necessary for us to "see" and "hear."

For the longest time, psychobiologists applied a similar common-sense approach to the study of how the brain processes visual and auditory input. It was thought that, starting with the retina in the eye and the tympanic membrane in the ear, sight and sound could be understood by way of the nerve-cell connections which terminate in the "visual and auditory areas" within the brain. There is much to recommend such a view. If a portion of the occipital lobes is destroyed in humans, blindness results. Soldiers with shrapnel wounds in this area provided poignant proof of the importance of the posterior part of the brain for visual function. From here, it was a short, logical step to conclude that if there was a "visual center," there must also be a "hearing center," a "touch center," and so on. Extension of this reasoning led to the "science of phrenology," in which all human attributes came to be associated with a unique location within the brain. Throughout this book we have stressed the narrowness of this localization theory of brain functioning. This doesn't imply, however, that every part of the brain is equal to every other part in every way. The principle of equipotentiality can only extend so far. The trick is to distinguish those brain functions which occur in comparatively localized brain regions from those which are diffusely, perhaps holographically, stored throughout the brain. One way of doing it comes from sharpening some of our concepts about human perception.

Consider the performance of two blind people. The first is totally cared for by his relatives, led around by a companion, and, for entertainment, spends the majority of his time listening to the radio. The second blind person lives alone, can skillfully negotiate the busiest streets with the help of a cane, and, for longer trips, takes advantage of the aid of a seeing-eye dog companion. The second man spends several hours a day "reading" books written in braille, and is presently an experimental subject in a project aimed at the development of ultrasonic equipment to help translate distance into patterns of sound.

Both men are "blind" in the legal and medical sense, but here the similarities end. The first man's contact with his environment is severely limited and he remains completely dependent on others for his continued survival. The second man, in contrast, has learned to use his sound and touch receptors as well as to integrate his body movements with the gentle nudges of a seeing dog. To an extent, he has enlarged his perceptual horizons through learning a form of *sensory substitution:* perceiving through sound and body movement those aspects of the world that sighted people perceive visually.

In recent years, scientists and inventors have applied their imagination to the development of new forms of sensory substitution. In the case of blindness the emphasis has been on an aid to increase mobility and facilitate reading. Until recently, however, little chance has ever been held out that a blind person could learn to "visualize" objects and people around him. But before such a seemingly fantastic hypothesis could even be explored, the knotty problem of definition had first to be resolved. How do we define vision? Most of us can see, but how do we describe the experience?

An operational definition of vision is helpful and goes something like this: If a person can provide detailed information about surrounding objects and people, can adequately describe his perceptions and respond appropriately to them, we can safely assume that the person possesses "vision." The issue then becomes one of determining which areas of the brain are absolutely essential for vision.

In lower animals, over 98 per cent of the visual cortex can be removed without interfering with the animal's ability to make complex visual-pattern discriminations. If these surgical alterations are produced in monkeys, the visual performance drops but by no means results in a blind animal. The monkeys continue to reach out and accurately snare small visually presented objects. When the monkeys are released from their cages, they show excellent spatial vision, limited principally by an inability to recognize particular objects in formal discrimination tasks. The explanation for the remarkable performances depends on the projection of impulses from the retina. While visual impulses in humans project almost entirely to the visual cortex, the projections in lower animals follow two pathways. Along with projections to the visual cortex, fibers originating in the retina also make synaptic contact in the midbrain. It is postulated that a laboratory animal's continued visual performance following the surgical removal of the visual cortex depends on the clues supplied by these fibers which connect with the midbrain. There is, in other words, a duplication in the animal's visual pathway that enables it to maintain vision after extensive damage to the visual centers in the cerebral cortex. Here, too, a form of visual substitution is taking place in which alternative brain mechanisms come into play to produce vision.

Does a comparable substitute mechanism exist in human beings? For years, such a system was considered impossible because visual cells in the cerebral cortex were thought to be responsible only for vision.

Experiments on the visual cortex of cats, however, provide proof that the brain cells in the "visual" areas may respond to any one of a number of different stimuli. In one study, sound and touch were found to activate almost half of the cells in a cat's visual cortex, demonstrating an associative or integrating role of at least some cells traditionally considered "visual."

From here, psychobiologists began searching for some way to translate one input signal into another. There are already reasons for thinking such transformations might ordinarily be taking place. Blind subjects, for instance, are capable of detecting the approach of a warm cylinder from three times the distance required by a sighted person. The usual explanation for such enhanced performance is that touch receptors become "overdeveloped" in order to compensate for the loss of vision. But an alternative, and even more exciting, possibility exists: The brain is able to convert touch impulses into vision!

Sensory information of all kinds reaches the brain in the form of nerve impulses, which differ according to their temporal and spatial patterning. Visual information, for instance, travels along the optic nerve according to patterns and nerve discharges. The actual optical images reach no further than the retinal receptor before they are converted into patterns of nerve-cell discharges that are interpreted by the brain as vision. The same holds true for hearing, smelling, tasting, and touching. In each case, the nerve impulse is the same; only coding and patterning are different. Thus, information transfer to the brain depends on the stimulation of a population of receptors in such a way as to produce coded nerve impulses which can be delivered to the brain for final interpretation.

In addition, there are many communication channels where different stimuli converge. The reticular formation performs its alerting function partly by providing a point of convergence for nerve impulses arriving from varied channels. Thus, impulses originating in the touch receptors of the skin synapse in the reticular formation along side cells, conveying impulses originating in the receptors of the retina. Associative links are thus formed in the lower brainstem which may be partly responsible for the integration of different sensory stimuli into unified perceptual experience. The work on microcircuits discussed in Chapter Nine lends further support to the concept that the brain works primarily via complex patterned events taking place in large and varied brain-cell populations. We no longer believe that the brain consists of centers responsible for the processing of the separate sensory impulses. Rather, all of the inputs are integrated throughout the system—from the primary receptors in the retina all the way up to the associative fibers in the cerebral cortex. Based on this, the brain should be capable, at least theoretically, of "interpreting" touch impulses as vision. But how to test such an unorthodox proposal?

Over the last ten years, Dr. Paul Bach-y-Rita and Dr. Carter C. Collins, of the Smith-Kettlewell Institute of Visual Science in San Francisco, have been developing a tactile vision substitution system which has revolutionized our ideas on how the normal brain processes information.

Starting with the theory that touch may be transformable into vision, Bach-y-Rita and Collins began by linking a small television camera to 400 small skin stimulators arranged in a 20 x 20 array and placed it on the back of a wheelchair. The blind subject sits in the wheelchair and leans back against the array while operating the hand-held TV camera. When the camera is directed at an object—say, a vase of flowers—each point of the television image is converted into a pattern of electrical stimulation delivered to the subject's back from some portion of the array. While the TV image consists of thousands of separate dots, Bach-y-Rita and Collins found that satisfactory resolution could be achieved through the use of only a 400-point array. With further experience, the stimulating array has been increased to 1,024 separate points laid out on a 32 x 32 grid.

At first, the subjects were taught to recognize horizontal, vertical, and curved lines. Later, they were able to identify geometrical forms such as triangles and squares. From here, a basic 25-word "vocabulary" was presented consisting of common objects such as telephones, chairs, or even a small model of a Volkswagen. With practice, recognition time drops from several minutes to only a few seconds. But to achieve that success, the subjects must operate the camera themselves. If a stationary TV monitor replaces the hand-held unit, object recognition fails.

Bach-y-Rita's system is based on the brain research we described in Chapter Six. The experiments with the active and passive kittens demonstrated the importance of active exploration. The kittens who walked around gained normal sensory co-ordination, while their passive counterparts remained helpless. Testing humans with prism goggles led to the same experimental finding: Active exploration is necessary for normal sensory integration. The blind subject's hand-operated TV can function as his visual receptor for conveying information to the skin where integration begins.

At first, perception takes place strictly on a trial-and-error basis. With increasing practice, however, the errors become fewer and the "visual" perceptions increasingly sophisticated. As the camera scans an object, an "image" is conveyed across the skin receptors where impulses are generated and forwarded to the brain for final interpretation. Even though the skin receptors originally carry impulses that are interpreted within the brain as "touch," the link-up with a primary visual substitute (the TV camera) enables the brain to reinterpret the impulses visually. At each step of the process, the active operation of the camera provides dynamic input concerning the object being scanned. The skin has, in a sense, been converted to an optical relay system from the artificial receptor (TV) to the brain.

Bach-y-Rita compares the experience of sensory substitution to learning how to drive a car. Initially, each movement has to be consciously and laboriously considered. Braking has to be hard enough to stop the car, but not so rapid as to jolt other passengers out of their seats. At first, full attention has to be given to the mechanism of turning the wheel precisely enough to maintain the center of the road, while at the same time retaining the flexibility needed for rapid lane changes. With practice, the adjustments become automatic and most drivers can carry out an animated conversation punctuated by periodic glances away from the road. With still more practice, the vehicle becomes an extension of the driver. The movements of the car as it dodges in and out of traffic can sometimes be anticipated down to a matter of inches. Obviously, when parking the car, the driver, for instance, cannot see his rear bumper or know that it is only two or three inches from the car behind him. Nevertheless, the skilled driver can sense the dimensions of his car with a precision that closely approximates the appreciation of his own body boundaries.

In a similar manner, blind subjects using sensory substitution start out with crude and clumsy efforts which are initially unsuccessful. With further practice, the input from the TV camera begins to integrate with the electrical stimulation administered to the skin surface. At a later stage, the integration becomes automatic and follows the same pattern encountered in normal vision. A telephone cord, for instance, picked up by a TV camera and relayed to the array in an excitation pattern, will provoke the response "a telephone." "A blind subject looking at a display of objects must initially consciously perceive each of the relative factors, such as the perspective of the table, the precise contour of each object, the size and orientation of each object, and the relative position of parts of each object and others nearby. With experience, information regarding several of these factors is simultaneously gathered and evaluated," says Bach-y-Rita.

The resulting images produced by the visual prosthesis can exhibit considerable detail. "Although tactile image recognition has not displayed the same level of performance as that of the eye, our subjects have learned to extract a surprising amount of pictorial information utilizing the tactile modality," says Dr. Carter Collins, a co-developer of the visual prosthesis. "Since there is no sign of saturation, this trend suggests that performance could theoretically be increased about four times if the skin of the entire trunk could be utilized."

Clearly the skin acts as an integrator of visual stimuli originating from the TV camera. The active manipulation of the camera seems to "program" the brain to operate in the visual mode. The light reflected from an object is recorded by the TV camera, which emits patterned impulses activating the vibrators. Each of the 1,024 vibrators corresponds to a segment of the image that orients a sighted person looking at the monitor. As the camera sweeps over the object, different vibrators are activated, allowing the "image" to move across the receptors on the skin.

Bach-y-Rita's substitution system works in a way that is dramatically different from the usual visual aids. A cane, for instance, is only an extension of the sense of touch which enables a blind person to detect an object before bumping into it. Other tactile displays are also extensions of the skin. The small cassette-sized Optacon device consists of an array of photo cells that are moved across the lines of print on a page. This is immediately converted to a tactile display produced by vibrating wire tips delivered to the surface of the subject's fingers. Despite the gain in reading efficiency (80 words per minute in some cases) the method does not involve sensory substitution. The skin is used to pick up tactile sensation and no conversion or substitution is carried out. In Bach-y-Rita's system, however, the skin takes over the role of an intermediary receptor between the eye (TV camera) and the brain.

With increasingly sophisticated technology, sensory substitution systems are becoming portable. In one system, a TV camera, weighing less than seven grams, is attached to a pair of spectacles. The circuitry for converting the camera image to the patterned electrical stimulus delivered by the matrix is housed in the pouch of a vest. Beneath the subject's shirt is the electrotactile stimulator. The entire system weighs about five pounds. As the apparatus becomes less bulky and conspicuous, more challenging experiments will soon become possible.

For instance, if several small letters are placed on a wall, a blind subject can enter the room, walk immediately to a letter, correctly identify it and, after some experience, pick it up. The whole performance sometimes takes less than fifteen seconds!

From here, Bach-y-Rita became interested in how his subjects would perform in more "natural" experiments. How would they react, for instance, if they were able to "view" their friends and relatives? To find out, Bach-y-Rita began substituting several *Playboy* pinups for the usual neutral photographs. He selected two blind college students as subjects and observed their reactions.

"We soon noted that, although they could describe much of the content of the photographs, the experience had no affectual component; no pleasant feelings were aroused," Bach-y-Rita recalls. "This greatly disturbed the two men, who were aware that similar photographs contain an affectual component for their normally sighted friends." After several repetitions of this experience, Bach-y-Rita discontinued use of the pinups. "We were not yet prepared to cope with the psychological aspects of the resulting situation."

But soon, even more distressing situations developed. Several of the subjects expressed their desire to turn the TV camera on their wives or girlfriends. When given the opportunity to do this, the resulting experience was often perceived as unpleasant. Bach-y-Rita recalls one subject who, after five minutes of manipulating the camera and "viewing" his girlfriend, turned away. Rather than a happy experience, the opportunity elicited feelings of sadness and disappointment.

At first glance, such responses seem surprising; but, as we have seen with Professor Gregory's patient, S.B., the recovery of sight may be deeply disturbing. Along with the perceptual modifications created by the newly acquired vision, the brain must also be capable of modifying a person's emotional responses. Paradoxically, this may be most difficult in those blind subjects who have made the best adjustment to their blindness!

In a study of emotional reactions to blindness, Dr. R. G. Fitzgerald discovered that it is easier to adjust to complete loss of vision than to only a partial loss. In Fitzgerald's words, the blind person must "die a sighted person in order to be reborn as a blind man." Acceptance of blindness and the willingness to acquire new skills turn out to be the two factors most critical for successful adjustment to blindness. With sight completely lost, the blind person must often struggle for years until he has formed for himself a new "sightless" identity. S.B. became a boot repairman, the head of a family, and a respected companion of his co-workers. But with the recovery of sight, all this vanished, leaving S.B. caught in a poignantly agonizing identity crisis: "Is sight possible? Will I have to restructure my whole life yet again?"

Bach-y-Rita and other psychobiologists now working in the area of sensory substitution are acutely aware of the ambivalence shared by many blind people regarding the prospect of acquiring vision. For this reason, sensory substitution is used as an adjunct in areas that seem important to their subjects. One recently married female subject lost interest in continuing with the substitution system only to enthusiastically return when the experimenters provided display objects which she could relate to her new life (e.g., kitchen utensils, etc.).

But in almost all cases, some benefits result. Few blind people, for example, have any appreciation of perspective and seem surprised, when "viewing" an object via sensory substitution, that the object's appearance changes when the camera photographs it from different angles.

Dr. Karl Frank, former director of the Neural Control Center of the National Institute of Health, recalls testing a patient with one of the early visual prosthesis devices. "The subject was skilled at immediately identifying common objects. Instantaneously, he correctly identified a cup which was placed alongside a telephone. When I turned the cup on its side, however, he was puzzled."

Despite similar perceptual changes, sighted people have no difficulty performing such a recognition task. As we discussed in Chapter Six, the brain assimilates the data, compares it with other sensory information, and concludes that a cup is unchanged and that only the angle of vision has been altered. After practicing with the sensory substitution system, most blind subjects are able to make similar deductions when confronted with objects presented from unusual perspectives.

The work of Piaget and Bower demonstrates that perspective is acquired in sighted people only through the interaction of the developing brain with its environment. Deprive the developing child of tactile and other sensory stimulation and it's likely that when tested years later, its performance would be similar to the blind person "seeing" the cup on its side for the first time via sensory substitution. It's Molyneux's question all over again: Will the blind person who acquires sight be able to distinguish the cube from the sphere? From S.B.'s experience, we know that he will not. After practice with sensory substitution, however, blind subjects also begin to make sophisticated deductions when confronting objects for the first time.

The results so far with the visual prosthesis have been gratifying. "Blind students have been trained to read graphs and meters, to analyze and debug electronic circuitry," says Collins. "A microscope-mounted TV camera has permitted blind students to study biological specimens, crystal structures, and microcircuits, and has permitted a blind worker to successfully complete industrial production, line assembly, and inspection jobs at the Hewlett-Packard Electric Plant in Palo Alto, comparing favorably in time and accuracy with other plant personnel." In a more recent development, tactile transformation of flight-instrument information enabled pilots to make satisfactory "landings" with a sophisticated aircraft flight simulator.

Along with tactile substitution devices, there have been attempts to provide vision by means of direct implantation of visual stimulators. In fact, the original visual prosthesis was envisioned as an internal device implanted within the brain. According to early speculations, stimulation of appropriate parts within the visual areas might produce a "visual" experience. Now, several years and millions of dollars later, most experts in the field are less confident that a working *internal* visual prosthesis will ever become a reality. For one thing, foreign materials are poorly tolerated within brain tissue. Up until now, most implants have stimulated scar formation within the brain, and sometimes the devices have even become ensnared within scar tissue. In addition, the electrodes, despite their tiny size—which sometimes approaches the limits of naked-eye visibility—are still too big. "There is a practical limit just how close the electrodes can be to each other," according to Dr. Karl Frank. "There is also a limitation on how frequently electrodes can be stimulated. These two limitations—electrode size and stimulus frequency—are major limitations that may never be completely overcome."

At the present time, workers at NIH are attempting to implant depth electrodes into the visual cortex. Instead of relying on superficial electrodes, the scientists hope, through depth electrodes, to achieve a breakthrough in the quality of image produced. Even if this is achieved, however, the feasibility of such an approach remains uncertain in light of the previously mentioned tendency of the brain to reject "foreign invaders." This, of course, provides a further stimulus for the development of external devices, such as the electrotactile visual prosthesis, capable of helping the brain to form visual images.

While the operative mechanisms responsible for the success of visual prostheses remain speculative, psychobiologists know a lot more about auditory prostheses. A single-channel electrode, for instance, when placed on the chorda tympani of the inner ear, can serve as a lipreading aid for the deaf. Ordinarily, lip reading proceeds via the rapid visual recognition of the lip and tongue movements employed in consonant formation. Vowels, in contrast, are not formed by the lips, but depend on the rate of movement of the vocal cords. Therefore, lip readers cannot discriminate one vowel from another. Despite this handicap, a skilled lip reader is usually able to fill in the vowels in the same way a reader can make sense out of sentences written with the vowels omitted.

410

"T ll th tr th" is quickly recognized as "tell the truth" both when written out and spoken in front of a skilled lip reader.

The auditory prosthesis picks up cadence information given off by the speaker's vocal cords and transforms it into an electronic signal, usually a beep or a buzz. The brain then combines the signal with the person's already acquired lip-reading skill to produce enhanced auditory receptive capabilities.

"Encouraged by the success of this single-channel prosthesis, we are now in the process of developing a multichannel electrode array that can be placed within the ear," says Dr. Terry Hambricht, Director of the NIH's Neural Control Division. "If we place enough electrodes along the cochlea to reproduce the frequency receptive capacities of the human ear, we will be able to combine multiple sounds, and eventually produce pitch information. Once a person achieves the ability to discriminate true pitch, then the rate of possible information transfer is increased tremendously."

When fully developed, an auditory prosthesis may enable profoundly deaf people to hear by literally feeling the words. Dr. David W. Sparks, of the Applied Physics Laboratory at the University of Washington, has developed such a system. Known by the acronym MESA (Multipoint Electrotactile Speech Aid), the system converts spoken syllables into pattern impulses that can be felt by the "listener" via a special belt.

In the initial step, a person talks into a microphone which feeds the signal into an electronic analogue computer that functions like the cochlea of the ear, breaking down the speech sounds into their pitch and loudness components. Next, the signals are registered on a matrix of 288 individual electrodes incorporating both pitch and loudness when held against the listener's abdomen. The pattern of the pulses is geared to the sound patterns of speech. So far, although the apparatus is still in the experimental stage, both lip readers and non-lip readers have demonstrated enhanced ability to follow speech.

Patients with strokes can also be aided by augmentation devices. Movement of a paralyzed hand, for instance, can be controlled by a tiny microprocessing unit strapped to the forearm to program the degree of pressure the hand must exert in order to carry out such acts as grasping and holding a cup of water. If the hand exerts too much pressure, the muscles can be made to relax. If the grip is too loose, tightening of the muscles can be brought about. But so far, such devices have suffered from the limitations of requiring visual monitoring on the patient's part. In the future, the goal is a more "natural" performance in which the person is free to concentrate on other matters while the hand carries the glass to the lips.

But other prosthetic devices are already in use in the management of strokes. Walking, for instance, can now be aided by the electrical stimulation of a nerve in the leg which raises the foot as the person walks. Ordinarily, a stroke victim cannot quite clear the ground with the front part of the foot and may stumble. The self-activated stimulator raises the foot at just the right moment. But first the walker must concentrate in order to properly time the nerve stimulation. After a while, the timing becomes almost as automatic as the act of walking itself.

Stimulators may also find application in states of chronic pain. Many patients who suffer from prolonged and intense pain are obtaining relief by activating a stimulating unit whose terminals are located within the depths of the brainstem. When the pain builds up, the patient simply pushes a button and activates the unit, which continues the stimulation until the pain goes away.

In each of the above examples, the prosthetic device is helping the brain to carry out functions it has been prevented from doing on its own. Along with this, however, the prostheses are also capable of temporarily augmenting the brain's performance in the event of dysfunction. Such devices are often informally referred to as "brain pacemakers," despite the fact that their method of operation differs considerably from other body pacemakers.

Cardiac pacemakers, for instance, maintain a regular heartbeat via periodic electrical stimulation to the heart. The cardiac pacemaker actually takes over the role of regulating the heart's rhythm, a task usually performed by intrinsic tissue located within the muscle of the heart wall. From the first prolonged clinical application, in 1952, cardiac pacemakers have grown into a hundred-million-dollar-a-year industry helping over 300,000 patients. In fact, it has been estimated that a cardiac pacemaker is implanted every five minutes somewhere in the world. Will "brain pacemakers" ever achieve this degree of success? There are several reasons why this is unlikely.

First of all, pacemakers for the control of pain are likely to be superseded over the next few years by new developments in the area of nonaddictive painkilling drugs. From the work on the endorphins described in Chapter Fifteen, breakthroughs are likely to come about in the control of pain. This alone will eliminate the largest single demand for "brain pacemakers."

Despite this, limited, but nevertheless lifesaving, applications will continue in the near future. For one thing, electrophrenic stimulation will be used to artificially ventilate patients with injuries in the spinal cord, as well as sufferers from the mysterious disease Ondone's curse, marked by an arrest of respiration accompanying sleep or unconsciousness. In other cases, prostheses will be used to empty the bladder of paralyzed patients. Finally, a prosthetic stimulator may be capable of preventing seizures and, in some cases, modifying behavior.

Although crystal-ball predictions are hazardous in any field, certain carefully qualified expectations seem reasonable regarding the future of brain pacemakers. In cases of blindness and deafness, they will offer the one and only hope for the restoration of sight and hearing. But, at the moment, the exact direction prosthetic research may take is still largely a matter of speculation. Nor should this be surprising. Since we are only beginning to understand the workings of the normal brain, we are at a distinct disadvantage when it comes to explaining what goes wrong to produce a dysfunctional brain. And, as a consequence, we are less prepared to provide prosthetic devices capable of correcting the dysfunction.

Brain research, however, is a two-way street. Just as greater understanding of the mechanisms of brain function lead to breakthroughs in the possible application of this knowledge, so, too, innovative applications such as the brain prosthesis can boost our confidence that our theories about brain function are sound. The brain prosthesis work provides proof that a holistic concept of brain function is undoubtedly correct. Certainly, according to traditional theories, vision based on a pattern of touch stimulation sounds like an unrestrained fantasy. But already progress in visual substitution systems indicates that greater advances can be expected in the near future. Will the blind be able to see and the deaf hear through advances in psychobiology? At this point, no one would dare to confidently predict such developments. Besides, the question may not be nearly as important as what precautions we will take in order to help the blind and deaf to adapt to their new-found "gifts." After less than ten years of active research in neural prostheses, the human response to newly acquired sight indicates the likelihood of profound, but not always happy, personality and behavioral alterations.

19
The Verdict

The verdict is not suddenly arrived at, the proceedings only gradually merge into the verdict.

FRANZ KAFKA

A teacher who hesitates to repeat shrinks from his most important duty, and a learner who dislikes to hear the same thing twice over lacks his most essential acquisition.

WILLIAM GOWERS

In this last chapter I would like to sum up some of the implications of psychobiology. As I mentioned in the introduction, I believe psychobiology can help us with many of the social problems we are now facing. For this reason, I've presented brain research from the point of view of its relevancy for social policy. Stated simply: The world we're in the process of creating for ourselves and future generations is the product of our brains; therefore, only by understanding how our brains work, can we hope to achieve true insights regarding individual and species motivations.

Already, there has been a profound change in our understanding of the importance of conscious experience as a determinant of why people and nations behave as they do. When behavior is "explained" in terms of stimulus/response or other strictly observable phenomena, dehumanization is inevitable. From the model of a rat in a cage pushing buttons to obtain rewards, it is only a short conceptual step to describing human behavior as differing only by the size of the cage and the complexities of the rewards.

But once it's realized that internal mental processes exert control over individual and collective behavior, certain consequences inevitably follow. Subjective feelings, values, and goals thus become the primary determinants of human action. In addition, as knowledge about brain function advances, there is an increasing need for approaches to human behavior based upon the psychobiological study of consciousness, its transformations, and especially its deviations. This places psychobiology in the unique position of bringing about unity between brain research and our highest humanistic and cultural aspirations. For the first time in history, the interest of science and the interest of the humanities coincide.

Since man is a product of his brain, self-understanding will ultimately depend on the degree and depth of understanding man achieves regarding his own brain. At the basis of our interest in the brain lies a curiosity regarding, in Paul MacLean's words, "why we are here, what we are doing here, and where we are going." Despite its reasonableness, few contemporary philosophical thinkers seem to have grasped this rather self-evident truth. The writings of most of them remind one of the productions of a medieval monk who sits complacently in his cell while self-indulgently spinning high-sounding theories about how the mind "must work." If nothing else, psychobiology in the last fifteen years has pointed up the intellectual hauteur and bad faith inherent in such a narrow approach. The question of human "thinking" cannot be examined in the same way a group of medieval theologians once argued the question as to whether angels, under certain conditions, were capable of lying.

Such easy-chair approaches are not appropriate when it comes to determining the relationship of the mind to the brain. We must get beyond formulations which postulate the brain as of interest only to untidy biologists who must occasionally enter messy laboratories in order to employ their trained powers of observation as, presumably, cumbersome and inconvenient checks on their imaginations. Fortunately, such attitudes are rapidly changing. From here on in, those who would expound on the mysteries of human thought will have to know as much about brain physiology as they do about Greek philosophy.

In the last fifteen years, progress in psychobiology has forced us to rethink some of our most fundamental philosophical positions. "It now becomes important in any instance to distinguish between pre-1965 and post-1965 versions of a given philosophic stance," says Roger Sperry in regard to just one aspect of psychobiological research, the split-brain findings.

The work of Paul MacLean encourages us to formulate a concept of human nature that correlates our brain functioning with that of our remote reptilian and early mammalian ancestors. Religious and moral leaders from Christ to Gandhi have insisted that all men are brothers. Others, such as St. Francis of Assisi and certain Eastern mystics, have gone even further, emphasizing the kinship of all living things. Now our knowledge of the triune brain, particularly the limbic system, establishes a scientific basis for these assertions. The limbic system is a repository of ancient emotional responses that are pre-logical—or "unreasonable," if you prefer. These are inherited by us from our less "advanced" ancestors and expressed through the action of those parts of our brain which we share with the reptiles and early mammals. This implies, in my opinion, that what has been referred to over the centuries as "human nature" is to a large extent genetically programed.

As a corollary to this, our emotions are probably untrustworthy when it comes to providing us with a basis for action. Fear and aggression, for instance, were useful during thousands of years of prehistory when our ancestors had to battle for survival against savage and cunning enemies. But today, these same emotions, when unrecognized and unchecked, lead to such dangerous acts as the relentless stockpiling of nuclear arms, or the unnecessary expansion of territorial borders. MacLean's explanation for this is summed up perfectly in his reference to our "paranoid streak."

To what extent are we able to free ourselves from the influence of our ancient animal forebears? Do we want to cut ourselves off from our biological origins? How can we retain what is good and eliminate or control the destructive consequences that may result from the unchecked expression of the limbic system? These are only three questions that psychobiology may help us to answer over the next several years.

But the answer to such questions requires, I believe, no less than a re-examination of what it is to be human: in essence, a restatement of the biological and psychological underpinnings of our "human nature." In doing this, we are moving well beyond the area of biology and psychology into the domain of the philosopher and theologian. In essence, can we establish a biological foundation for ethics?

"Value systems are probably influenced to an unknown extent by emotional responses programed in the limbic system in the brain," according to Edward Wilson. In the last part of the twentieth century, the task of psychobiology will be to help us define these constraints imposed on us by our brains.

One of the steps toward accomplishing this involves a radical re-evaluation of our present explanations as to why people think and act the way they do. Behavioral psychology as a legitimate explanation of human behavior is a failure and should be quietly buried at a small private funeral limited to B. F. Skinner and other members of the immediate family. In its place, we must develop a psychology that is relevant to the everyday world we all inhabit. Most of us don't spend our lives in laboratory cages, and an explanation of the subtleties of human behavior based on such a model is ludicrous.

We behave the way we do because we are endowed with a brain which enables us to have certain experiences, think certain thoughts, and carry out certain actions that seem desirable. From the study of illusions and those behavior patterns that have a strong genetic component, we're beginning to realize some of the limitations imposed on us by our brains. The question "What is the nature of reality?" is turning out to be too ambitious. "What is the nature of the reality that our brains are capable of experiencing?" is more appropriate. We already know that our brain is capable of perceiving certain parts of the "real world" and not others. In addition, as seen from Kinsbourne's study of left-hemisphere activation, the questioning process itself favors the perception of some aspects of the world to the exclusion of others which are perhaps equally important.

A physicist who studies subatomic particles must remind himself constantly that, according to the Heisenberg principle, the method of observation can influence the results obtained. So, too, our brain, while it provides us with an exquisitely sensitive instrument for gaining knowledge about the world, also contains inherent limitations. We are bound by such constraints as our need for logical consistency, purpose, and intelligibility. These are important attributes for us, but, so far, there are no indications that the physical universe must conform to them. The world existed for millions of years before our brains developed. Why, then, should we presume that the universe must somehow conform to mental operations imposed on us by the organization of our brain? Yet, what other course is open to us than to base our philosophy on what we can learn about how our brain works? Our dilemma was anticipated by the philosopher Democritus, who had the senses address the inquiring mind: "Wretched mind, do you, who get your evidence from us, yet try to overthrow us? Our overthrow will be your downfall." Truly, the only way we can deepen our self-understanding is by learning as much as we possibly can about how our brain works. This will not eliminate the need for the accurate observation and interpretation of behavior, but it will put behavior in the correct context: the manifestation of the working brain.

Along with these theoretical concerns, it seems to me that psychobiology implies the adoption of certain practical measures. As things now stand, the typical graduate of most major universities knows next to nothing about the workings of his own brain. This is particularly ludicrous when you consider that all mental life is directly dependent on the integrity and optimal functioning of the

419

brain. How absurd that a person, in order to understand the world, should study history, sociology, even psychology, yet possess no information at all about the regulator of *all* human activities. As I have tried to show throughout this book, if our brains were different, the history of our species would be different as well. Certainly when it comes to the development of individual brains, this is clearly true. Malnutrition in infancy, brain damage due to trauma or vascular disease, the mysterious brain-cell deaths that result in senility—all effect the individual in ways that are so fundamental that sometimes it requires a continuing moral effort on our part to recognize, in the worst cases, the continued humanity of the afflicted individual.

In the past several years, there has been an awakening on the part of large numbers of people that they should bear some responsibility for their own health. Although it is probably unrealistic to envision a time when everyone is his own doctor, the principle behind such efforts is sound. Knowledge about our health and how to preserve it is personally liberating, since, for one thing, when we learn about our bodies we no longer remain helpless and dependent on health care "experts" for every illness, however trivial. I believe that if this attitude can be applied to the brain, the results are likely to be even more beneficial. Along with understanding the harmful effects on the brain of such factors as poor diet and high blood pressure, our understanding of how the brain works offers opportunities for self-realization and growth. As psychobiologists establish increasingly accurate correlations between behavior and brain functioning—as they are already doing and can be expected to do in the future —further opportunities will be available to apply this knowledge toward increasing personal as well as social potential. The concept of disregulation introduced by Gary Schwartz—to take just one example—provides a conceptual framework for evaluating our response to stress, the degree of physical activity we require for optimal health, even ways of combating certain illnesses.

Along with these indications of future progress, I think caution in our efforts to understand the human brain is in order. As Chomsky emphasized, explanation itself is the product of our brain and is probably as limited as our other powers. We cannot fly or live forever (at least in the physical sense of never dying), and our bodies impose severe limitations on how fast we can run, reproduce, and even think. Our comprehensive powers, too, are no less limited, which means that the brain may or may not be accessible to our understanding.

As I progressed with this book, my own attitude toward psychobiology underwent a change. Although I still think that this is the most exciting area of current human intellectual endeavor, I have, nonetheless, become less confident that the mysteries of the human brain are going to be totally explained by the investigations of scientists. This is not quite the same thing as believing in a "spiritual" substance; nor is it an excuse for foraging in the labyrinths of mentalism. It's based on how psychobiology has progressed in the past.

The history of psychobiology, as with all the sciences, has passed through phases of explanation which depended for their credibility on ideas that were popular at the time. When Aristotle, for instance, postulated that the heart was the organ responsible for the mind, his explanation was based on the fact that human dissection had revealed the importance of the pumping action of the heart; the brain still awaited an anatomist observant enough to realize that the brain is much more than an elaborate refrigerator to cool the blood.

We must remember that our concept of a physical explanation changes over time. Today we accept without question the idea of gravitational field as a physical explanation. We are also comfortable with massless particles and antimatter. These concepts would be logically untenable to a medieval logician to whom the existence of a particle would logically contradict the absence of its mass. In a similar way, our concept of physical explanation can be expected to change. At least consider the possibility as to whether, in Chomsky's words, "the evolution of human mentality can be accommodated within the framework of physical explanation as it is presently conceived, or whether there are new principles, now unknown, that must be invoked, perhaps principles that emerge only at higher levels of organization than can now be submitted to physical investigation."

I'm suggesting that the mysticism of one generation is often the scientific demonstration of the next. Perhaps this is what Oscar Wilde meant when he defined science as "the record of dead religions."

Despite these cautionary notes, I think there is room for a lot of optimism. From studies on malnutrition, we can now state with almost absolute certainty that a "good" brain requires a minimum of food received in a stimulating and enriching environment. Today, the brains of over three hundred million children throughout the world are developing below their potential because of inadequacies in diet and their social and cultural environments. What will the rest of us do about this? We can ignore it, as we've done in the past, and retreat into the privacies of our reptilian R-complexes; or, preferably, we can operate on the cortical level and reason out systems of social equity. In either case, psychobiology is forcing on us some hard facts as replacements for centuries of empty speculation. If an inadequate diet in deprived communities spawns unhealthy brains, then we must take direct responsibility for our unwillingness to try to improve this aspect of those people's lives. If we're not exactly our brothers' keepers, it seems that we are, at least, the keepers of our brothers' brains.

In addition, psychobiological research has something useful to tell us about the brain development of prematures. The newborn nursery experience suggests that special kinds of stimulation involving movement and physical closeness are needed to guarantee the brain's normal development. Nurseries are not just warehouses for children unfortunate enough to be born before their expected time. In less than ten years, we have progressed from a bleak, depressing picture of brain development in prematures toward a hopeful optimism that their brains, too, can develop normally when provided with proper stimulation. As with many of the things we've discussed in this book, attitude changes follow directly from psychobiological research. I expect that, during the next quarter of a century, similar findings regarding senescence will foster new attitudes toward aging. Senility and mental deterioration will be understood and prevented, at least in part. I believe that most of the present pessimism and neglect about the aged spring from our inability to do anything about brain aging. From work on slow viruses, arteriosclerosis, and hereditary forms of mental deterioration—to mention just three promising research areas—brain scientists may eventually be able to prevent the more severe forms of senility.

Premature infants and tottering octogenarians, at first sight, seem to share very little except their humanity. But they also share the danger that the rest of us may define them as nonpersons. Euthanasia and abortion are two of the most controversial medical social issues of our time. I think psychobiological research on human development does away with the idea that, at a certain point, we can define a fetus as a "person"; or, at another point, require a ninety-year-old person to surrender his humanity like an outdated passport. Brain development progresses in stages in which the total is always greater than the sum of its parts. Psychobiologists cannot tell us exactly when consciousness appears.

Alexander Luria's work in the steppes of Central Asia proves that consciousness varies among individual societies in different historical periods. For this reason, it's unlikely that there will ever be a scientific technique for measuring the brain that will be sensitive enough to detect the presence of "personality" or anything as elusive as "consciousness."

Some people place great emphasis on EEG. But in the absence of a recordable EEG, or brain wave, in a fetus, what conclusion should be drawn? Does it indicate that there is no brain activity? or only that our measurements are insufficiently sensitive? Cosmologists have dealt with similar kinds of questions for centuries. Every astronomer who ever proclaimed that his telescope could envision the "outer limits" of the universe, became an object of ridicule to the developers of the next most powerful telescope.

Since we cannot say for certain when the brain is capable of "human performance," it seems to me we should stop deceiving ourselves about abortion and euthanasia. These are not issues that can be settled by reference to studies on early brain development. Nor is there any indication that we will ever be able, in the foreseeable future, to state with assurance that at a certain point the fetal brain assumes "human" dimensions. All along the continuum from birth to death the human brain's performance, at any point, is a trade-off between its potential and the uses to which it is actually put. Psychobiology, thus, has little to offer, other than caution, on the subjects of abortion and euthanasia. If we can't decide when consciousness emerges in a person, or even a society, how can we possibly justify destruction of an undeveloped or a senile brain? Besides, since every one of us possesses a triune brain inherited from our prehistoric ancestors, who can even begin to define a "human level" of brain performance?

The history of Western philosophy can be considered a record of our speculations about our place in the universe, how we came to be, the origin of the world and its "purpose." Since we seem to be the only creatures on earth concerned with these questions, they take on a particularly human quality. Implicit in philosophical speculation, however, are certain assumptions that may reflect more on the nature of the human brain than the structure and reality of the universe. Our brain perceives, processes information, and regulates our behavior according to certain relatively fixed principles of operation. Added to this, our brain, in many cases, may not utilize the best strategies when it is attempting to arrive at the truth. A particularly striking example of this involves how we calculate probabilities.

Most of us live in highly predictable environments. Our jobs, our friends, even our experiences, change very little from day to day. Most of the choices and decisions we make involve very little in the way of probability measurements. Accidents, divorces, the loss of a job—these are less predictable but, except for life insurance policies, are rarely taken into account when we plan our futures. This predictability of our environment shields us from the glaring inadequacies we all feel when we try to calculate probabilities.

Imagine a hypothetical situation where I have just introduced you to a solemn, rather pompous-looking man in his mid-forties who dresses in tweeds and speaks with a terribly proper and cultured Boston accent. Is he a laborer or a professor? Most people would reply at once that he's a professor, ignoring probability considerations: The total number of laborers in the world far exceeds the number of professors. The reason for this choice is, of course, that most people have quite definite ideas about the kind of person who teaches at a university and how he is likely to differ in appearance and behavior from someone who digs foundations for buildings.

Based on observations of dress, speech, and manner, the brain ignores strict probability considerations and decides that solemnity, tweeds, and a cultured Boston accent are more likely to belong to an academic than a laborer. Such mental operations are performed every day and, in fact, have to be if we are to survive in the "real world" of practical observations and decisions. Probability considerations are routinely preempted all the time by "knowledge" gained from experience. Even if I were to somehow locate a laborer who speaks and acts like a professor, you would be wise to ignore this exception and, given the identical situation, choose on the basis of your "knowledge of the world" rather than on probability considerations alone. In essence, the human brain, except under highly artificial conditions, doesn't deal with probabilities at all, but with "rough and ready" generalizations which, in ordinary life, are correct 99 per cent of the time.

Based on considerations such as these, it's difficult—for me at least—to arouse any enthusiasm for the prospect of our brains' arriving at anything other than equally "rough and ready" philosophical explanations. When it comes to speculation on the likelihood of an afterlife, for example—to take an extreme instance of speculative thought—it seems to me that the only truly honest answer is, "We don't know." There is nothing in our genetic or personal histories that provides a basis for anything other than a wild guess about such matters. The human brain is simply not capable of estimating the probabilities of such a likelihood. Such questions involve judgments of a type where common knowledge of our "everyday world" is of little help. Nor is there any reason to think that a deeper understanding of our brain, no matter how detailed, will enable us to determine these probabilities. Even asking such questions sets in motion certain brain events that prejudice the types of answers we are capable of arriving at. I'm reminded, once again, of Kinsbourne's studies on how verbal performance can distract from our perceptions of the left visual field. Does a similar situation exist in our attempts to understand meaning? "Why am I here?" "What is life's purpose?" There is little evidence that our brains are capable of answering these questions.

Since the brain is unlike any other structure in the known universe, it seems reasonable to expect that our understanding of its functioning—if it can ever be achieved —will require approaches that are drastically different from the way we understand other physical systems. This is not a call for nihilism, but humility. Our most striking recent conceptual advances in psychobiology—microcircuitry and holography—are the result of having the courage to relinquish traditional and no longer rewarding viewpoints in favor of bold, imaginative, and innovative conceptual schemes. By the turn of the century, I suspect our approach to the brain will again be radically different from our present orientation. These changes in our understanding will probably closely parallel the changes that have already taken place, thanks to nuclear physics, in our understanding of the physical universe. Nor should we be surprised or discomforted by such developments. Why should we expect the world to conform to our notions? In the words of V. B. Mountcastle, the dean of living psychobiologists: "Each of us believes himself to live directly within the world that surrounds him, to sense objects and events precisely, and to live in real and current time. I assert that these are perceptual illusions. Contrarily, each of us confronts the world from a brain linked to what is 'out there' by a few million fragile sensory nerve fibers, our only information channels, our lifelines to reality."

Is there, therefore, any more reasonable response to such considerations than humility? After all, are not our brains a part of the same physical universe whose essential nature remains, after thousands of years of speculation, essentially mysterious?

Notes

CHAPTER ONE
The Philosopher's Myth

The chess computations are taken from "The Glorious and Bloody Game" in Arthur Koestler's *The Heel of Achilles,* essays 1968–73 (London: Hutchinson & Co., 1974).

The example of the eight-year-old boy is modified from Gilbert Ryle's "Descartes' Myth" in *The Concept of Mind* (London: Hutchinson & Co., 1949).

CHAPTER TWO
Dr. Punishberg and Dr. Rewardnick

The observation of the behavior of laboratory rats is taken from the work of Professor S. A. Barnett carried out over the past twenty-five years. An excellent summary of Dr. Barnett's findings is found in "Activity, Exploration, Curiosity and Fear: An Ethological Study," *Interdisciplinary Science Reviews,* Vol. I, No. 1 (1976), pp. 43–62.

The limitations of behaviorism are described in *Behaviorism and the Limit of Scientific Method* by Brian D. MacKenzie (London: Routledge and Kegan Paul, 1977).

The dynamic observations of behavior (contrasted with behaviorism's static and unimaginative explanations for the same observations) were suggested by the work of the late Russian psychologist Alexander Luria. The best summary of Luria's thinking is contained in "Neuropsychology: Its Sources, Principles and Prospects," in *The Neurosciences: Paths of Discovery*. This essay, as well as Luria's *The Working Brain,* provide the inspiration for Chapter Three.

CHAPTER FOUR

Answers to Questions Which Have Not Yet Been Raised

The Paul MacLean material is from interviews carried out in 1976–77 as well as the following papers and books: "A Triune Concept of the Brain and Behavior," University of Toronto; *Zygon/Journal of Religion and Science,* Vol. VIII, No. 2 (June 1973); "On the Evolution of Three Mentalities," *New Dimensions in Psychiatry: A World View,* Vol. II, ed. by Dr. Silvano Arieti and Dr. Gerard Chrzanowki (New York: John Wiley & Sons, 1977); "The Paranoid Streak in Man," in *Beyond Reductionism* (London: Hutchinson & Co., 1969); "Cerebral Evolution and Emotional Processes: New Findings on the Striatal Complex," *Annals of the New York Academy of Sciences,* Vol. 193 (August 25, 1972), pp. 137–49. An excellent account of MacLean's work was published in 1976 by the National Institute of Mental Health as an NIMH program report, entitled the "Archeology of Affect," by Anne H. Rosenfeld.

CHAPTER FIVE

The Murder of a Child

The Edward O. Wilson quotes are from an interview on October 27, 1977 at the Museum of Comparative Zoology, Harvard University. Quotes are also taken from Wilson's *Sociobiology: The New Synthesis* (Cambridge, Mass.: Harvard University Press, 1975) and from "The Social Instinct," *Bulletin of the American Academy of Arts and Sciences,* Vol. 30 (1:11–25), 1976.

The Wallace–Darwin exchange is from a personal communication from Arthur Koestler, November 1977.

The Harry J. Jerison quotes are from "Evolution of the Brain" in *The Human Brain* by M. C. Wittrock, et al. (Englewood Cliffs, New Jersey: Prentice Hall, 1977), and from "Paleoneurology and the Evolution of Mind," *Scientific American,* Vol. 234, No. 1 (January 1976), pp. 90–101.

CHAPTER SIX
Mirror of Reality

The D. H. Lawrence quote is from *The Complete Poems of D. H. Lawrence* (London: William Heinemann), reprinted in *Imagist Poetry,* ed. by Peter Jones (New York: Penguin Books, 1972).

The Hubel and Wiesel work was originally published as "Receptive fields, binocular interaction and functional architecture in a cat's visual cortex," *Journal of Physiology,* 160 (London, 1962), pp. 106–54, and "Shape and arrangement of columns in a cat's striate cortex," *Journal of Physiology,* 165 (London, 1962–63), pp. 559–68.

The Jerome Lettvin work is from J. Y. Lettvin, et al., "What the Frog's Eye Tells the Frog's Brain," *Proceedings of the Institute of Radio Engineers,* 47, pp. 1940–51.

The Colin Blakemore experiments were described in "Development of the Brain Depends on the Visual Environment" by C. Blakemore and G. Cooper, in *Nature,* 228 (1970), pp. 477–78. A review of this and similar material was written by Blakemore in "Developmental Factors in the Formation of Feature Extracting Neurons," which was published as Chapter 10 in *The Neurosciences, Third Study Program* (1974), pp. 105–13.

The Richard Gregory material is from an interview on August 2, 1977, at St. Catherine's College, Oxford, England. Material is also taken from *The Intelligent Eye* by Richard Gregory (London: Weidenfeld, and Nicolson, 1970); *Eye and Brain: The Psychology of Seeing* by Richard L. Gregory (New York: McGraw-Hill, 1966); "Recovery from Early Blindness: A Case Study," in *Concepts and Mechanisms of Perception* by Richard L. Gregory (London: Gerald Duckworth & Co., 1974), pp. 65–129.

The Cree Indian Study is from R. C. Annis and B. Frost, "Human Visual Ecology and Orientation Antistropies in Acuity," *Science,* 182 (1973), pp. 729–31.

The T. G. R. Bower material is from "Repetitive Processes in Child Development," *Scientific American,* Vol. 235, No. 5 (November 1976), pp. 38, and *The Perceptual World of the Child,* T. G. R. Bower (Cambridge, Mass.: Harvard University Press, 1977).

The R. L. Fanz work is described in "The Origin of Form Perception," *Scientific American,* Vol. 204, p. 66 (1961).

The Gibson "Visual Cliff" experiments were described in "Visual Cliffs," *Scientific American* (April 1960).

The Stratton experiments were originally described in "Some Preliminary Experiments on Vision," *Psychol. Rev.,* No. 3, 611 (1896); "Vision without inversion of the retinal image," *Psychol. Rev.,* Vol. 4, 341 (1897).

The Held and Hein experiments were described in "Movement produced stimulation and the development of visually guided behavior," *Journal of Comparative and Physiological Psychology,* R. Held and A. Hein, 56, pp. 872–76 (1963).

CHAPTER SEVEN
Children of the Moon

The waterbed material is from "Effects of Flotation on Premature Infants: A Pilot Study," *Pediatrics,* Vol. 56, No. 3 (September 1975), by Anneliese F. Korner, Ph.D., et al.; and from "Reduction of Apnea in Premature Infants Through Oscillating Waterbeds," a paper presented at a meeting of the American Psychological Association in San Francisco, August 30, 1977, by Anneliese F. Korner, et al.

Material on the visual cortex of premature infants is

taken from "Morphogenesis of the Visual Cortex in the Preterm Infant" by Dominick P. Purpura, in *Growth and Development of the Brain,* ed. by Mary A. B. Brazier (New York: Raven Press, 1975).

The effects of enriched environments on brain development are described in "Brain Changes in Response to Experience" by Mark R. Rosenzweig, Edward L. Bennett, and Marian Cleeves Diamond in *Scientific American* (February 1972).

The Albert Globus and J. Altman studies are described in "Early Sensory Influences on Regional Activity of Brain ATPases in Developing Rats" by Esmail Meisami, in *Growth and Development of the Brain: Nutritional, Genetic and Environmental Factors* by Mary A. B. Brazier (New York: Raven Press, 1975).

The material on the effects of prematurity is from "The Consequences of Prematurity: Understanding and Therapy" by Arnold J. Sameroff and Lauren C. Abbe, in *Psychology: From Research to Practice* (Washington, D.C.: American Psychological Association). The effects of nutritional deprivation on brain growth was developed at the International Conference on Behavioral Effects of Energy and Protein Deficits, held November 30 through December 2, 1977, in Washington, D.C.; and by interviews with Merrill S. Read (Growth and Development Branch of the National Institute of Child Health and Human Development), Dr. Janina R. Galler (Department of Child Psychiatry, Boston University Medical Center), and Robert B. Livingston (Neurosciences Department, Institute for Information Systems, University of California, San Diego).

CHAPTER EIGHT
Cain's Curse

The material on the effects of movement on the development of violence is based on interviews with Dr. James W. Prescott, of the National Institute of Child Health and Human Development, in January 1978. The

433

material is also developed in "Phylogenetic and Ontogenetic Aspects of Human Affectional Development," in *Selected Proceedings of the 1976 International Congress of Sexology,* Robert Gemme and Connie C. Wheeler, eds. (New York: Plenum, 1977); "Early Somatosensory Deprivation as an Ontogenetic Process in the Abnormal Development of the Brain and Behavior," in *Medical Primatology* (1970); Second Conference exp. Med. Surg. Primates, New York, 1969; "Developmental Sociobiology and the Origins of Aggressive Behavior," a paper presented at the XXIst International Congress of Psychology, July 18–25, 1976, Paris; "Somatosensory Deprivation and Its Relationship to the Blind," in *The Effects of Blindness and Other Impairments on Early Development,* Zofja S. Jastrzemska (The American Foundation for the Blind, 1976).

The Robert Heath material is from "Modulation of Emotion with a Brain Pacemaker: Treatment for Intractable Psychiatric Illness," *Journal of Nervous and Mental Disease,* Vol. 165, No. 5 (1977), pp. 300–17.

The effects of early maternal infant experiences on later activity are described in "Effects of Maternal Mobility on the Development of Rocking and Other Behaviors in Rhesus Monkeys: A Study with Artificial Mothers" by William A. Mason and Gershon Berkson, in *Developmental Psychobiology,* 8 (3) (New York: John Wiley & Sons, 1975), pp. 197–211; and in "Looking Behavior in Monkeys Raised with Mobile and Stationary Artificial Mothers," Robert F. Eastman and William A. Mason, in *Developmental Psychobiology,* 8 (3) (New York: John Wiley & Sons, 1975), pp. 213–21.

CHAPTER NINE
Jumping Frogs and Purple People

The story of the development of lithium is set forth in "The Story of Lithium" by John F. Cade, in *Discoveries in Biological Psychiatry,* Frank J. Ayd, Jr., ed. (Philadelphia: J. B. Lippincott, 1970).

The rivalry between Ramon J. Cajal and Camillo Golgi regarding the ultimate nature of the neuron is described in "Neuroanatomical Basis of Epilepsy: The Neuron Doctrine," in *Science and Epilepsy: Neuroscience Gains in Epilepsy Research* by James L. O'Leary and Sidney Goldring (New York: Raven Press, 1976).

The discovery of the electrical nature of the nerve impulse is described in "From Animal Electricity to Electrophysiology" by O'Leary and Goldring, ibid., pp. 61–84.

An excellent up-to-date review of the present understanding of brain chemistry and behavior is found in "Behavioral Neurochemistry: Neuroregulators of Behavioral States" by Jack D. Barchas, et al., *Science,* Vol. 200 (May 26, 1978), pp. 964–73.

The story of the development of our present theories about "second messengers" in the brain is described in "Second Messengers in the Brain" by James A. Nathanson and Paul Greengard, *Scientific American* (August 1977), pp. 108–19.

The comparison between brain "states" and neutrinos in the universe was made by Dr. Kenneth Boulding, President of the American Association for the Advancement of Science, at its annual meeting on February 7, 1978, in Washington, D.C. Similar material was developed in a letter to the author, April 10, 1978.

The brain's microcircuits are described in "Electrotonic Processing of Information by Brain Cells: Recent research augurs an important role for neuronal local circuits in higher brain function," Frances O. Schmitt, Parvati Dev, and Barry H. Smith, *Science,* Vol. 193 (July 9, 1976), pp. 114–20; and in "Microcircuits in the Nervous System" by Gordon M. Shepherd, *Scientific American* (February 1978), pp. 93–103.

CHAPTER TEN

The Jekyll and Hyde Solution

The story of Dr. A. L. Wigan's discovery of a patient with a single cerebral hemisphere is taken from "Hemispheric Specialization: Implications for Psychiatry" by David Galin, in *Biological Foundations of Psychiatry,* R. G. Grenell and S. Gabay, eds. (New York: Raven Press, 1976). Galin's speculation on the role of the right hemisphere in "unconscious processes" is also contained in this paper.

The Mary Hastings description was taken from a patient described by Dr. M. Critchley in his book *The Parietal Lobes* (New York: Hafner, 1953).

The Roger Sperry material is taken from "Lateral Specialization in the Surgically Separated Hemispheres," in *The Neurosciences,* Third Study Program (1974); "Changing Concepts of Consciousness and Free Will," in *Perspectives in Biology and Medicine* (Autumn 1976), 9–19; and "In Search of Psyche," in *Neurosciences: Paths of Discovery* by Frederick G. Worden, et al. (Cambridge, Mass.: M.I.T. Press, 1975).

The Marcel Kinsbourne material was gathered during interviews in Toronto, Canada, in June 1977, and in August of the same year in Oxford, England. Material was also taken from "The Neuropsychological Analysis of Cognitive Deficit," in *Biological Foundations of Psychiatry* (1976), pp. 527–89; "Direction of Gaze and Distribution of Cerebral Thought Processes," in *Neuropsychologia,* Vol. 12 (1974), pp. 279–81; and "The Evolution of Language in Relation to Lateral Action," in *The Asymmetrical Function of the Brain,* Marcel Kinsbourne, ed. (Cambridge Univ. Press, 1977).

The differences in the two hemispheres are described in "Right-Left Asymmetries in the Brain: Structural Differences Between the Hemispheres May Underlie Cerebral Dominance" by Albert M. Galaburda and Norman Geschwind, in *Science,* Vol. 199 (February 24, 1978); and "The Fundamental Nature of Human Infant Brain Asymmetry" by Juhn A. Wada and Alan Davis, *Le Journal Canadien des Sciences Neurologiques* (August 1977), pp. 203–07.

The studies on early differences in behavior based on brain developmental differences is described in "The Development of Lateral Differences in the Human Infant" by Gerald Turkewitz, in *Lateralization in the Nervous System,* Stevan Harnad, et al., eds. (New York: Academic Press, 1977).

The study on hemisphere specialization in nonhuman primates is from "An Assessment of Hemispheric Specialization in Monkeys" by Charles R. Hamilton, in "Evolution and Lateralization of the Brain," Stewart J. Dimond and David A. Blizard, eds., in the *Annals of the New York Academy of Sciences,* Vol. 299 (September 30, 1977).

The studies by Raquel and Ruben Gur are published in "Correlates of Conjugate, Lateral Eye Movements in Man," in *Lateralization in the Nervous System,* Stevan Harnad, ed.; and in "Classroom Seating and Functional Brain Asymmetry," *Journal of Educational Psychology,* Vol. 67, No. 1 (1975), pp. 151–53.

The Jerre Levy material was obtained during an interview in June 1977. Similar material was also developed in "The Origins of Lateral Asymmetry," in *Lateralization in the Nervous System* (New York: Academic Press, 1977); and in "Perception of Bilateral Chimeric Figures Following Hemispheric Deconnection" by J. Levy et al., in *Brain,* Vol. 95 (1972), pp. 61–78.

The material on aesthetic preference is from "Lateral Dominance in Aesthetic Preference" by Jerre Levy, in *Neuropsychologia,* Vol. 4 (1976), pp. 431–45.

The sex differences material is from *Explorations in Sex Differences* by Lloyd and Archer, eds. (New York: Academic Press, 1976); "Sex Differences in Finger Tapping: A Developmental Study" by Peter H. Wolff and Irving Hurwitz, in *Neuropsychologia,* Vol. 14 (1976), pp. 35–41; and "Human Intelligence: Sex Differences" by Lissy F. Jarvik, in *Acta Geneticae Medicae et Gemel-*

lologiae, Vol. 24 (1975), pp. 189–211. The material was developed from a paper presented by Karl H. Pribram and Diane McGuiness ("The Object-Person Dichotomy and Information Acquisition: Implications for Science Education") given at the annual meeting of the American Association for the Advancement of Science, Washington, D.C., February 17, 1978; also, I developed much of this material from a discussion with Dr. Pribram in Washington, D.C., in February 1978. A discussion of the differential effects of brain damage on male and female brains can be found in "Sex Differences in the Cerebral Organization of Verbal Functions in Patients with Unilateral Brain Lesions" by Jeannette McGlone, in *Brain,* Vol. 100, Part 4 (December 1977), pp. 775–93.

CHAPTER ELEVEN
The Princess and the Philosopher

The Gilbert Ryle quote is from *Concept of Mind* (London: Hutchinson & Co., 1949).

The dialogue between Descartes and Princess Elizabeth of Bohemia is described in *Body and Mind* (Walton Hall, Bletchley, Buckinghamshire, England: The Open University Press, 1973).

The Wilder Penfield and Herbert Jasper studies of the exposed cerebral cortex are described in *Epilepsy and the Functional Anatomy of the Human Brain* by Wilder Penfield and Herbert Jasper (Boston: Little, Brown, 1954).

The Karl Pribram experiments on monkeys are described in *Languages of the Brain* by Karl H. Pribram (Englewood Cliffs, New Jersey: Prentice Hall, 1971).

The Edward Evarts work on monkeys is described in "Brain Mechanisms in Movement," *Scientific American* (July 1973); "Changing Concepts of Central Control of Movement: The Third Stevenson Lecture," in *Canadian Journal of Physiology and Pharmacology,* Vol. 53, No. 2 (April 1975); "Representation of Movements and Muscles by Pyramidal Tract Neurons of the Precentral Motor

Cortex," in *Neurophysiological Basis of Normal and Abnormal Motor Activities*, M. D. Yahr and D. Purpura, eds. (New York: Raven Press, 1967).

Experiments on the "readiness potential" are described in *The Self and Its Brain* by Karl R. Popper and John C. Eccles, Springer International (1978), pp. 283–86; "Cerebral Cortex Cerebellum and Basal Ganglia: An Introduction to Their Motor Functions" by H. Kornhuber, in *The Neuro Sciences: Third Study Program*, pp. 267–80.

Walter B. Weimer's ideas on structural ambiguity are developed in "Manifestations of Mind: Some Conceptual and Empirical Issues," in *Consciousness and the Brain: A Scientific and Philosophical Inquiry*, Gordon Globus, et al., eds. (New York: Plenum Press, 1976).

Karl Pribram's ideas on the influence of structure on brain processes were developed during interviews in February 1978 in Washington, D.C. Similar material is also developed in *Languages of the Brain* and in *The Holographic Hypothesis of Memory Structure in Brain Function and Perception* by Karl H. Pribram, et al.; "Holonomy and Structure in the Organization of Perception," in *Images, Perception and Knowledge*, John M. Nicholas, ed. (Dordrecht, Holland: D. Reidel Pub. Co., 1977). Further material on holography can be found in *A Guide to Practical Holography* by Christopher Outwater and Eric Van Hamersveld (Beverly Hills, Calif.: Pentangle Press, 1974).

CHAPTER TWELVE
The Soul of a Peacock

The material on states of consciousness is from "Ongoing Thought: The Normative Baseline for Alternate States of Consciousness" by Jerome L. Singer, in *Alternate States of Consciousness*, Norman Zinberg, ed. (New York: The Free Press, 1977).

The Piaget material is from *The Essential Piaget: An Interpretative Reference and Guide*, Howard E. Gruber and J. Jacques Vonèche, eds. (New York: Basic Books, 1977).

The Alexander Luria material on the historical development of consciousness is developed in *Cognitive Development: Its Cultural and Social Foundations* by Alexander Luria (Cambridge, Mass.: Harvard University Press, 1976).

The McGill University experiments on sensory deprivation were described in "Cognitive Effects of a Deceased Variation in the Sensory Environment" by W. Heron, et al., *American Journal of Psychology,* Vol. 8, No. 366 (1953).

Experiments on sensory deprivation by John C. Lilly were described in "Mental Effects of Reduction of Ordinary Levels of Physical Stimuli on Intact Healthy Persons," in *Psychia. Res. Rep.,* 5 (1956), 13, 1–9.

The philosophical implications of split-brain work were developed in *The Journal of Medicine and Philosophy,* Vol. 2, No. 2 (June 1977), which is entitled *Mind-Body Quandaries.* I am also grateful to Dr. H. Tristram Engelhardt, Jr., of the Georgetown University Institute of Bioethics, for his help in the preparation of this chapter.

The Roger Sperry quotes are from "Forebrain Commissurotomy in Conscious Awareness," in *The Journal of Medicine and Philosophy,* Vol. 2, No. 2 (June 1977). The Jerome Shaffer quotes are also from that volume, in a paper entitled "Personal Identity: The Implications of Brain Bisection and Brain Transplants."

CHAPTER THIRTEEN
Window on the Mind

The evoked-potential work is described in "Evoked Potentials in Man" by Charles Shagass, in *Biological Foundation of Psychiatry,* R. G. Grenell and S. Gabay, eds. (New York: Raven Press, 1976), pp. 199–253; "Event-related Slow Potentials in Psychiatry" by M. Dongier, et al., in *Psychopathology and Brain Dysfunction,* C. Shagass, et al., eds. (New York: Raven Press, 1977), pp. 339–52; and "Twisted Thoughts, Twisted Brain Waves?" by C. Shagass, in the same volume, pp. 353–78.

P300 work is described in "Brain Electrical Correlates of Pattern Recognition" published in *Signal Analysis and Pattern Recognition in Biomedical Engineering,* G. F. Inbar (Ed.), John Wiley and Sons, New York, and Israeli University Press, Jerusalem, 1975. The quotes from Professor Donchin are taken from that work. The drawing on p. 258 is adopted from a diagram in the Inbar volume.

The ongoing work with event-related potentials was taken from an interview with Marta Kutas of the Cognitive Psychophysiology Laboratory of the University of Illinois, June 1977, in Miami, Florida.

Edward C. Beck's work on event-related potentials in children and adults is described in "The Evoked Response: Its Use in Evaluating Brain Function of Children and Young Adults" by R. E. Dustman, and E. C. Beck, in *Mental Health in Children,* Vol. 2, D. V. Siva Sankar, ed. (Westbury, New York: PJD Publ. Ltd., 1976); and "Electrophysiology and Behavior," *Annual Review of Psychology,* Vol. 26 (1975), pp. 233–62.

The study of environmentally deprived children is from "Developmental Electrophysiology of Brain Function as Reflected by Changes in the Evoked Response" by Edward C. Beck and Robert E. Dustman, in *Brain Function and Malnutrition,* James W. Prescott, et al., eds. (New York: John Wiley & Sons, 1975).

The use of event-related slow potentials in mental illness is described in "Event-Related Slow Potentials in Psychiatry" by M. Dongier, et al., in *Psychopathology and Brain Dysfunction,* C. Shagass, et al., eds. (New York: Raven Press, 1977), pp. 339–52.

The use of event-related potentials in sociopathy is described in "Psychophysiology of Sociopathy: Electrocortical Measures" by Karl Syndulko, et al., in *Biological Psychology,* Vol. 3 (1975), pp. 185–200.

The use of event-related potentials in multiple sclerosis is described in "Pattern Reversal Evoked Visual Potential in the Diagnosis of Multiple Sclerosis" by W. B. Matthews, et al., in *British Journal of Neurology, Neurosurgery and Psychiatry,* Vol. 40, No. 10 (October 1977), pp. 1009–14; and "Evoked Potentials, Saccadic Velocities and Computerised Tomography in the Diagnosis of Multiple Sclerosis" by F. L. Mastaglia, et al., in *British Medical Journal,* Vol. 1 (1977), 1315–17.

The use of brainstem-evoked responses in the clinical diagnosis of death is described in "Detection and Localization of Occult Lesions with Brain Stem Auditory Responses" by James J. Stockard, et al., in *Mayo Clinic Proceedings* (December 1977), pp. 761–69.

The E. Roy John material was developed in an interview on April 28, 1978, at the Brain Research Laboratories of New York University Medical Center. Similar material was also developed in "A Model of Consciousness" by E. Roy John, in *Consciousness and Self-Regulation,* Vol. 1; Gary Schwartz and David Shapiro, eds. (New York: Plenum, 1976), pp. 1–50; and "Neurometrics" by E. Roy John, et al., in *Science,* Vol. 196, No. 4297 (June 24, 1977), pp. 1393–1410.

CHAPTER FOURTEEN
Put on a Happy Face

The Gary Schwartz material was developed during an interview on December 20, 1976, at Yale University Medical Center.

The material on facial muscle patterning was published as "Facial Muscle Patterning to Affective Imagery in Depressed and Non-depressed Subjects" by Gary E. Schwartz, et al., in *Science,* Vol. 192 (April 30, 1976), pp. 489–91. A survey of Schwartz's work as well as an overview of biofeedback can be found in "Psychosomatic Disorders and Biofeedback: A Psychobiological Model of Disregulation" by Gary Schwartz, in *Psychopathology: Experimental Models,* Jack D. Maser and Martin E. P.

Seligman, eds. (San Francisco: W. H. Freeman & Co., 1977), pp. 270–307.

The Barbara Brown material is from *New Mind, New Body: Biofeedback, New Directions for the Mind* by Barbara Brown (New York: Harper & Row, 1974).

The role of learning in biofeedback is described in "Learning in the Autonomic Nervous System" by Leo V. DiCara, *Scientific American,* Vol. 222, No. 1 (January 1970), pp. 30–39.

The implications of biofeedback for psychotherapy and behavioral modification were developed from the interview with Gary Schwartz, also from his paper "Psychobiological Foundations of Psychotherapy and Behavior Change," in *Handbook of Psychotherapy and Behavior Change* by S. L. Garfield and A. E. Bergin (New York: John Wiley & Sons, 1977). Similar material was also developed in the interview with Dr. E. Roy John.

CHAPTER FIFTEEN
A Treat Instead of a Treatment

Many of the concepts developed in this chapter were treated in a slightly different manner in my "The Brain Makes Its Own Narcotics," *Saturday Review* (March 5, 1977), pp. 6–11. An excellent review of the same material is found in "The Opiate Receptor and Morphine-like Peptides in the Brain" by Solomon H. Snyder, in *The American Journal of Psychiatry,* Vol. 135, No. 6 (June 1978), pp. 645–52; and "Neurotransmitter Activity in the Brain: Focus on the Opiate Receptor" by Solomon H. Snyder, in *Interdisciplinary Sciences Review,* Vol. 3, No. 1 (1978), pp. 46–54.

The material on Nathan Kline's treatment of mental illness with endorphins is developed in "Doctors Debate Brain Hormone Dilemmas," in *Medical World News* (January 9, 1978), pp. 86–96.

CHAPTER SIXTEEN
The Dancing Bees

Much of this material was taken from an interview with Noam Chomsky at the Massachusetts Institute of Technology in October 1977. Thanks are also due Professor Chomsky for reviewing this chapter and making many helpful commentaries. Quotes are also taken from Chomsky's *Language and Mind* (New York: Harcourt Brace Jovanovich, 1968–72); and *Reflections on Language* (New York: Pantheon Books, 1975).

CHAPTER SEVENTEEN
Because You Asked Me To

The Roger Schank material was developed from interviews at the Yale University Artificial Intelligence Laboratory in December 1977. Quotations are also taken from *Scripts, Plans, Goals and Understanding: An Inquiry into Human Knowledge Structures* by Roger Schank and Robert Abelson (Hillsdale, New Jersey: Lawrence Erlbaum Associates, 1977).

The Joseph Weizenbaum material was developed from interviews in October 1977. Dr. Weizenbaum also kindly reviewed this chapter and made many helpful comments. Quotations are also taken from Weizenbaum's *Computer Power and Human Reason: From Judgement to Calculation* (San Francisco: W. H. Freeman & Co., 1976).

Quotes are also taken from *Understanding Natural Language* by Terry Winograd (New York: Academic Press, 1972) and *The Thinking Computer: Mind Inside Matter* by Bertram Raphael (San Francisco: W. H. Freeman & Co., 1976).

CHAPTER EIGHTEEN
A Seductive Analogy

One of the major sources for this chapter is *Brain Mechanisms in Sensory Substitution* by Paul Bach-y-Rita (New York: Academic Press, 1972).

I am also grateful to Drs. Karl Frank and Terry Hambricht, both of the Neural Control Laboratory of the National Institutes of Health. Another major resource was Dr. Carter C. Collins, of the Department of Visual Sciences, University of the Pacific, San Francisco. Quotations were taken from "Studies of Intuitive Education of the Blind with a Tactile Seeing Aide" and "Electrotactile Visual Prosthesis," both of which were prepared for conferences on "Systems and Devices for the Disabled," June 1–3, 1977, Seattle, Washington.

Bibliography

Atkinson, Richard C. *Psychology in Progress Readings from Scientific American* (San Francisco: W. H. Freeman & Co., 1975).

Ayer, A. J. *The Central Questions of Philosophy* (New York: Penguin Books, 1976).

Benson, D. Frank; and Blumer, Dietrich. *Psychiatric Aspects of Neurologic Disease* (New York: Grune and Stratton, 1975).

Blakemore, C. *Mechanics of the Mind* (New York: Cambridge University Press, 1977).

Bolinger, David. *Aspects of Language* (New York: Harcourt Brace Jovanovich, 1975).

Bower, T. G. R. *Perceptual World of a Child* (Cambridge, Mass.: Harvard University Press, 1977).

Brazier, Mary. *Growth and Development of the Brain: Nutritional, Genetic, and Environmental Factors* (New York: Raven Press, 1975).

Brown, Hugh. *Brain and Behavior* (New York: Oxford University Press, 1976).

Calder, Nigel. *The Human Conspiracy: The New Science of Social Behavior* (New York: Viking Press, 1976).

————. *The Mind of Man* (New York: Viking Press, 1970).

Chomsky, Noam. *Reflections on Language* (New York: Pantheon Books, 1975).

————. *Language and Mind* (New York: Harcourt Brace Jovanovich, 1972).

————. *Problems of Knowledge and Freedom: The Russell Lectures* (New York: Vintage Books, 1971).

Cohen, David. *Psychologists on Psychology* (London: Routledge and Kegan Paul, 1977).

447

De Chardin, Teilhard. *The Phenomenon of Man* (New York: Harper & Row, 1959).

Eccles, John. *The Understanding of the Brain* (New York: McGraw-Hill, 1973).

Engelhardt, H. Tristram, Jr. "Mind-Body Quandries," *The Journal of Medicine and Philosophy*, Vol. II, No. 2 (June 1977).

Flew, Anthony. *Body, Mind and Death* (New York: Macmillan, 1974).

Fodor, Jerry A. *The Language of Thought* (New York: Thomas Y. Crowell, 1975).

Gazzaniga, Michael S. *The Bisected Brain* (New York: Appleton-Century-Crofts, 1970).

Globus, Gordon G.; Maxwell, Grover; and Savodnik, Irwin, eds. *Consciousness and the Brain: A Scientific and Philosophical Inquiry* (New York: Plenum, 1976).

Granit, Ragnar. *The Purposive Brain* (Cambridge, Mass.: M.I.T. Press, 1977).

Gregory, Richard. *Concepts and Mechanisms of Perception* (London: Gerald Duckworth, 1974).

————. *The Intelligent Eye* (London: Weidenfeld and Nicolson, 1970).

Grenell, Robert G.; and Gabay, Sabit, eds. *Biological Foundations of Psychiatry*, Vols. I and II (New York: Raven Press, 1976).

Griffin, Donald R. *The Question of Animal Awareness: Evolutionary Continuity of Mental Experience* (New York: Rockefeller University Press, 1976).

Hanfling, Oswald. *Body and Mind* (Walton Hall, Bletchley, Buckinghamshire, England: The Open University, 1973).

Harnad, Stevan, et al. *Lateralization in the Nervous System* (New York: Academic Press, 1977).

Hilgard, Ernest R.; Atkinson, Roger C.; and Atkinson, Rita L. *Introduction to Psychology* (New York: Harcourt Brace Jovanovich, 1975).

Kinsbourne, Marcel; and Smith, W. Lynn, eds. *Hemispheric Disconnection and Cerebral Function* (Springfield, Ill.: Charles C. Thomas, 1974).

Koestler, Arthur. *The Heel of Achilles* (London: Hutchinson & Co., 1974).

Lausch, Erwin. *Manipulation: Dangers and Benefits of Brain Research* (New York: Viking Press, 1972).

Luria, Alexander. *Cognitive Development: Its Cultural and*

Social Foundations, Michael Cole, ed. (Cambridge, Mass.: Harvard University Press, 1976).

————. *The Working Brain: An Introduction to Neuropsychology* (New York and London: Penguin Press, 1973).

Maser, Jack D.; Seligman, Martin E. P., eds. *Psychopathology: Experimental Models* (San Francisco: W. H. Freeman & Co., 1977).

Neisser, Ulric. *Cognition and Reality: Principles and Implications of Cognitive Psychology* (San Francisco: W. H. Freeman & Co., 1976).

Ornstein, Robert E., ed. *The Nature of Human Consciousness: A Book of Readings* (San Francisco: W. H. Freeman & Co., 1973).

Penfield, Wilder; and Jasper, Herbert. *Epilepsy and the Functional Anatomy of the Human Brain* (Boston: Little, Brown, 1954).

Popper, Karl R.; and Eccles, John C. *The Self and Its Brain: An Argument of Interactionism* (Springer International, 1977).

Phillips, John L., Jr. *The Origins of Intellect: Piaget's Theory* (San Francisco: W. H. Freeman & Co., 1975).

Prescott, James W.; Read, Merrill S.; and Coursin, David B., eds. *Brain Function and Malnutrition: Neuropsychological Methods of Assessment* (New York: John Wiley & Sons, 1975).

Raphael, Bertram. *The Thinking Computer: Mind Inside Matter* (San Francisco: W. H. Freeman & Co., 1976).

Restak, Richard M. "Brain 'Pacemakers' Would Watch More Than Brain," *The New York Times—Ideas and Trends* (July 1974).

————. "Jose Delgado—Exploring Innner Space," *The Saturday Review* (August 9, 1975).

————. *Premeditated Man: Bioethics and the Control of Future Human Life* (New York: Viking Press, 1975).

————. "Psychiatry in Search of Identity," *The New York Times—Ideas and Trends* (January 12, 1975).

————. "The Brain Makes Its Own Narcotics!," *The Saturday Review* (March 5, 1977).

————. "The Hemispheres of the Brain Have Minds of Their Own," *The New York Times—Ideas and Trends* (January 25, 1976).

————. "The Promise and Peril of Psychosurgery," *The Saturday Review* (September 25, 1976).

————. "Some Drugs Are Clarifying the Mind," *The New York Times—Ideas and Trends* (December 12, 1976).

————. "What Happens in Brainwashing Is Only Vaguely Understood," *The New York Times—Ideas and Trends* (February 22, 1976).

Rosenfeld, Albert, ed. *Mind and Super Mind* (Holt, Rinehart and Winston, 1977).

Ryle, Gilbert. *Concept of Mind* (London: Hutchinson & Co., 1949).

Schank, Roger; and Abelson, Robert. *Scripts, Plans, Goals, and Understanding: An Inquiry into Human Knowledge Structures* (Hillsdale, New Jersey: Lawrence Erlbaum Associates, 1977).

Shagass, Charles; Gershon, Samuel; and Friedhoff, Arnold J., eds. *Psychopathology and Brain Dysfunction* (New York: Raven Press, 1977).

Schmitt, Frances O.; and Worden, Frederick G., eds. *The Neurosciences, Third Study Program* (Cambridge, Mass.: M.I.T., 1974).

Teyler, Timothy. *A Primer of Psychobiology: Brain and Behavior* (San Francisco: W. H. Freeman & Co., 1975).

Valenstein, Elliot S. *Brain Control* (New York: John Wiley & Sons, 1973).

Waddington, C. H. *Tools for Thought* (London: Jonathan Cape, 1977).

Weiner, Herbert, *Psychobiology and Human Disease* (New York: Elsevier North-Holland, 1977).

Weizenbaum, Joseph. *Computer Power and Human Reason: From Judgement to Calculation* (San Francisco: W. H. Freeman & Co., 1976).

Whitaker, Haiganoosh; and Whitaker, Harry, eds. "Perspectives in Neurolinguistics and Psycholinguistics," *Studies in Neurolinguistics,* Vol. III (New York: Academic Press, 1977).

Wilson, Edward O. *Sociobiology: The New Synthesis* (Cambridge, Mass.: Harvard University Press, 1975).

Worden, Frederick G.; Swazey, P.; and Adelman, George, eds. *The Neurosciences: Paths of Discovery* (Cambridge, Mass.: M.I.T. Press, 1975).

Zinberg, Norman E. *Alternate States of Consciousness: Multiple Perspectives on the Study of Consciousness* (New York: The Free Press, 1977).

Index

456

458

BOOKS ON HEALTH
FROM WARNER BOOKS

HOW TO LOOK TEN YEARS YOUNGER
by Adrien Arpel　　　　　　　　　(L97-823, $8.95)
Adrien Arpel, president of her own multi-million-dolla
cosmetics company, tells you how to give yourself a 10
years-younger image in just one day via her 5-step re
packaging system; how to lift and firm your body in 1
minutes a day; how to follow the 24-hour emergency diet
She will also show you how to dress to hide body flaw
as well as how to use make-up and win the big six ski
battles to be wrinkle-free.

THE 15-MINUTE-A-DAY NATURAL FACE LIFT
by M. J. Saffon　　　　　　　　　(L97-788, $3.95)
Now you can give yourself all the beautifying effects o
a face lift—safely and naturally without surgery—throug
a unique series of exercises created by an internationa
beauty expert. In just minutes a day, you can smooth th
forehead, banish frown lines, round out hollow cheeks
prevent puffy eyelids, erase crow's feet, remove a doubl
chin, tighten neck muscles, and much more.

THE EYE/BODY CONNECTION
by Jessica Maxwell　　　　　　　　(L87-950, $6.95)
Your eyes forecast the onset of disease—and your eye:
reveal the effects of stress, diet, and heredity on you
body. This book presents 59 eye photographs and thei
readings that will tell you what to look for in your own
eyes. The charts enable you to pinpoint vital areas b
matching flaws in your iris with points on the diagrams
This is the first book on this subject for laymen and wi
provide you with a valuable diagnostic tool to the earlies
signs of physical disorder.

JAMES BEARD'S FISH COOKERY
by James Beard　　　　　　　　　(L33-060, $2.95)
America's master of the culinary arts gives you all the
information you'll ever need about choosing and cook
ing fish. He gives you recipes for salt-water fish such as
halibut and pompano, for fresh water fish like trout and
buffalo fish, for shellfish of all kind—and delicacies such
as terrapin and tortoise.

IMPROVE YOUR HEALTH
WITH WARNER BOOKS

LOW SALT SECRETS FOR YOUR DIET
by Dr. William J. Vaughan   (L37-223, $3.95)
Not just for people who must restrict salt intake, but for everyone! Forty to sixty million Americans have high blood pressure, and nearly one million Americans die of heart disease every year. Hypertension, often called the silent killer, can be controlled by restricting your intake of salt. This handy pocket-size guide can tell you at a glance how much salt is hidden in more than 2,600 brand-name and natural foods.

EARL MINDELL'S VITAMIN BIBLE
by Earl Mindell   (L30-300, $3.75)
Earl Mindell, a certified nutritionist and practicing pharmacist for over fifteen years, heads his own national company specializing in vitamins. His VITAMIN BIBLE is the most comprehensive and complete book about vitamins and nutrient supplements ever written. This important book reveals how vitamin needs vary for each of us and how to determine yours; how to substitute natural substances for tranquilizers, sleeping pills, and other drugs; how the right vitamins can help your heart, retard aging, and improve your sex life.

To order, use the coupon below. If you prefer to use your own stationery, please include complete title as well as book number and price. Allow 4 weeks for delivery.